SELF-ORGANIZING METHODS IN MODELING

STATISTICS: Textbooks and Monographs

A SERIES EDITED BY

D. B. OWEN, Coordinating Editor
Department of Statistics
Southern Methodist University
Dallas, Texas

OTHER VOLUMES IN PREPARATION

SELF-ORGANIZING METHODS IN MODELING

GMDH Type Algorithms

edited by

Stanley J. Farlow

The University of Maine
Orono, Maine

MARCEL DEKKER, INC.   New York and Basel

Library of Congress Cataloging in Publication Data

Main entry under title:

Self-organizing methods in modeling.

(Statistics, textbooks and monographs ; v. 54)
Bibliography: p.
Includes index.
1. GMDH algorithms. 2. Regression analysis--Data
processing. 3. Self-organizing systems--Data processing.
I. Farlow, Stanley J., [date]. II. Series.
QA278.2.S45 1984   519.5   84-7061
ISBN 0-8247-7161-3

MARCEL DEKKER, INC.
270 Madison Avenue, New York, New York 10016

Current printing (last digit):
10 9 8 7 6 5 4 3 2 1

PRINTED IN THE UNITED STATES OF AMERICA

This book is dedicated to Professor Alexey Grigoryevich Ivakhnenko on his seventieth birthday.

Alexey Grigoryevich Ivakhnenko

Alexey Grigoryevich Ivakhnenko was born near Poltava, USSR, in March 1913. He graduated from the Leningrad Institute of Electrical Engineering in 1938. In 1943 he became a candidate in technical sciences at the Physics Institute of Moscow University, and in 1944 he joined the Ukrainian Academy of Sciences in Kiev. He wrote his doctoral dissertation on invariance theory in 1953.

In 1944 he joined the Polytechnic Institute of Kiev, first as Lecturer and later as Professor in the Department of Electrical Engineering. He is now director of the Combined Control System Department, Institute of Cybernetics, Ukrainian Academy of Sciences, and chief editor of the journal Avtomatika. He is the author of several books, including Electroavtomatika, Engineering Cybernetics, Self-Organizing Systems with Positive Feedback, Cybernetic Predicting Devices, Cybernetic Systems with Combined Control, and The Heuristic Self-Organizing System in the Engineering Cybernetics and Inductive Method of Model Self-Organization of Complex Systems. In addition, he had published over 200 research papers on automatic control and engineering cybernetics (specializing in the modeling of complex systems). For the past fifteen years he has introduced and developed the special method of modeling known as the group method of data handling (GMDH). Since 1969 his papers and books have been devoted to this method.

Preface

In statistics nowadays there is a distinguishable trend away from the restrictive assumptions of parametric analysis and toward the more computer-oriented area of what is generally known as nonparametric data analysis. One of the more fascinating concepts from this new generation of research is what is known as the GMDH algorithm, which was introduced and is currently being developed by the Ukrainian cyberneticist and engineer A. G. Ivakhnenko.

What is known these days as a heuristic, the GMDH algorithm constructs high-order regression-type models for complex systems and has the advantage over traditional modeling in that the modeler can more-or-less throw into the algorithm all sorts of input/output types of observations, and the computer does the rest. The computer self-organizes the model from a simple one to one of optimal complexity by a methodology not unlike the process of natural evolution.

It is the purpose of this book to introduce to English-speaking people the basic GMDH algorithm, present variations and examples of its use, provide computer programs and, last but not least, list a bibliography of all published work in this growing area of research. The book is aimed at a fairly wide variety of readers. Those persons interested would most likely be modelers in economics, biology, ecology, and other unstructured areas, statisticians interested in nonparametric regression, and some computer scientists. The book could be used as a reference source for researchers or as a textbook for specialized courses and seminars in modeling, applied mathematics, and applied statistics.

A suggested study guide for this book might be:

Chapter 1 → Chapter 5 → application chapters (the Ivakhnenko course)

Chapter 1 → Chapter 3 → application chapters (self-organizing methods in regression course)

Chapter 1 → Chapters 2, 4 → application chapters (adaptive learning networks approach to GMDH course)

Chapter 1 → application chapters (the quick and dirty course)

The editor would like to thank the authors for contributing to what he feels is a valuable resource work, as well as the staff at Marcel Dekker, Inc. for their professional assistance in making a very complicated process seem simple.

Stanley J. Farlow

Contributors

ANDREW R. BARRON, M.S.* Consultant, Adaptronics, Inc., Subsidiary of Flow General, Inc., McLean, Virginia

ROGER L. BARRON, B.S.E., S.M.[†] Principal Scientist, Systems Technologies Division-East, General Research Corporation, Subsidiary of Flow General, Inc., McLean, Virginia

HUGH A. BROOKS[‡] Department of Fisheries, Virginia Institute of Marine Science, Gloucester, Virginia

FRANCIS J. COOK, Ph.D.[§] Vice President, Adaptronics, Inc., Subsidiary of Flow General, Inc., McLean, Virginia

JOSEPH NELSON CRAIG, Ph.D.** Director, Research and Product Development Division, Adaptronics, Inc., Subsidiary of Flow General, Inc., McLean, Virginia

STANLEY J. FARLOW, Ph.D. Professor of Mathematics, The University of Maine, Orono, Maine

SHIN-ICHI FUJITA, M.S. Mathematical Engineer, Research Planning Section, Environmental Pollution Control Center of Osaka Prefectural Government, Osaka, Japan

---

Present affiliations:

*Information Systems Laboratory, Department of Electrical Engineering, Stanford University, Stanford, California
[†]Barron Associates, Inc., Annandale, Virginia
[‡]Systems Development Corporation, Research Triangle Park, North Carolina
[§]Senior Research Scientist, Advanced Technologies Division, General Research Corporation, Subsidiary of Flow General, Inc., McLean, Virginia
**Manager, Adaptive Systems Group, General Research Corporation, Subsidiary of Flow General, Inc., McLean, Virginia

CHARLES E. HUTCHINSON, Ph.D. Professor, Department of Electrical and Computer Engineering, University of Massachusetts, Amherst, Massachusetts

SABURO IKEDA, Ph.D. Associate Professor, Institute of Socio-Economic Planning, University of Tsukuba, Sakura, Ibaraki, Japan

A. G. IVAKHNENKO, Ph.D. Director, Institute of Cybernetics, Ukrainian Academy of Sciences, Kiev, Ukraine, Union of Soviet Socialist Republics

HIROSHI KOI, M.S. Environmental Engineer, Life Environment Department of Osaka Prefectural Government, Osaka, Japan

TADASHI KONDO, Ph.D.* Department of Precision Engineering, Osaka University, Osaka, Japan

WILLIAM M. LEBOW, M.S. Research Mathematician, Control Sciences Division, Scientific Systems, Inc., Cambridge, Massachusetts

JAMES M. MALONE II, Ph.D. Chief Consultant, Innovative Information Technology, Schenectady, New York

RAMAN K. MEHRA, Ph.D. President, Scientific Systems, Inc., Cambridge, Massachusetts

ANTHONY N. MUCCIARDI, M.S.E., Ph.D.† President, Adaptronics, Inc., Subsidiary of Flow General, Inc., McLean, Virginia

JUNJI NOMURA, B.S.E. Assistant Supervisor, Research and Development Laboratory, Matsushita Electric Works, Ltd., Osaka, Japan

KENICHI OHASHI, Ph.D. Economist, Investment Department, The World Bank, Washington, D.C.

MICHAEL H. PRAGER, B.S. N.S.F. Graduate Fellow, Graduate School of Oceanography, University of Rhode Island, Narragansett, Rhode Island

---

Present affiliations:

*Research Engineer, Heavy Apparatus Engineering Laboratory, Toshiba Corporation, Tokyo, Japan
†Deputy Director, Advanced Technologies Division, General Research Corporation, Subsidiary of Flow General, Inc., McLean, Virginia

THOMAS H. PROBERT, Ph.D.* Computer Scientist, Science and Technology Division, MITRE Corporation, McLean, Virginia

HARBERT RICE, Ph.D.† Vice President, Energy Group, Strategic Information, Burlington, Massachusetts

SAUL B. SAILA, Ph.D. Professor, Graduate School of Oceanography, University of Rhode Island, Narragansett, Rhode Island

DONALD E. SCOTT, Ph.D. Associate Professor, Department of Electrical Engineering, University of Massachusetts, Amherst, Massachusetts

HIROYUKI TAMURA, Ph.D. Associate Professor, Department of Precision Engineering, Osaka University, Osaka, Japan

PAUL M. TOLDALAGI, Ph.D.‡ Senior Research Engineer, Scientific Systems, Inc., Cambridge, Massachusetts

---

Present affiliations:

*Director, Computer and Software Engineering Division, Institute for Defense Analyses, Alexandria, Virginia
†Consultant, Digital Signal Processing Division, Analog Devices, Inc., Norwood, Massachusetts
‡Senior Staff Engineer, Digital Signal Processing Division, Analog Devices, Inc., Norwood, Massachusetts

Contents

1

The GMDH Algorithm

STANLEY J. FARLOW / The University of Maine, Orono, Maine

I. INTRODUCTION AND OVERVIEW
OF THE GMDH ALGORITHM

Imagine if you will a mathematical model of a complex biological organism
that is constructed in much the same way as evolution constructs the organism itself. That is, starting with a few basic primeval forms or equations,
we construct a new generation of more complex offsprings and then allow a
survival-of-the-fittest principle to determine which new equations live and
which equations die. This new generation of hardy equations will then be
better suited to describe our biological organism than was the original. If
we now combine all the new equations of our new generation, we will get
yet another generation of more complex equations. Continuing this process
for many generations, we will "grow" a progeny of mathematical models
that will behave more and more like the real system. One can even stop the
process when the mathematical equations overspecialize the real world. In
other words, one truly gets a model of optimal complexity—not too simple
and more important, not too complex.

You say: "That sounds like a pretty good idea—maybe I should try it
myself"? Well, you do not have to; we are talking about the group method
of data handling (GMDH) algorithm of Ivakhnenko. In 1966, a Ukrainian
cyberneticist, A. G. Ivakhnenko, was discouraged by the fact that many
types of mathematical models require the modeler to know things about the
system that are generally impossible to find. In fisheries modeling, for
example, the modeler may be required to know the migration patterns of
certain species of fish, or the fertility levels of certain age groups. If modelers are forced to make wild guesses at these variables, they can hardly
expect to produce a model with a great deal of reliability as to prediction.

1

In other words, Ivakhnenko felt that many of the highly structured models in existence were starting to crumble under their own weight. It was time for researchers to stop putting their prejudices into models and to start building purely objective ones: in other words, models that look only at the data and nothing else.

Those primitive forms or equations that we have been discussing are just the input variables of our system—x_1, x_2, ..., x_m. It is combining them to get the next generation that is interesting. We start by computing the regression equations

$$y = A + Bx_i + Cx_j + Dx_i^2 + Dx_j^2 + Fx_i x_j$$

for each pair of input variables x_i and x_j and the output y (see Fig. 1). This will give us $m(m - 1)/2$ higher-order variables for predicting the output y in place of the original m variables x_1, x_2, ..., x_m.

After finding these regression equations (from a set of input/output observations), we then find out which ones to save. This will give us a collection of quadratic regression models (say, m_1) which best predict y (note that each prediction depends on two independent variables). We now use each of the quadratic equations that we have just computed and generate new independent observations (which will replace the original observations

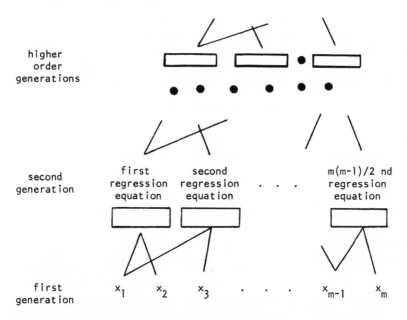

Fig. 1 Computation of higher-order variables from simple ones.

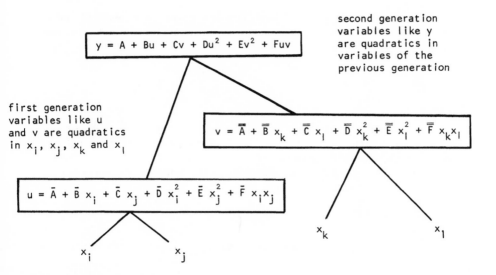

second generation
variables like y
are quadratics in
variables of the
previous generation

$$y = A + Bu + Cv + Du^2 + Ev^2 + Fuv$$

first generation
variables like u
and v are quadratics
in x_i, x_j, x_k and x_l

$$v = \bar{\bar{A}} + \bar{\bar{B}} \, x_k + \bar{\bar{C}} \, x_l + \bar{\bar{D}} \, x_k^2 + \bar{\bar{E}} \, x_l^2 + \bar{\bar{F}} \, x_k x_l$$

$$u = \bar{A} + \bar{B} \, x_i + \bar{C} \, x_j + \bar{D} \, x_i^2 + \bar{E} \, x_j^2 + \bar{F} \, x_i x_j$$

x_k x_l

x_i x_j

zeroth generation
variables are the
original inputs
x_i, x_j, x_k, x_l

Fig. 2 Basic scheme of propagation of variables.

of the variables x_1, x_2, $\ldots$, x_m). From these new independent variables
we will combine them exactly as we did before. That is, we compute all of
the quadratic regression equations of y versus these new variables (two at
a time) (see Fig. 2).

This will give us a new collection of $m_1(m_1 - 1)/2$ regression equations
for predicting y from the new variables, which in turn are estimates of y
from the previous equations. Essentially what we have is a collection of
polynomials of degree 4 in four variables. The reader no doubt is starting
to get the picture of how this algorithm is proceeding. We now merely
select the best of these new estimates, generate new independent variables
from the selected equations to replace the old, and combine all pairs of
these new variables. We continue this process until the regression equations
begin to have a poorer predictability power than did the previous ones. This
will happen since we will test the regression equations against a new independent set of observations. In other words, the model will start to become
overspecialized. After stopping we pick the best of the quadratic polynomials in that generation. What we have is an estimate of y as a quadratic of
two variables, which are themselves quadratics of two more variables,
which are themselves quadratics of two more variables, $\ldots$, which are

quadratics in the original variables. In other words, if we were to make the necessary algebraic substitutions, we would arrive at a very complicated polynomial of the form

$$y = a + \sum_{i=1}^{m} b_i x_i + \sum_{i=1}^{m} \sum_{j=1}^{m} c_{ij} x_i x_j + \sum_{i=1}^{m} \sum_{j=1}^{m} \sum_{k=1}^{m} d_{ijk} x_i x_j x_k + \cdots$$

which is known as the Ivakhnenko polynomial. For example, if we started with 10 input variables x_1, x_2, $\ldots$, x_{10} and our process went through eight generations, one would have an Ivakhnenko polynomial in x_1, x_2, $\ldots$, x_{10} of degree 256. To get a healthy respect for the complexity of such a model, one should realize that terms like

$$x_1^2 x_3^4 x_4^6 x_6^{11} x_7^9 x_{10}^3$$

would be commonplace and that the polynomial would have thousands of such terms.

Our goal now is to explain the GMDH algorithm in minute detail so that the reader will be able to carry out the algorithm's computations on a computer.

II. THE GMDH ALGORITHM IN DETAIL

We begin by collecting regression-type data: that is, n observations of the form shown in Fig. 3. The reason we subdivided the observations into two distinct subsets will become clear shortly.

Step 1 (constructing new variables z_1, z_2, $\ldots$, $z_{\binom{m}{2}}$): The first step is very simple. We merely take all of the independent variables (columns of X) two at a time and construct the regression polynomial

$$y = A + Bu + Cv + Du^2 + Ev^2 + Fuv$$

that best fits the dependent observations y_i in the training set. In other words, each of the $\binom{m}{2} = m(m-1)/2$ regression polynomials would be computed from the observations $\{(y_i, x_{ip}, x_{iq}) : i = 1, \ldots, nt\}$ for different p, q = 1, $\ldots$, m. These $m(m-1)/2$ regression surfaces are illustrated in Fig. 4.

Now, for each of these regression surfaces, evaluate the polynomial at all n data points. For example, in the first surface shown above we would evaluate

$$y = A + Bx_1 + Cx_2 + Dx_1^2 + Ex_2^2 + Fx_1 x_2$$

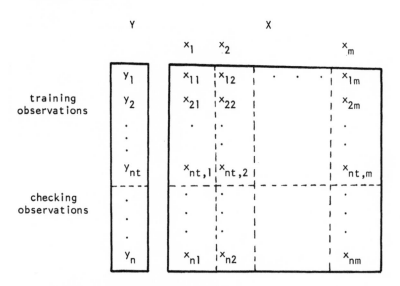

Fig. 3 Input to the GMDH algorithm.

at the n data points (x_{11}, x_{12}), (x_{21}, x_{22}), ..., (x_{n1}, x_{n2}) and store these n new observations (new generation of variables) in the first column of a new array Z (see Fig. 5).

The resulting columns of z are computed in a similar manner. One should interpret these new variables as new improved variables that have better predictability powers than those of the original generation x_1, x_2,

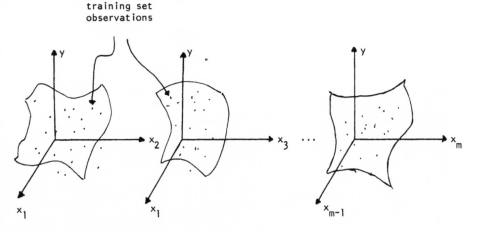

Fig. 4 Construction of the new variables.

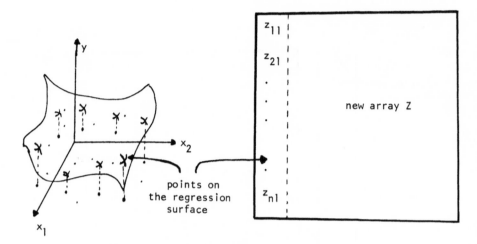

Fig. 5 Construction of the new array Z.

..., x_m. The object is to keep only the <u>best</u> of these new variables (survival of the fittest), and this is where the observations in the <u>checking set</u> come into play (so far we have not used these observations).

Step 2 (screening out the least effective variables): This step replaces the columns of X (old variables) by those columns of Z (new variables) that best estimate the dependent variable y <u>in the checking set</u>. That is, for each column j of Z we compute the <u>root mean square</u> (also called the <u>regularity criterion</u>) r_j

$$r_j^2 = \frac{\Sigma_{i=nt+1}^{n} (y_i - z_{ij})^2}{\Sigma_{i=nt+1}^{n} y_i^2} \quad j = 1, 2, \ldots, \binom{m}{2}$$

(or some other measure of association), order the columns of Z according to increasing r_j, and then pick those columns of Z satisfying $r_j < R$ (R some prescribed number chosen in advance) to replace the original columns of X (those columns of Z will be the new X). One should note that the number of variables saved, say m_1, may be less than or greater than the original number m. Also note that the test of goodness of fit r_j was summed over the observations in the <u>checking set</u>.

Step 3 (test for optimality): From step 2 we find the <u>smallest</u> of the r_j's and call it RMIN. We then plot it on a graph as shown in Fig. 6. If the value of RMIN is less than the value of RMIN from the previous generation, we go back and repeat steps 1 and 2. If the value of RMIN is greater than the preceding value, we assume that the "RMIN curve" has reached its minimum and stop the process (and use the results from the preceding generation).

From experimental work it has been verified that the RMIN curve has the general shape shown in the figure. For the specific RMIN curve shown in Fig. 6, one would stop the process after four iterations (or generations). The first column of the array Z will contain the values $\bar{y}_i$ of the Ivakhnenko polynomial:

$$\bar{y}_1 = a + \sum_{i=1}^{m} b_i x_{1i} + \sum_{i=1}^{m} \sum_{j=1}^{m} c_{ij} x_{1i} x_{1j} + \cdots$$

$$\bar{y}_2 = a + \sum_{i=1}^{m} b_i x_{2i} + \sum_{i=1}^{m} \sum_{j=1}^{m} c_{ij} x_{2i} x_{2j} + \cdots$$

$$\vdots$$

$$\bar{y}_n = a + \sum_{i=1}^{m} b_i x_{ni} + \sum_{i=1}^{m} \sum_{j=1}^{m} c_{ij} x_{ni} x_{nj} + \cdots$$

evaluated at the original n data points. In other words, the first column of Z is the predictor of the dependent variable. To find the coefficients a, b_i, c_{ij}, d_{ijk}, $\cdots$ in the Ivakhnenko polynomial, one must save all the quadratic regression surfaces

$$y = A + Bu + Cv + Du^2 + Ev^2 + Fuv$$

that were computed at each iteration and systematically evaluate this tree

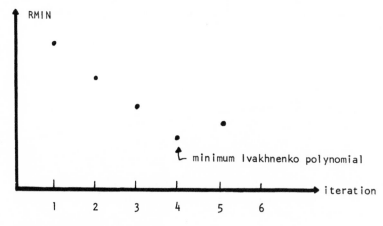

Fig. 6 Stopping criterion.

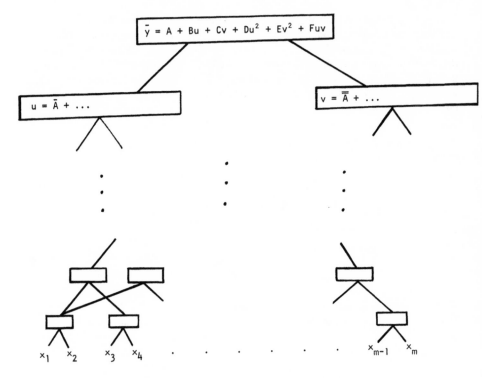

Fig. 7 Tree of polynomials.

of quadratics until one arrives at the high-order polynomial in the original variables x_1, x_2, $\cdots$, x_m. From a computational point of view, once we have all the quadratic regression polynomials stored in the computer, it is possible to compute the prediction $\bar{y}$ of the output y from these quadratics (see Fig. 7).

III. COMMENTS ON THE BASIC ALGORITHM

There are a number of aspects of this algorithm that the reader should observe.

A. Rationale for Subdividing the Observations into Two Parts

The reason for subdividing the n data points into the training and checking sets can easily be understood from the following analogy. If one wished to find the least-squares polynomial

$$y = a_0 + a_1 x + a_2 x^2 + \cdots + a_m x^m$$

of some degree m that best fits a set of data points $(x_1, y_1), \ldots, (x_n, y_n)$, one can of course make the polynomial pass exactly through each point by picking m = n - 1 (see Fig. 8). The problem is, however, that if we now test this polynomial against a new set of independent observations (checking set), we would get a large sum of squares, since the model is overspecialized.

There are, of course, other statistical techniques, such as the PRESS statistic (prediction sum of squares) and others that can be used for determining models of optimal complexity, and many authors have started to use these as an alternative to subdividing the data into two parts.

B. Computation of the Ivakhnenko Polynomial

After each iteration the values of the Ivakhnenko polynomial will be stored in the first column of Z and the degree of the polynomial itself doubles at each iteration. After the first iteration the degree is 2, then 4, 8, and so on.

C. The GMDH as a Predictor and System Identifier

The Ivakhnenko polynomial can be used both as a predictor for estimating the output of complex systems and for identification purposes in determining which inputs are the most important in the model. Both types of analysis are present in this book.

D. Comparison of GMDH with High-Order Regression

To illustrate some advantages of the GMDH algorithm over standard high-order regression, suppose that one wishes to find a regression polynomial

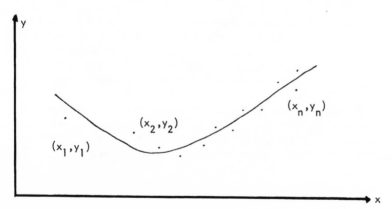

Fig. 8 Least-squares polynomial fit.

in m = 4 variables x_1, x_2, x_3, and x_4 of degree p = 4; that is,

$$y = r_0 + r_1x_1 + r_2x_2 + r_3x_3 + r_4x_4 + r_5x_1^2 + r_6x_2^2 + r_7x_3^2 + r_8x_4^2$$
$$+ r_9x_1x_2 + r_{10}x_1x_3 + r_{11}x_1x_4 + r_{12}x_2x_3 + r_{13}x_2x_4 + r_{14}x_3x_4 + r_{15}x_1^3$$
$$+ \cdots + r_{69}x_1x_2x_3x_4$$

This equation has 1 constant term, 4 first-order terms, 10 second-order terms, 20 third-order terms, and 35 fourth-order terms, for a total of 70 coefficients. If we wished to find these coefficients r_0, r_1, $\cdots$, r_{69} by the usual regression techniques, we would have to solve a system of 70 ill-conditioned linear equations in 70 unknowns. Although this is not difficult by today's standards, we should keep in mind that we are talking about a very simple problem. The computational problems one faces in finding the Ivakhnenko polynomial consist of solving six regression equations in six unknowns A, B, C, D, E, and F (many times).

The real advantage of the GMDH algorithm comes about when the modeler works with large, complex systems. For example, if the model contains m input variables, then the pth-order regression polynomial would contain a total of $(m + 1)(m + 2) \cdots (m + p)/m!$ terms. Even if m = 10 and p = 8, one would have 43,758 terms in the regression polynomial. To find these coefficients by solving the normal equations is out of the question by anyone's standards. The GMDH algorithm, on the other hand, is able to find this regression polynomial (or what we call the Ivakhnenko polynomial) by repeatedly solving regression equations of order 6. If we kept all of the quadratic polynomials for each of three generations, we would have to solve a total of

$$\left[\binom{10}{2} = 45 \right] + \left[\binom{45}{2} = 990 \right] + \left[\binom{990}{2} = 487,555 \right] = 490,590$$

systems of order 6. Of course, in practice we keep far less than all of these polynomials, so the number would be much less. One should note that it is possible by means of the GMDH algorithm to construct a model containing hundreds of parameters from only a few data points. This is in contrast to classical regression, where one needs at least as many observations as parameters.

The reason we do not run up against ill-conditioned normal equations as severely when using the GMDH algorithm as we often do when using high-order regression is that the GMDH algorithm constantly throws out variables along the way, and this helps to eliminate linear dependence.

E. Variations of GMDH

There are many variations of the basic GMDH algorithm. One should realize that the GMDH algorithm is what one might call a <u>heuristic method</u>. In other words, we cannot prove that the method yields a given amount of accuracy as we can in standard regression theory—only that the method works and makes a lot of intuitive sense. There is no reason why we cannot change many of the specific steps to arrive at modified algorithms. The changing of the quadratic polynomials to different types of functions and the use of different decision rules rather than the statistic r_j (called the regularity criterion) for selecting the partial descriptions at each level are only two common variations. One should take a look at Chap. 12 in this book, together with the article by Ikeda et al. (1976) (and, of course, Chap. 5), for some useful variations.

F. Objective Nature of the GMDH Algorithm

The GMDH algorithm allows one to model complex systems without having specific knowledge of the system or massive amounts of data. The modeler need only observe a relatively small number of input/output observations; the computer does the rest. The philosophy is to let the computer do more of the thinking so that the researcher can do more data collecting.

IV. OTHER CRITERIA FOR SELECTING VARIABLES

We have seen how the basic GMDH algorithm uses the regularity criterion, r_j, for selecting models at each iteration (as well as telling us when to stop the iterations). The disadvantage of this criterion is that it is a poor selector in the presence of noisy data and in the problem of medium- and long-range prediction (when the observations might be consecutive values in a time series). For these reasons we will describe two other selectors used by Ivakhnenko.

A. The Unbiased Criterion
(Constructs Models Insensitive to Noise)

The unbiased criterion (sometimes called the minimum-of-bias criterion) requires that the model selected be insensitive to the data on which it is built. For example, suppose that our data set is as shown in Fig. 9 and that we subdivide it into two parts, A and B. If we have a time series where successive data points represent observations at consecutive values in time, data sets A and B might be chosen as alternating observations (A could be the even years and B, the odd ones). From the observations in A we construct $m_1 = m(m - 1)/2$ regression models of the usual form,

$$y = A + Bu + Cv + Du^2 + Ev^2 + Fuv$$

(for all pairs of independent variables x_i and x_j) and then evaluate each of

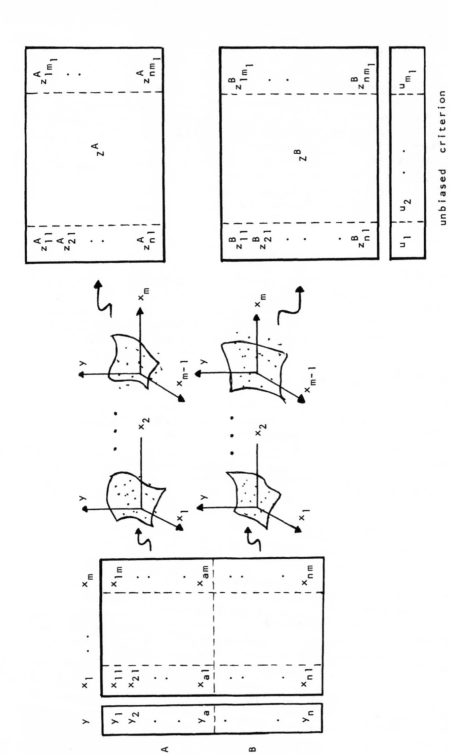

Fig. 9 Construction of the unbiasedness criterion.

them at <u>all n data points</u> (these evaluated numbers are then stored in the columns of Z_A). We then do the same for the observations in B. This gives us the two $n \times m_1$ arrays Z_A and Z_B, from which we can compute the statistics

$$u_j^2 = \frac{\sum_{i=1}^{n} (z_{ij}^A - z_{ij}^B)^2}{\sum_{i=1}^{n} y_i^2} \qquad j = 1, 2, \ldots, m_1$$

which is known as the <u>unbiased criterion</u> (there are other versions of this statistic as well).

We now order the computed values of u_j in ascending order and select those variables that satisfy $u_j < U$ (U some predefined constant). By doing this we will have selected a model that is insensitive to the observations on which it was built. If a given selected variable has $u_j = 0$, the corresponding regression equations are the same (the same computed values, anyway) for both sets of observations A and B. Calling UMIN the smallest of the u_j^s, we then stop our iterations when the UMIN curve reaches its minimum (as we did in the RMIN curve). The unbiased criterion is useful for short-range predictions in the case in which the observations are time series and successive observations represent consecutive points in time.

The weakness of the unbiasedness criterion is the fact that it does not choose models that have the outstanding predictability properties of the regularity criterion r_j. For that reason we might wish to use the <u>combined criterion</u>.

B. The Combined Criterion
    (Combination of the Regularity and Unbiased Criteria)

We have seen that the regularity criterion r_j picks models that have good predictability properties (but might be sensitive to noisy data), whereas the unbiased criterion u_j is insensitive to noisy data (but might not have good predictability properties). For that reason the best criterion combines these two ideas in a single <u>combined criterion</u>. Here we subdivide our observations into <u>three</u> subsets, A, B, and C (generally 40%, 40%, and 20% in each of the three subsets), as seen in Fig. 10. We now construct matrices Z_A and Z_B as we did in Fig. 9 and also construct matrix Z as we did in Fig. 5, using the combined observations A and B for our training set (C would act as our checking set). From these arrays we compute the regularity criterion r_j and the unbiased criterion u_j as we did before, but now we compute the <u>combined criterion</u> c_j:

$$c_j = (r_j^2 + u_j^2)^{\frac{1}{2}} \qquad j = 1, 2, \ldots, m_1$$

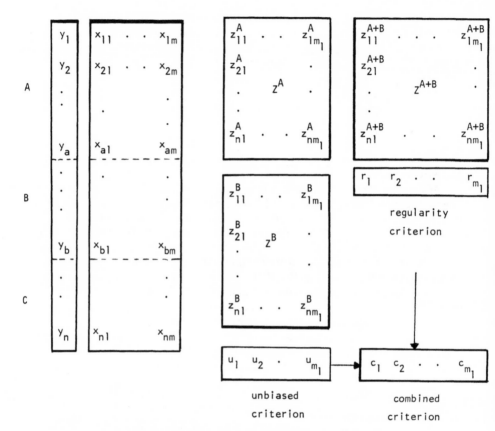

Fig. 10 Construction of the combined criterion.

Now we will order these c_j^s in ascending order and pick those models that
satisfy $c_j < C$ (C some predefined number). If we call CMIN the smallest of
these c_j^s, we stop our iteration when the CMIN curve reaches its minimum
(as we did in the RMIN and UMIN curves). This criterion will then select
models that have good predictability properties but are also insensitive to
noise.

V. LONG-TERM PREDICTIONS BY MEANS OF
THE BALANCE-OF-VARIABLES CRITERION

The idea here is to use a criterion (the balance-of-variables criterion) that
allows us to construct a model over the interpolation region of the data
points that is most likely to hold over the extrapolation area. In this way

we can build models that hold in the distant future. We now explain how this technique works.

Suppose that we have m variables x_1, x_2, ..., x_m as before, but now let us not make any assumptions as to which of them are inputs and which are outputs. In other words, all we know is that we have m interrelated variables, and we would like to predict any or all of them at future values of time (see Fig. 11). We explain this prediction in two steps.

Step 1 (construction of the balance-of-variables criterion): For simplicity, let us assume that we have only m = 3 variables, x_1, x_2, and x_3. The general case can easily be extended by the reader. We first subdivide our observations (which are assumed to be a time series) into two parts, A and B. Then, starting with the observations in A, we find the least-squares estimates of x_1 (in terms of x_2 and x_3) in the form of the rational functions in Table 1 (or some other class of functions). After finding these 63 rational functions, we evaluate each of them at all n observations and store the evaluated functions in the first 63 columns of the matrix Z^A (see Fig. 12). We now find the 63 least-squares estimates of the second variable x_2 (in terms of x_1 and x_3) and store these evaluated functions in the next 63 columns of Z^A. Finally, the last 63 columns of Z^A contain the evaluated least-squares estimates of x_3 (in terms of x_1 and x_2).

We now do exactly the same for the observations in B and store the results in the n × (3)(63) matrix Z^B. Now for each of the (3)(63) columns of Z^A and Z^B, we compute the unbiased criterion (slightly different from the earlier version)

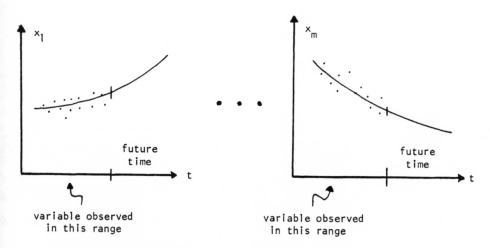

Fig. 11 Long-range prediction of variables.

Table 1 Rational Functions

$$z_1 = a_0 + a_1 x$$

$$z_2 = a_0 + a_1 y$$

$$z_3 = a_0 + a_1 x + a_2 y$$

$$z_4 = a_0 + a_1 xy$$

$$z_5 = a_0 + a_1 x + a_2 xy$$

$$z_6 = a_0 + a_1 y + a_2 xy$$

$$z_7 = a_0 + a_1 x + a_2 y + a_3 xy$$

$$z_8 = \frac{a_0}{1 + b_1 x}$$

$$z_9 = \frac{a_0 + a_1 x}{1 + b_1 x}$$

$$z_{10} = \frac{a_0 + a_1 y}{1 + b_1 x}$$

$$z_{11} = \frac{a_0 + a_1 x + a_2 y}{1 + b_1 x}$$

$$z_{12} = \frac{a_0 + a_1 xy}{1 + b_1 x}$$

$$z_{13} = \frac{a_0 + a_1 x + a_2 xy}{1 + b_1 x}$$

$$z_{14} = \frac{a_0 + a_1 y + a_2 xy}{1 + b_1 x}$$

$$z_{15} = \frac{a_0 + a_1 x + a_2 y + a_3 xy}{1 + b_1 x}$$

$$z_{16} = \frac{a_0}{1 + b_1 y}$$

$$z_{17} = \frac{a_0 + a_1 x}{1 + b_1 y}$$

$$z_{18} = \frac{a_0 + a_1 y}{1 + b_1 y}$$

$$z_{19} = \frac{a_0 + a_1 x + a_2 y}{1 + b_1 y}$$

$$z_{20} = \frac{a_0 + a_1 xy}{1 + b_1 y}$$

$$z_{21} = \frac{a_0 + a_1 x + a_2 xy}{1 + b_1 y}$$

$$z_{22} = \frac{a_0 + a_1 y + a_2 xy}{1 + b_1 y}$$

$$z_{23} = \frac{a_0 + a_1 x + a_2 y + a_3 xy}{1 + b_1 y}$$

$$z_{24} = \frac{a_0}{1 + b_1 x + b_2 y}$$

$$z_{25} = \frac{a_0 + a_1 x}{1 + b_1 x + b_2 y}$$

$$z_{26} = \frac{a_0 + a_1 y}{1 + b_1 x + b_2 y}$$

$$z_{27} = \frac{a_0 + a_1 x + a_2 y}{1 + b_1 x + b_2 y}$$

$$z_{28} = \frac{a_0 + a_1 xy}{1 + b_1 x + b_2 y}$$

$$z_{29} = \frac{a_0 + a_1 x + a_2 xy}{1 + b_1 x + b_2 y}$$

$$z_{30} = \frac{a_0 + a_1 y + a_2 xy}{1 + b_1 x + b_2 y}$$

$$z_{31} = \frac{a_0 + a_1 x + a_2 y + a_3 xy}{1 + b_1 x + b_2 y}$$

$$z_{32} = \frac{a_0}{1 + b_1 xy}$$

$$z_{33} = \frac{a_0 + a_1 x}{1 + b_1 xy}$$

$$z_{34} = \frac{a_0 + a_1 y}{1 + b_1 xy}$$

$$z_{35} = \frac{a_0 + a_1 x + a_2 y}{1 + b_1 xy}$$

$$z_{36} = \frac{a_0 + a_1 xy}{1 + b_1 xy}$$

$$z_{37} = \frac{a_0 + a_1 x + a_2 xy}{1 + b_1 xy}$$

$$z_{38} = \frac{a_0 + a_1 y + a_2 xy}{1 + b_1 xy}$$

$$z_{39} = \frac{a_0 + a_1 x + a_2 y + a_3 xy}{1 + b_1 xy}$$

$$z_{40} = \frac{a_0}{1 + b_1 x + b_2 xy}$$

$$z_{41} = \frac{a_0 + a_1 x}{1 + b_1 x + b_2 xy}$$

$$z_{42} = \frac{a_0 + a_1 y}{1 + b_1 x + b_2 xy}$$

$$z_{43} = \frac{a_0 + a_1 x + a_2 y}{1 + b_1 x + b_2 xy}$$

$$z_{44} = \frac{a_0 + a_1 xy}{1 + b_1 x + b_2 xy}$$

$$z_{45} = \frac{a_0 + a_1 y + a_2 xy}{1 + b_1 x + b_2 xy}$$

$$z_{46} = \frac{a_0 + a_1 y + a_2 xy}{1 + b_1 x + b_2 xy}$$

$$z_{47} = \frac{a_0 + a_1 x + a_2 y + a_3 xy}{1 + b_1 x + b_2 xy}$$

$$z_{48} = \frac{a_0}{1 + b_1 y + b_2 xy}$$

$$z_{49} = \frac{a_0 + a_1 x}{1 + b_1 y + b_2 xy}$$

$$z_{50} = \frac{a_0 + a_1 y}{1 + b_1 y + b_2 xy}$$

$$z_{51} = \frac{a_0 + a_1 y}{1 + b_1 y + b_2 xy}$$

$$z_{52} = \frac{a_0 + a_1 xy}{1 + b_1 y + b_2 xy}$$

$$z_{53} = \frac{a_0 + a_1 x + a_3 xy}{1 + b_1 y + b_2 xy}$$

$$z_{54} = \frac{a_0 + a_1 y + a_2 xy}{1 + b_1 y + b_2 xy}$$

$$z_{55} = \frac{a_0 + a_1 x + a_2 y + a_3 xy}{1 + b_1 y + b_2 xy}$$

Table 1 (Continued)

$$z_{56} = \frac{a_0}{1 + b_1 x + b_2 y + b_3 xy}$$

$$z_{57} = \frac{a_0 + a_1 x}{1 + b_1 x + b_2 y + b_3 xy}$$

$$z_{58} = \frac{a_0 + a_1 y}{1 + b_1 x + b_2 y + b_3 xy}$$

$$z_{59} = \frac{a_0 + a_1 x + a_2 y}{1 + b_1 x + b_2 y + b_3 xy}$$

$$z_{60} = \frac{a_0 + a_1 xy}{1 + b_1 x + b_2 y + b_3 xy}$$

$$z_{61} = \frac{a_0 + a_1 x + a_2 xy}{1 + b_1 x + b_2 y + b_3 xy}$$

$$z_{62} = \frac{a_0 + a_1 y + a_2 xy}{1 + b_1 x + b_2 y + b_3 xy}$$

$$z_{63} = \frac{a_0 + a_1 x + a_2 y + a_3 xy}{1 + b_1 x + b_2 y + b_3 xy}$$

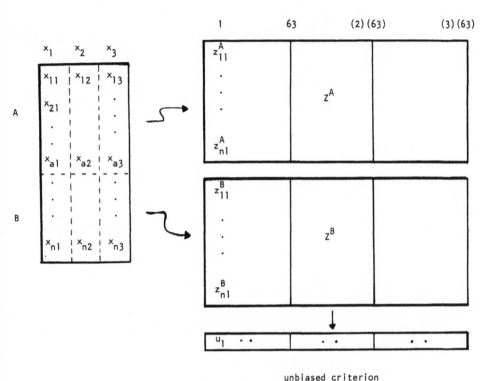

Fig. 12 Construction of the balance-of-variables criterion.

$$u_j^2 = \frac{\sum_{i=1}^{n} \left(z_{ij}^A - z_{ij}^B \right)^2}{\sum_{i=1}^{n} \left(z_{ij}^A \right)^2 + \left(z_{ij}^B \right)^2} \qquad j = 1, 2, \ldots, (3)(63)$$

and pick the three minimums of the u_j corresponding to each of the three variables. This will give us the most unbiased estimates of each of the three variables as rational functions of the other two (we may now wish to go back and reestimate the three rational functions <u>from all n observations</u>). What we now have is three rational estimates:

$$x_1 = f_1(x_2, x_3)$$
$$x_2 = f_2(x_1, x_3)$$
$$x_3 = f_3(x_1, x_2)$$

of each of our variables in terms of the others. We now solve for the
inverse functions. That is, we solve for

x_1 from equation 2 (call it $\bar{x}_1$)

x_2 from equation 3 (call it $\bar{x}_2$)

x_3 from equation 1 (call it $\bar{x}_3$)

or

$$\bar{x}_1 = f_2^{-1}(x_2, x_3)$$

$$\bar{x}_2 = f_3^{-1}(x_1, x_3)$$

$$\bar{x}_3 = f_1^{-1}(x_1, x_2)$$

The balance-of-variables criterion is then computed by the formula

$$B = \frac{\sum_{i=1}^{N'} (x_1 - \bar{x}_1)^2}{\sum_{i=1}^{N'} x_i^2} + \frac{\sum_{i=1}^{N'} (x_2 - \bar{x}_2)^2}{\sum_{i=1}^{N'} x_i^2} + \frac{\sum_{i=1}^{N'} (x_3 - \bar{x}_3)^2}{\sum_{i=1}^{N'} x_i^2}$$

where the summation is over the extrapolation region and where N' is the
number of data points in this extrapolation (or prediction) region. This criterion probably seems a little obscure at the moment, but after the second
step is explained and an example is presented, everything will clear up.

Step 2: We now attempt to predict the three variables x_1, x_2, and x_3
as functions of time t in the form of the polynomials shown in Table 2 (or
some other class of functions). Which of the polynomials we pick will depend on the balance-of-variables criterion. We now compute for each variable x_1, x_2, and x_3 (each column of the data matrix) all 31 least-squares
polynomial estimates of the form shown in Table 2—and then for each of the
$(31)^3$ combinations compute the balance-of-variables criterion B. The specific
combination $x_1^*(t)$, $x_2^*(t)$, and $x_3^*(t)$ that gives the smallest B is chosen as
the best long-range estimate of the three variables. The fact that the balance-
of-variables criterion B should theoretically be zero over the prediction
range, combined with the assumption that our system is "constant" over
time, gives reasons to pick the model that makes B a minimum. We present
an example taken from Ivakhnenko's 1975 paper that will clear up any mys-
teries of the process.

A. Prediction of the World's Population (and Related Variables)

The three variables considered were

x_1 = world population

x_2 = energy consumption

x_3 = natural resources

Table 2 Functions for Long-Range Predictions

$$y_1 = a_0 + a_1 t$$

$$y_2 = a_0 + a_1 t^2$$

$$y_3 = a_0 + a_1 t + a_2 t^2$$

$$y_4 = a_0 + a_1 t^3$$

$$y_5 = a_0 + a_1 t + a_2 t^3$$

$$y_6 = a_0 + a_1 t^2 + a_2 t^3$$

$$y_7 = a_0 + a_1 t + a_2 t^2 + a_3 t^3$$

$$y_8 = a_0 + a_1 t^4$$

$$y_9 = a_0 + a_1 t + a_2 t^4$$

$$y_{10} = a_0 + a_1 t^2 + a_2 t^4$$

$$y_{11} = a_0 + a_1 t + a_2 t^2 + a_3 t^4$$

$$y_{12} = a_0 + a_1 t^3 + a_2 t^4$$

$$y_{13} = a_0 + a_1 t + a_2 t^3 + a_3 t^4$$

$$y_{14} = a_0 + a_1 t^2 + a_2 t^3 + a_3 t^4$$

$$y_{15} = a_0 + a_1 t + a_2 t^2 + a_3 t^3 + a_4 t^4$$

$$y_{16} = a_0 + a_1 t^5$$

$$y_{17} = a_0 + a_1 t + a_2 t^5$$

$$y_{18} = a_0 + a_1 t^2 + a_2 t^5$$

Table 2 (Continued)

$$y_{19} = a_0 + a_1 t + a_3 t^2 + a_3 t^5$$

$$y_{20} = a_0 + a_1 t^3 + a_2 t^5$$

$$y_{21} = a_0 + a_1 t + a_2 t^3 + a_3 t^5$$

$$y_{22} = a_0 + a_1 t^2 + a_2 t^3 + a_3 t^5$$

$$y_{23} = a_0 + a_1 t + a_2 t^2 + a_3 t^3$$

$$y_{24} = a_0 + a_1 t^4 + a_2 t^5$$

$$y_{25} = a_0 + a_1 t + a_2 t^4 + a_3 t^5$$

$$y_{26} = a_0 + a_1 t^2 + a_2 t^4 + a_3 t^5$$

$$y_{27} = a_0 + a_1 t + a_2 t^2 + a_3 t^4 + a_4 t^5$$

$$y_{28} = a_0 + a_1 t^3 + a_2 t^4 + a_3 t^5$$

$$y_{29} = a_0 + a_1 t + a_2 t^3 + a_3 t^4 + a_4 t^5$$

$$y_{30} = a_0 + a_1 t^2 + a_2 t^3 + a_3 t^4 + a_4 t^5$$

$$y_{31} = a_0 + a_1 t + a_2 t^2 + a_3 t^3 + a_4 t^4 + a_5 t^5$$

and data were taken over the years from 1953 to 1974 (22 observations). Here the data sets A and B were chosen to be consecutive observations in the time series (A was the even years and B the odd years). The three least-squares relationships (so-called direct functions) between x_1, x_2, and x_3 were found to be

$$x_1 = f_1(x_2, x_3) = \frac{5.05}{1 - 666x_2x_3 + 6.4x_3}$$

$$x_2 = f_2(x_1, x_3) = \frac{0.04}{1 - 0.99x_1}$$

$$x_3 = f_3(x_1, x_2) = 1.57 - 2.09x_1$$

and the inverse functions (solve them yourself)

$$\bar{x}_1 = f_2^{-1}(x_2, x_3) = \frac{x_2 - 0.004}{0.99x_2}$$

$$\bar{x}_2 = f_1^{-1}(x_1, x_3) = \frac{5.05 + (1 + 6.4x_3)x_1}{666.5x_1x_3}$$

$$\bar{x}_3 = f_1^{-1}(x_1, x_2) = \frac{5.05 + x_1}{(666.5x_2 - 6.43)x_1}$$

Note that we have solved for $\bar{x}_2$ from the first equation since the third equation does not contain x_2. Ivakhnenko then found that the balance-of-variables criterion B (B is a function of time since x_1, x_2, and x_3 are now functions of time) was a minimum (summed over N' = 5 years 1975-1980 in the prediction interval) for the estimates

$$x_1^*(t) = 6036.8 - 96.4t + 2.8t^2 - 0.004t^3 \quad \text{(world population)}$$

$$x_2^*(t) = 3206.9 - 41.9t + 0.005t^3 \quad \text{(energy consumption)}$$

$$x_3^*(t) = 734,008 - 8978t + 201t^2 \quad \text{(natural resources)}$$

VI. OTHER READING

The next step in making a thorough study of GMDH algorithms might be to read Chaps. 2, 3, and 5 (a complete bibliography generated from the INSPEC data base is given at the end of the book). In addition, Chap. 5 by Ivakhnenko lists some important papers. A more complete description of the balance-of-variables criterion as applied to long-range prediction can be found in the three papers by Ivakhnenko et al. (1974-1976). Another nice general-purpose reference is the 1979 article by Ivakhnenko et al.

REFERENCES

Ikeda, S., Ochiai, M., and Sawaragi, Y. (1976). Sequential GMDH Algorithm and Its Application to River Flow Prediction. IEEE Trans. Syst. Man Cybern. SMC-6(7):473-479.

Ivakhnenko, A. G., and Ivakhnenko, N. A. (1974). Long-Term Prediction by GMDH Algorithms Using the Unbiased Criterion and the Balance-of-Variables Criterion. Sov. Autom. Control 7(4):40-45.

Ivakhnenko, A. G., and Ivakhnenko, N. A. (1975). Long-Term Prediction by GMDH Algorithms Using the Unbiased Criterion and the Balance-of-Variables Criterion, Part 2. Sov. Autom. Control 8(4):24-38.

Ivakhnenko, A. G., Ivakhnenko, N. A., Vysotskiey, V. M., and Cheberkus, V. I. (1976). Long-term Prediction by GMDH Algorithms Using the Unbiased Criterion and the Balance-of-Variables Criterion, Part 3. Sov. Autom. Control 9(2):28-42.

Ivakhnenko, A. G., Krotov, G. I., and Visotsky, V. N. (1979). Identification of the Mathematical Model of a Complex System by the Self-Organization Method. Theoretical Systems Ecology: Advances and Case Studies, E. Halfon (Ed.). Academic Press, New York, Chap. 13.

2

Adaptive Learning Networks: Development and Application in the United States of Algorithms Related to GMDH

ROGER L. BARRON* / General Research Corporation, Subsidiary of Flow General, Inc., McLean, Virginia

ANTHONY N. MUCCIARDI,[†] FRANCIS J. COOK,[†] JOSEPH NELSON CRAIG,[†] and ANDREW R. BARRON[‡] / Adaptronics, Inc., Subsidiary of Flow General, Inc., McLean, Virginia

I. INTRODUCTION

The GMDH family of modeling algorithms used today discovers the structure (functional form) of empirical models as well as performing the traditional task of fitting model coefficients to bases of observational or postulated data. Forty years ago scientists began seeking such inductive algorithms in their quest for underlying principles governing the activity of the central nervous system. It was believed—with good reason—that a grasp of these principles would yield improvements in feedback control systems and in the design of automatic calculating machines.

In the United States, algorithms related to GMDH are traceable directly to studies in the 1940s of the behavior of neurons and neuron aggregates. In the USSR, more emphasis was directed (particularly in the 1960s) toward the mathematics of cybernetic systems than toward the emulation of neurons. Both lines of development began to flow together for the 1970s, and in the past five years, scientists in Japan have also had an impact on the course of GMDH work.

Present affiliations:

*Barron Associates, Inc., Annandale, Virginia

[†]General Research Corporation, Subsidiary of Flow General, Inc., McLean, Virginia

[‡]Information Systems Laboratory, Stanford University, Stanford, California

Recent U.S. activity has emphasized use of a predicted squared error criterion for prevention of model overfitting. Using this criterion, it is not necessary to divide the data base into groups: all available data can be used for model fitting, and overfitting is prevented at each stage of model synthesis, including establishment of the structures of individual elements within the network (polynomial model). Because the data base is not grouped into subsets, much of the current U.S. work in the field of this book cannot strictly be said to be GMDH activity. Nevertheless, the GMDH imprint has been indelible, and the authors feel it appropriate to outline the principal U.S. line of development—generally referred to as the adaptive learning network (ALN) approach—in a volume devoted substantially to the Soviet treatment, known as GMDH.

This chapter will trace the origins and development of the ALN method in relation to GMDH, summarize the predicted squared error criterion and its use in ALN synthesis, outline other principal distinctions between ALN and GMDH algorithms, and report on representative applications of the ALN method. No attempt will be made to describe details of the GMDH algorithm; these are treated elsewhere in this volume, as is the derivation of the predicted squared error criterion. In some cases the authors of the present chapter quote at length from earlier papers that are not available in readily accessible publications.

II. BACKGROUND: DEVELOPMENT OF ADAPTIVE LEARNING NETWORK TECHNIQUES PRIOR TO 1971

In the 1940s, the physical and information sciences had already felt the impact of probabilistic models replacing many of the earlier deterministic representations of natural phenomena. The mathematics of these probabilistic models appealed to interdisciplinary scientists seeking to explain the actions of the human nervous system and apply their findings to the improvement of communication and control processes. Wiener et al. [1] in their 1943 paper "Behaviour, Purpose, and Teleology" showed the "possibility of treating such teleological notions as 'goal,' 'purpose,' 'evolution,' etc. in a quantitative manner. Philosophers had been debating these terms for centuries, and it was quite startling in the context of the times to be shown that a quantitative treatment was indeed possible. As an outgrowth of this way of thinking, Wiener in 1948 founded the science of Cybernetics [2], meaning, in his words, the study of control and communication in animal and machine [3].

Continuing from Ref. 3:

Wiener's development of this theme was couched in a highly abstract form for the times and not easily accessible to the engineer. His ideas were put at the disposal of the engineer by the British psychiatrist, W. R. Ashby [4,5] . . . [who] demonstrated . . . that it is possible, working from purely cybernetic ideas, to develop machines that would show such peculiar animal-like characteristics as purpose, goal, and survival potential in an a priori unknown environment. It is important to note that a brain was not being constructed, but rather a machine which exhibited behavior which is presumed to be caused, in animals, by a nervous system.

Ashby amplified the definition of Cybernetics as follows: "Many a book has borne the title 'Theory of Machines,' but it usually contains information about things, about levers and cogs. Cybernetics, too, is a 'theory of machines' but it treats not things but ways of behaving. It does not ask, 'What is this thing?' but, 'What does it do?' It is thus functional and behavioristic." . . . The point of Cybernetics, then, is the explicit possibility of quantitatively treating such behavioral parameters independently of the mechanisms presumed to give rise to them.

A number of scientific people . . . were caught up in the ferment of these ideas. The U.S. Department of Defense became interested, and in the period 1958-60, through one of its elements, the U.S. Air Force, initiated the Bionics (another word for applied Cybernetics) program. We will not go into detail on the various types of learning and self-organizing machines studied since this period. One type, known as a probability state variable (PSV) machine, was ultimately selected for intensive development. . . .

The concepts for a PSV machine grew out of the work of R. J. Lee in the 1950s and early 1960s on artificial neurons and neuron networks [6-9]. Theoretical work began in 1961 on the use of PSV machines for control system applications [10, 11], and a laboratory prototype of a PSV controller was first constructed in 1964 [12]. Successful flight testing of an elementary PSV controller took place in 1969 [13, 14].

A PSV controller is self-directed toward a performance goal, using internal "reward" and "punishment" (selective reinforcement) actions to influence its behavior. These rewards and punishments are interpreted by the controller in the context of the prior decisions that produced them, and the controller modifies the statistics of its internal states accordingly.

To acquire sufficient feedback information for the purposes of identification and control of a plant, PSV controllers must interact with that plant. This interaction takes the form of small experiments conducted at a rate that is generally comparable to the bandwidth of the closed-loop system. There are many interesting theoretical and practical topics associated with this class of controller; unfortunately, these are outside the scope of this paper.

References 10 to 33 detail many of the aerospace applications of PSV self-organizing controllers (SOCs). Other uses of PSV control have been made, chiefly in systems that must continuously reallocate resources in the context of rapid environmental changes; there is, however, little in the open literature on these other applications.

U.S. cybernetics research in the 1960s tended to focus on self-organizing control processes, but in 1963, attention began to be directed toward aspects of empirical modeling (a design process). Although self-organizing control systems cannot have an explicit "teacher" and must rely on self-assessment of performance, empirical modeling processes usually can be guided by an explicit, stored data base and gauge their performance by means of a goodness-of-fit criterion. In 1963 it was not known how to discover the structure of an empirical model from the data base, except to the extent that identification of parameter values within an assumed overall structure might cause certain terms in the model to become negligible. The structure was postulated by the analyst, and a multiparameter numerical search was used to find the values of the model parameters [34,35].

The first empirical modeling application addressed from a cybernetics viewpoint appears to have been the high-speed prediction (in 1963) of trajectories of atmospheric ballistic reentry vehicles [34,35]. A network of 72 algebraic elements was prestructured using multilinear, two-input elements of the form

$$y = w_0 + w_1 x_i + w_2 x_j + w_3 x_i x_j \tag{1}$$

where x_i and x_j represent the inputs; $w_0, \ldots, w_3$ are constants; and y is the output of the element. A "guided" random search was employed to find the values of the 288 parameters in this network. The performance objective for the search was the minimization of average absolute error on all points in the training data set. After training, an independent testing set was used to verify that the model performed properly on new data. A typical data base consisted of radar tracking data for 50 trajectories, with half the data used in the training set and the balance in the testing set. Prediction accuracies for these models were comparable to those usually obtained through serial integration of differential equations of motion, and solution speeds were several orders of magnitude faster than via integration.

From the perspective afforded by the 1960s trajectory prediction model synthesis work, L. O. Gilstrap, Jr. wrote as follows in 1971 [36]:

> One of the more critical problems in the design of intelligent machines is how to construct a large enough space of possible transformations or mappings in the performance unit of a learning machine. If a learning system is to find a suitable mapping of inputs into outputs, that mapping must be within its range of possible mappings.
>
> This problem is most acute in systems with many interacting variables. One important solution to this problem is provided by a method for approximating nonlinear hypersurfaces. Just as arbitrary (but reasonably well-behaved) functions of one variable can be approximated by a polynomial, so can arbitrary functions of many variables be approximated by a suitable high-degree multinomial.* Constructing a space of

---

*Kolmogorov-Gabor polynomial.

hypersurfaces must be done indirectly, however. To show this we define first multi-linear, homogeneous, and complete multinomials.

A <u>multi-linear multinomial</u> is a polynomial in m variables in which all possible product pairs, product triples, ..., and the m-way products appear, but no variable appears to a degree higher than first.

As an example, the multi-linear multinomial in three variables is

$$y = a_{000} + a_{100}x_1 + a_{010}x_2 + a_{001}x_3 + a_{110}x_1x_2 + a_{101}x_1x_3 + a_{011}x_2x_3$$
$$+ a_{111}x_1x_2x_3$$

Note that $\partial y/\partial x_1$ is not a function of x_1, and similarly for the other variables in a multi-linear multinomial. Also, if none of the coefficients is zero, there are $N_M = 2^m$ terms in a multi-linear multinomial in m variables.

A <u>homogeneous multinomial</u> of degree d in m variables is a polynomial such that the exponents of all the variables that make up each term sum to d.

As an example, the homogeneous multinomial of degree two in three variables is

$$y = a_{200}x_1^2 + a_{020}x_2^2 + a_{002}x_3^2 + a_{110}x_1x_2 + a_{101}x_1x_3 + a_{011}x_2x_3$$

Note that if none of the coefficients is zero, the number of terms in a homogeneous multinomial of degree d in m variables is

$$N_H = \frac{(d + m - 1)!}{d!(m - 1)!}$$

A <u>complete multinomial</u> of degree n in m variables is the sum of all homogeneous multinomials from zero[th] degree through n[th] degree.

The number of terms in a complete multinomial of degree n in m variables, provided none of the coefficients is zero, is

$$N_C = \frac{(n + m)!}{n!m!}$$

From the magnitudes of N_M, N_H, and N_C for even relatively small values of n and m, it is apparent that it is not practical to use multinomials directly to approximate nonlinear hypersurfaces. For example, in the case of m = 24 variables, for which a fifth degree (n = 5) surface is to be fitted, $N_C = (24 + 5)!/24!5! = 118,755$ coefficients would be required to specify the surface. If a multi-linear multinomial in these

Barron et al.

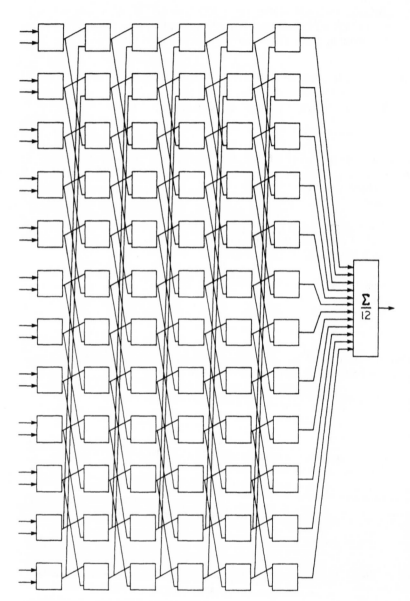

Fig. 1 Uniform spiral 72–element network.

24 variables were desired, then $N_M = 2^{24} \approx 1.6 \times 10^7$ coefficients would have to be specified.

However, high degree multinomials in many variables can be generated using a basic building block element that computes a multi-linear multinomial in two variables:

$$y = w_0 + w_1 x_1 + w_2 x_2 + w_3 x_1 x_2$$

where x_1 and x_2 are inputs, y is the output, and the w's are arbitrary coefficients. . . .

If three of these scaled units are connected into a triangular network . . ., all the terms in a multi-linear multinomial in the four input variables, $x_1, \ldots, x_4$, appear in the output, y_3. However, it should be noted that the coefficients appearing in y_3 are not all independent. As stated earlier, a multi-linear multinomial in four variables requires $2^4 = 16$ independent coefficients, while y_3 has only 12 independent coefficients (the four w's in each of the three units, ... While this [lack of independence for all coefficients in the multinomial] might appear to be a defect in the generation of larger multinomials, it turns out to be the key factor in the practical realization of high degree multinomials in many variables. Although physical systems in many variables can exhibit strong interactions in these variables, the interactions are constrained in many respects, and a multinomial with relatively few degrees of freedom in the coefficients can be used to describe the interactions.

The degree of the multinomial produced by a network is dependent on the connectivity pattern of the building block elements. Figure 1 shows a rectangular network of 72 [multi-linear, two-input] elements connected in a feedforward fashion. In a rectangular net, the highest degree term in any one variable can be as high as one less than the number of columns in the net. The cross-product terms will appear to higher degree; e.g., in the net in Fig. 1, cross products with 12 variables appear in the net output, so that the multinomial produced by the net is a sum of two complete multinomials of degree five in 12 variables and two multi-linear multinomials of degree six.

It is important to obtain complete mixing of inputs in a net if there is no prior knowledge as to which cross-product terms in the multinomial are needed. When the products of all pairs of inputs to a net appear in the net output, the net is said to have <u>sufficiently rich</u> connectivity. The multinomial produced by a sufficiently rich connectivity generally contains all the zeroth, first, and second degree terms, i.e., is at least a quadratic multinomial. The net shown in Fig. 1 does not satisfy the sufficiently rich condition. It is a cylindrically connected net with a uniform spiral of pitch two; i.e., the outputs from any row element are connected to the elements in the same row and to the elements two rows lower. The last two elements are connected back to the first two

rows, much as if the net were wrapped around a cylinder and then connected according to a uniform pattern. Rectangular nets with 12 rows
and six columns with an alternating pitch of one and two do satisfy the
sufficiently rich condition, although nets of pitch one or two alone do
not. Note also that a sufficiently rich rectangular net with 72 elements
and 24 inputs similar to that of Fig. 1 would actually produce a multinomial with more than the 118,755 terms in a fifth degree multinomial
in 24 variables, and it is apparent why no effort was made to write out
the output of the net of Fig. 1. . . .

There are several questions that might be posed of this method of
approximating multinomials, even assuming that the correct coefficients
can be found, such as:

1. How good is the approximation to an arbitrary multinomial?
2. Is the set of network coefficients unique?

The first question is difficult to answer theoretically, but, on purely
practical grounds, various empirical data bases containing from eight
to 24 variables have been satisfactorily approximated. Since it is
impractical to realize fifth-degree functions of as many as 24 variables,
there is nothing to compare results with. About the best that we can do
is to compute the distribution of errors over the hypersurface. If the
distribution of errors is acceptable, then the network approximation
can be used, much as any function-generator or multi-variable table
can be.

In general, the set of network coefficients is not unique, but uniqueness
of the coefficients is important only when the coefficients have some
physical meaning, and it is the values of the coefficients that are desired, rather than the approximation to the multinomial. Since network
coefficient values are partly a result of the connectivity pattern of the
net, they cannot have much physical meaning, and the question of
uniqueness is somewhat irrelevant.

Having indicated that it is possible to generate multinomials, the next
problem is how to find the four coefficients in each of the network elements. Classical matrix inversion is clearly of no value here, since
the coefficients of the multinomial terms are nonlinear combinations
of the coefficients in the multi-linear multinomials in the elements of
the net. (In the network of Fig. 1, there are four coefficients per element and 72 elements in the net, for a total of 288 coefficients to be
found.) The most satisfactory procedure for finding these coefficients
should be independent of the initial values used in the search and should
converge quickly; such a procedure is provided by . . . guided random
search. . . .

Search processes can be divided into two groups, deterministic and
random. The latter group appears to be the most suitable for finding

multinomial coefficients. R. L. Barron [14] has summarized a comparison of a guided random search with the gradient search (one of the deterministic searches) as follows:

1. The guided random search converges faster for spaces of high dimensionality.
2. The guided random search is effective for multi-modal surfaces (the gradient search is basically suited only for unimodal surfaces).
3. It [the guided random search] is an effective search for time-varying surfaces.
4. The guided random search is more effective in coping with sensor or measurement noise than the gradient search.
5. The guided random search can be used to search all parameters simultaneously.
6. The guided random search can be mechanized simply, and in software versions generally requires less storage and less computation time than the gradient search.

In view of the advantages of guided random searches, we will discuss only the random searches in what follows. Much of the discussion has been excerpted from an earlier paper by R. L. Barron [37].
Let $X = (x_1, x_2, \ldots, x_n)$ be a point in an n-dimensional space being searched. Associated with each point in space is some type of index or score, $S(X)$, which is a measure of the value or utility of that point. Typically, the objective of any search is to find the point, X_m, which yields the maximum (or minimum) value of S. Frequently, there are constraints on the permissible choices for X; among the simplest of these constraints is an upper and lower bound on each of the variables being searched. One consequence of a constraint of this type is that the maximum or minimum found within a permissible region may not correspond to the theoretically optimum point, since that point might lie outside the permissible region. Also, there is, in general, no guarantee that the point, X_m, in a bounded space is unique, but uniqueness is not always an essential characteristic of a solution to a search problem.
The rule for halting the search process, called the stop rule, may depend on such practical matters as having a fixed time in which to perform the search, or may be determined from the score function itself. For example, if a minimum value of the score function is sought and if that minimum value is zero, then the search can be halted if a . . . selected set of coefficients produces a score that is arbitrarily close to zero.
In an unguided (or parallel) search, the sequence of trial points, X^1, $X^2, \ldots, X^k$, is selected according to a fixed formula or algorithm which does not take into account the results of each trial point. In a guided search, the score from one trial point is used to guide the

selection of the next trial point, i.e.,

$$x^{k+1} = x^k + \Delta x^{k+1}$$

where Δx^{k+1} is selected on the basis of S^k, the score obtained on the kth trial, and the search becomes an iterative procedure.

Most of the guided random searches are modifications to the basic, unguided random search. In the unguided random search, points are selected at random from the total space being searched; scores are noted for all of the points and, after k trials, the best estimate for X_m is simply the point corresponding to the maximum score obtained during the k trials. Brooks [38] has shown that the probability, $p(f)$, that X_m lies in a certain fraction, f, of the total space after k trials is

$$p(f) = 1 - (1 - f)^k$$

which approaches unity as k increases indefinitely. The expected number of trials to achieve a given level of confidence that the maximum lies in a fraction, f, of the total space can be obtained from the above equation:

$$k = \frac{\log\,[1 - p(f)]}{\log\,(1 - f)}$$

In the basic, unguided random search, the sampling of points is obtained from a rectangular distribution, and no use is made of information gained on prior trials.

Although the unguided random search is slow compared to all guided searches, it is independent of the modality of S, i.e., it can be used to find the global maximum of S and is not subject to "trapping" by local maxima. Because of this desirable feature of the unguided random search, several modifications to the algorithm have been devised to improve the rate at which it converges. The simplest of these modifications is the change from sampling from a rectangular distribution for each variable to sampling from a normal distribution centered about the point corresponding to the maximum score obtained from the beginning of the search to the current trial. This search is guided, but it makes minimal use of prior information. Additional modifications to this algorithm include reversal, hill climbing, acceleration, and smoothing of the terminal search. These modifications assume continuity of S but are so designed that the search reverts to an unguided search if continuity does not hold and no correlations can be found in the accumulated information.

Reversal is based upon the principle that the opposite of downhill is actually uphill. Hence, if a given step, Δx^k, produces a worsening of

the score, ΔX^{k+1} is set to $-\Delta X^k$; if this step also produces no improvement in score, then the step ΔX^{k+2} is taken at random.
Once a direction is found that produces an improvement in score, either by a random trial or by reversal, the continuation of trials in the same direction is . . . hill climbing. Although an uphill direction selected at random will not, in general, coincide with the direction of the maximum slope, improvement in performance may be noted for several steps in any uphill direction. Since the expected number of experiments [at the beginning of the search] required to find an uphill direction (not necessarily the maximum slope), is about one-half [the number of experiments performed], then hill climbing is seen to be a means for exploiting this limited information acquired by random trials.
The information as to whether a given direction is uphill or downhill is of limited value, and it should be exploited as quickly and efficiently as possible. This can be done by lengthening the average step size as long as performance continues to improve. Both arithmetic and geometric progressions of step sizes have been used in random searches during the hill climbing phase. This increase in step size each step is called acceleration of the search. In a geometric acceleration, e.g., doubling the step size each step, large overshoots can occur, and Matyas [39] has employed a deceleration to come as close as possible to the highest point in the given random direction. . . . Matyas also employed a bound on the largest possible step size. . . .
Although bounding the maximum possible step size does not speed up the random search, control of the average step size does appear to provide some improvement in the terminal search. Scaling of the average step size as a function of the score provides smoothing of the search in the region near the maximum as well as improving speed of search. In searching for minima using, for example, least squares score functions, average step size can be set proportional to the best-to-date score to provide automatic and continuous scaling of step size. The constant of proportionality can be adjusted for each problem. In searching for maxima, scaling can be inversely proportional to the score or can be any convenient, monotonically decreasing function of score.
More recently, Mucciardi [40] has described further refinements of the guided random search that improve the ability to shift modes.

A bibliography on random search is presented in Ref. 41. Further summaries of the pre-1971 development of ALN techniques in the United States are contained in Refs. 42 and 43.
In 1968, one of the authors of this chapter had the privilege of meeting with A. G. Ivakhnenko in the USSR while attending technical conferences in that country. Translations of several of Ivakhnenko's works on the theory of self-organizing systems and his earliest writings on GMDH had just begun to appear in the United States, although not in publications generally

accessible to the scientific community. Ivakhnenko was invited to submit a paper on GMDH for publication in the Transactions on Systems, Man, and Cybernetics of the Institute of Electrical and Electronics Engineers, Inc. The paper he produced, "Polynomial Theory of Complex Systems" [44], greatly stimulated interest in GMDH outside the USSR following its appearance in 1971.

The writings of Ivakhnenko on GMDH, which have appeared primarily in the bimonthly Kiev journal Avtomatika, are extensive.* Reference 45 is a succinct presentation of GMDH research by Ivakhnenko and his associates through the late 1970s.

III. APPLICATION AND REFINEMENT OF GMDH: DEVELOPMENT OF ADAPTIVE LEARNING NETWORK TECHNIQUES FROM 1971 TO 1978

The greatest significance of GMDH is in its capacity for "discovery" of the functional forms of empirical models. This capacity greatly lessens the need for analyst involvement in the model synthesis process and reduces the time (calendar and computer) that must be expended.

In 1970, Armco Steel Corporation became interested in the application of ALN techniques to processes in the steel industry. Their sponsorship of basic and applied research and development (R&D) in ALN areas was a significant factor in the development of GMDH empirical modeling in the United States, because the industrial emphasis of the Armco support demanded improvement in cost-effectiveness of the ALN methodology. Predicated on Ivakhnenko's paper [44], a "Polynomial Network Training Routine (PNETTR), Version I" was programmed and evaluated. Following on its success, Mucciardi formulated in 1971-1972 a Version II algorithm that incorporated several procedures that enhanced the basic GMDH. Version II was used extensively for approximately eight years, and proved itself in a great variety of applications [46].

The primary characteristics of PNETTR II not found in prior GMDH programs were:

1. The Mucciardi-Gose clustering procedure [47] was used to ensure the representativeness of the data groups (subsets), as discussed further below.
2. Three independent data subsets were employed—Fitting (F), sometimes called the Training subset; Selection (S), also referred to as the Testing subset; and Evaluation (E), used after model synthesis to predict the accuracy of the model.

*These works are available in the English-language translation of Avtomatika, titled Soviet Automatic Control, available through Scripta Publishing Company, 7961 Eastern Avenue, Silver Spring, MD 10910.

3. The original candidate inputs considered in layer 1 of the model were reintroduced as candidate inputs to each following layer (together with the outputs of the immediately preceding layer), thereby enlarging the combinational possibilities during evolution of the model.

4. The first minimum in the S subset error rate was used as the stopping criterion in model synthesis unless a significantly lower minimum was reached within a predetermined number of additional layers.

5. After preliminary synthesis of the model was obtained substantially in the manner taught by Ivakhnenko, the coefficients (weights) within the entire model were optimized (keeping the structure fixed) by means of a multiparameter search routine.

6. Except in instances which suggested that overfitting would have resulted, multiple, parallel subnetworks were created, with their outputs combined by summation, each subnetwork having been trained and tested on the error residuals (in F and S, respectively) from the aggregate of all lower-order subnetworks.

The use of a clustering algorithm was found to be valuable in applying the GMDH cross-validation procedure. It was also found that the identified cluster structure was highly useful in subsequent data screening to ensure that the model was being interrogated under conditions for which it had been trained [48]. It came to be more fully recognized than before that data cluster structures are, in themselves, valuable models if the behavior of the modeled system can be unambiguously correlated to the various input (observational) data clusters. The utility of a known prior cluster structure for data screening and data modeling was found to stem in part from the finite boundaries of data clusters—these boundaries signal immediately if the model is being called upon to extrapolate (risky) or interpolate (usually safe) [49]. Finally, it was found that the identification of cluster membership for an unknown input data point can sometimes be used as a "pointer" to lead the decision process to ALN-type models tailored for the particular clusters.

It is emphasized that data clusters are inherently bounded regions, whereas polynomial networks synthesized by GMDH and ALN methods extend to infinity along each axis of the modeled spaces. Because of their importance to the subject of empirical modeling and the discovery of data structures, attention is directed here to some of the details of cluster analysis.

The main idea in cluster analysis is to "discover" the groups, or clusters, of points that lie close together in the data space. Closeness is defined by a distance measure. The most suitable clustering algorithms are those that make the weakest assumptions about the statistical structure of the data, are order independent (i.e., the data may be introduced in any arbitrary sequence without materially altering the inferences made by the algorithm about the structure of the data base), and are recursively updatable (i.e., each time an additional input data point is introduced, the structure can be updated without calling forth the prior data except in terms of their statistical properties).

After a clustering analysis has been employed to examine the geometric (spatial) interrelationships of observed data, the results obtained include:

1. Structure of the Data Space: the number of distinct clusters (classes) that are discriminable based on a set of N measured parameters
2. Identification of Noisy Data: clusters or isolated points that are far removed from the main data body
3. Detection of Nonstationary Conditions: consistent observations of new data, which fall in the periphery or just outside existing clusters, whose effect is to cause cluster migration
4. Discovery of New Operating Regions: new data that form clusters which do not overlap with existing clusters (i.e., are statistically distinct via a multivariate F-test), but are not far enough removed to be considered noise
5. Avoidance of Extrapolation Error: screening new observations before interrogating a model to ensure that new data fall within or near those regions of N-space for which the model was synthesized
6. Establishment of Model Confidence Regions: modeling the error of a model to assign probabilities of error to future model outputs on new observations

A clustering analysis can therefore be used as an information filter [48] to detect the foregoing process descriptors from experimental data. These six aspects of data structural analysis are discussed below; further details can be found in the quoted references.

The Mucciardi-Gose CLUSTR algorithm [47] will now be outlined (refer to Fig. 2). The first data sample (a vector observation of N components) is introduced and the first cell (cluster) is centered at this point. The cells are hyperellipsoids, and their initial radii (principal axes) are preselected. The birth of each cell defines a new cluster in the space. The next sample is presented and it either falls within the boundary of the existing cell, within a "guard zone" surrounding the cell, or outside the guard zone so that a second cell is generated and centered at this point. Similarly, all succeeding points either fall within the cells in existence at that time, within their guard zones, or determine the generation of new cells. When a point falls within a cell, the location of the cell (its mean) and radii are changed to accommodate this new point. The cells thus locate themselves at the dense regions (modes) of the data and assume shapes that conform to the spread of data about these modes.

The CLUSTR algorithm requires the following parameters for each cell, which control the birth and growth rate of the cells: (1) shape factors, σ_i, proportional to the distances from the cell center to the cell boundaries in each dimension (these describe the N-dimensional shape of the cell); (2) a cell size factor, τ, equal to the radius of a hypersphere containing the

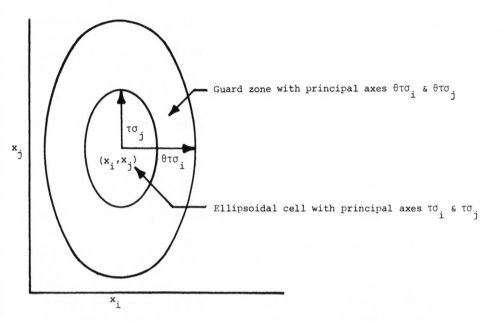

Fig. 2 Cell descriptors.

same hypervolume as the cell; and (3) a <u>guard zone size factor</u>, θ, which is the ratio of distances from the cell center to the outer and inner boundaries of the guard zone (see Fig. 2). The centers and shapes of the cells are adaptively updated as the data are sequentially introduced, and a provision is made for merging cells containing few points with their nearest large neighbor (if sufficiently close) after all the data have been examined.

Points falling within cell m satisfy

$$\sum_{i=1}^{N} \left[\frac{x_i - \bar{x}_{mi}(t)}{\sigma_{mi}(t)} \right]^2 \leq \tau^2$$

where $\bar{x}_{mi}(t)$ and $\sigma_{mi}(t)$ are the center coordinate and shape factor of the cell in dimension i at some time t, and x_i is the ith component of the input vector X. Points X satisfying the inequality are likely to lie inside or on the boundary of a hyperellipsoid in N dimensions with principal axes $\tau\sigma_{mi}$.

Cells are prevented from overlapping, at least during the initial part of the growth phase, by the use of a guard zone. A point falling in the guard zone is not allowed to generate a new cell. All such points are temporarily stored and tagged for later processing. If the size of the guard zone is carefully chosen, it will prevent the birth of new cells at points which are close enough to current cells so that the cells would be likely to cover

common regions of space after attaining their full growth. The reason for
temporarily storing these points is that the closest cell may grow due to
the influence of newer data. Thus, at some later time, a point that was
originally in the guard zone of a cell may be contained inside the cell. If
this does not happen after all the data have been examined, each point remaining in storage is used to update the cell nearest it. Mucciardi and
Gose [47] provide techniques for computing the cell descriptors as a function of dimension, N. The results reported by the CLUSTR algorithm are:

> The number of clusters (i.e., cells)
> Their (N-dimensional) location
> Their (N-dimensional) shape
> The identity of the data points in each cell
> The amount of overlap existing (if any) between cells

The CLUSTR program also computes the probability of observing new data
in the overlap regions. One of the uses of this information is for cluster
merging. When all clusters have been computed, a multidimensional F-test
is used in CLUSTR to determine if some of them are not statistically distinct. The generalized multivariate F-test for two N-dimensional distributions containing n_1 and n_2 points and with mean vectors and covariance
matrices X_1, X_2 and S_1, S_2, respectively, is

$$F = \frac{n_1 + n_2 - N - 1}{N(n_1 + n_2 - 2)^2} \frac{n_1 n_2}{n_1 + n_2} (X_1 - X_2)^T (S_1 + S_2)^{-1} (X_1 - X_2)$$

This statistic may be compared to $F_{\alpha; N, n_1 + n_2 - N - 1}$ at the $\alpha = 0.05$ level of
significance. If F is less than the tabular value, the hypothesis that the two
distributions are not statistically distinct is accepted. Any clusters that are
not distinct are merged automatically. The final result is a parsimonious
description of the data base structure. This description is useful in establishing an unbiased partitioning of data bases, in defining the effective
boundaries of prior data regions, and in determining exact relationships
between a new data point and prior data clusters, regardless of the dimensionality of the space.

The structure of the multivariate data space can be inferred from the
results given above. For example, one cluster containing the majority of
the data, surrounded by clusters containing a few points each, is the usual
result for a unimodal data structure. A bimodal data structure produces
either two clusters containing all the data, or two clusters containing a
majority of the data, surrounded by smaller satellite clusters, and so on.

If the data are from K classes and are to be used for synthesizing a
pattern recognition system, it is very helpful to perform K cluster analyses,
one for the features of each class. The degree of overlap between classes
is a measure of the irreducible error based on the parameters measured,

and this information can be used both for parameter selection and the design of the classifier.

The CLUSTR algorithm plays an important role in determining the quality and consistency of the observations in a multivariate data base. Sensor information might be degraded due to equipment failure or to temporary interruptions, such as dust or steam shielding an optical sensor. Therefore, it is very important to screen data in the input vector. Screening can be performed in the following way. A clustering analysis is first conducted on a set of data free of fault conditions to find the regions of the operating space in which the valid sensor data are clustered. Then, in an operational mode (as new data are observed), the cluster to which a new input vector is "closest" is determined (see below). The input vector is accepted as legitimate if it falls within the "nearest" cluster. If not within, this event signals either a fault condition or a time shift of the process.

Once a prior data cluster has been found, a normalized metric, D^2, may be computed for the distance between any new multidimensional point, X, and the mth cluster by the following equation:

$$D^2(m, X) = \sum_{i=1}^{N} \left(\frac{x_i - \bar{x}_{mi}}{\sigma_{mi}} \right)^2$$

where $\bar{x}_{mi}$ is the mean value of x_i for the mth cluster and σ_{mi} is the length of the ith principal axis of this cluster. D^2 is computed from each cluster, and thus the value of D^2 is found for that cluster, m^*, for which the normalized distance to point X is a minimum. If $D^2(m^*, X)$ is less than unity, X is inside cluster m^*; if this distance equals unity, X is on the m^* boundary; and if X is greater than unity, X is outside cluster m^*.

A decision to classify an input vector as a noisy, or fault, condition would probably be rendered if the data point is far from the nearest cluster. This is reasonable, since a distant point can result from one or more of its components taking on an extreme value with respect to the main portion of the data.

On the other hand, new observations which are just outside or in peripheral portions of clusters which represent the past observed operating regions probably suggest that a time shift is occurring. Depending on the physical process involved, the time shift may take the form that (1) all operating regions are moving simultaneously at the same rate, or (2) all operating regions are moving simultaneously but at different rates (some of which may be zero). The CLUSTR program can be used to detect which of the two types of time shift is taking place and to determine the rate of movement of each operating region (cluster) with respect to the coordinate axes of the data space.

If the frequency of data observations far removed from the main clusters increases and if a check confirms that the sensors are reporting properly, the possibility that new operating region(s) are being observed has to

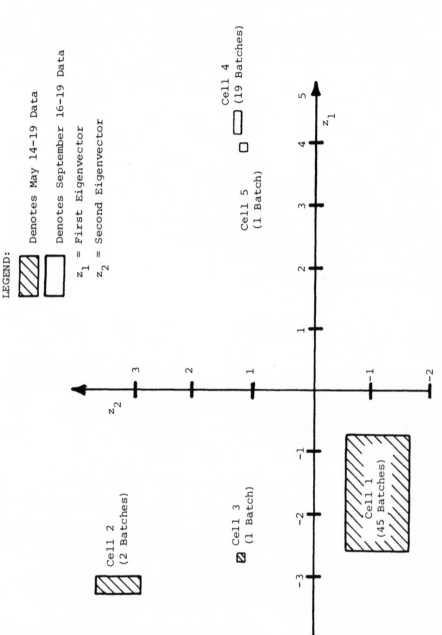

Fig. 3 Data structure for an industrial process. (From Ref. 49.)

be considered. New regions could signal, for example, previously unob-
served states of equilibrium in a chemical process. Incoming data points
should be screened to avoid extrapolation errors. Usually, the points that
are interior to prior data regions can be safely admitted for model interro-
gation, while points that are exterior should be kept in memory for use in
updating the model.

To aid the analyst in visualization of cluster structures of high dimensionality, one may reduce the displayed dimensionality of the cluster structure (with attendant loss of completeness) by computing eigenvectors (the
principal components z_1, z_2, $\cdots$, z_n) of the entire prior-available data
base, then transforming the boundaries of each X-space cluster into the
corresponding Z-space boundaries. The kth eigenvector parameter is derived as a linear transformation of the x's:

$$z_k = \sum_{i=1}^{N} u_{ki}(x_i - \bar{x}_i)$$

where u_{ki} is the ith coefficient of the kth eigenvector of the $N \times N$ covariance matrix and $\bar{x}_i$ is the mean value of the ith parameter. The clusters
may now be plotted as they appear when projected onto the z_1-z_2 plane,
and, as a further convenience, cluster ellipses may be drawn as rectangles.
The width of the jth cluster expressed as a rectangle along the kth z axis
is given by

$$z_{k,\text{lower}} = \sum_{i=1}^{N} u_{ki}[(x_i - s_{ji}) - \bar{x}_{ji}]$$

and

$$z_{k,\text{upper}} = \sum_{i=1}^{N} u_{ki}[(x_i + s_{ji}) - \bar{x}_{ji}]$$

where $\bar{x}_{ji}$ and s_{ji} are the mean and standard deviation values of the jth
cluster along the ith coordinate.

The first two eigenvectors (z_1 and z_2) are often adequate to reveal sepa-
rations between the data clusters, because these eigenvectors account,
cumulatively, for much of the variance in the data base (usually more than
half). An example is an industrial crystallization process for which data
clusters are drawn in Fig. 3. Data on initial conditions of 48 product batches
from this process were recorded during the interval May 14-19, 1975; these
data clustered as shown in the crosshatched rectangles. Data on another 20
batches were then recorded during the period September 16-19, and these
later data were found to cluster as shown in the open rectangles. From
Fig. 3 it is readily concluded that the process operating conditions, meas-

uring instruments, and/or the process itself were sufficiently different in September to be far removed in data space from the situations observed in June. This conclusion was not so immediately obvious to the analyst when inspecting a 30-dimensional data file!

The CLUSTR data screening procedure, using the D^2 distance metric, also detected that the September data from this process did not belong to the June data distribution. Accordingly, to avoid extrapolation, models trained on June data were not interrogated with the September data until model updating had been performed.

Once the statistical structure of the data base has been established as described above, the data base can be rationally divided into two or more parts for GMDH model synthesis (PNETTR II). The first part, the design data, consisting of F and S subsets, is used for model synthesis (i.e., model structure and coefficient determination). The second part, the evaluation data, used after the model has been found, is employed in a final test to ensure that the error rate that was obtained on the design data will occur approximately for all future data within the same regions in the N-dimensional parameter space. Notice that the X vectors are clustered as one group and then divided into subsets by selection (typically at random) within each cluster.

Once a model has been found, its output is an estimate of the dependent quantity:

$$\hat{y} = f(x_1, \ldots, x_N)$$

whose true value is y. For each observation j, let Δy_j denote the error committed by the model:

$$\Delta y_j = y_j - \hat{y}_j$$

The X vectors can now be divided according to their associated errors. That is, label as class 1 the set of X vectors for which the model produces the lowest error:

$$x_j \in \{X\}_1 \quad \text{if} \quad 0 \leq |\Delta y_j| \leq \epsilon_1$$

and so on until the class k set has been established as comprising the remaining observation vectors for which the model is least accurate (i.e., has the largest error):

$$x_j \in \{X\}_k \quad \text{if} \quad \epsilon_{k-1} < |\Delta y_j| \leq \epsilon_k$$

The data base has thus been divided into sets based on the modeling error, and now the model error can in turn be modeled as a multiclass pattern recognition problem. That is, given an X, what is the probability that the estimate of the model will be in error between ϵ_{k-1} and ϵ_k? This is expressed as

$$P(\epsilon_{k-1} < |\Delta y_m| \le \epsilon_k | X_m) = P(X_m | \epsilon_{k-1} < |\Delta y_m| \le \epsilon_k) \times P(\epsilon_{k-1} < |\Delta y_m| \le \epsilon_k)$$

The conditional probability distribution function (PDF) for the kth class, $P(X_m | \epsilon_{k-1} < |\Delta y_m| \le \epsilon_k)$, can be found from a k-class cluster analysis in which each of the classes is clustered separately [50].

The PDF for class k can be approximated by fitting a Gaussian distribution to each of the C_k major clusters for that class, using the data within the jth such region ($1 \le j \le C_k$) to estimate the mean and covariance matrix of that region and creating a weighted sum of the statistically distinct regions:

$$P(X_m | \epsilon_{k-1} < |\Delta y_m| \le \epsilon_k) \cong \sum_{j=1}^{C_k} w_{kj} A_{kj} \exp\left[-\frac{1}{2}(X_m - X_{kj})^T S_{kj}^{-1}(X_m - X_{kj}) \right]$$

where w_{kj} is the fraction of class k data in the jth region and A_{kj} is a constant inversely proportional to the square root of the determinant of S_{kj}.

As each new data point is observed and an estimate of y_i is rendered, the confidence in the estimate can be assigned via the last two equations above. This process can be easily made adaptive if the cluster structure is updated as soon as the true value of the error for X_i is available [51]. If Δy_i fell within the kth error band that was predicted, the regions of class k in which X_i falls is modified by updating its mean and covariance matrix due to the influence of this new point. If, on the other hand, Δy_i was not in the predicted kth error band, but fell instead in the nth error band, a new operating region for class n has been found. In this way, the probability of correctly classifying X—that is, the probability of assigning the correct model confidence—is made an adaptive process.

IV. POST-1978 DEVELOPMENT OF
     ADAPTIVE LEARNING NETWORKS

The utility of PNETTR II concealed, for a time, its deficiencies, which were:

1. Strong tendency to overfit the design data
2. Poor correlation between model structure and underlying "physics" of the modeled process

3. Need to partition the data base, which reduced the quantity of data available for model fitting
4. Limitation of model elements ("partial descriptors"—[44]) to bivariate, quadratic form
5. Limitation of model to a single output variable (requiring creation of multiple, unrelated models for the case of multiple outputs)
6. Inadequate mechanisms for influencing model synthesis and behavior on the basis of estimated errors in old and new observational data

In the 1970s, the work of Akaike [52,53] in Japan influenced additional developments in ALN synthesis algorithms. Akaike introduced an information criterion that incorporated two fundamental types of terms: one that signified the error performance of a model and a second term that conveyed a measure of the complexity of that model. Akaike suggested that the optimum model was that which, in any given instance, produced the minimum sum of error and complexity terms.

A. R. Barron ([54] and in Chap. 4) has carried the reasoning further. He suggests that two related questions have had a major role in the evolution of adaptive learning network synthesis programs. They are:

1. What is the criterion for determining ALN structure?
2. What is the expected performance of the ALN when it is presented with new data?

Ideally, there is one answer that resolves both of these questions; for if we know that a first structure will perform better on new data than another, we should adopt the first. The performance measure recommended by A. R. Barron is the expected squared error on new observations. Two estimates of the expected squared error, derived by him, are discussed below.

A natural way to estimate the expected squared error is by withholding a subset of observations during ALN training and evaluating the (empirical) average squared error on this subset. If the evaluation subset is kept independent of the creation of the ALN, and is representative of the universe of potential observations, it then provides a feasible measure of the performance of the model. However, if this subset is used, as in the GMDH cross-validation treatment, to help select the structure of the ALN, the withheld subset no longer provides an independent measure of future performance. In fact, the selection subset used in GMDH and in PNETTR II influences the future performance through the ALN it selects. Thus a selection data subset is not ideal for its original purpose—selecting structure according to an independent measure of performance on as yet unseen data. In PNETTR II a third group of data, the evaluation subset, did not participate in creation of the ALN and was used to estimate the expected squared error of the final model.

The use of two or three subsets of data permits using cross-validation to avoid overfitting and also (using three subsets) provides an estimate of the expected squared error. However, use of the cross-validation technique requires attention to the partitioning of the original observations into representative data groups and reduces by a factor of 2 to 3 the quantity of data available for model synthesis.

A. R. Barron has shown that the expected squared error on new data can be evaluated analytically with only mild statistical assumptions. The result is a criterion that can be applied to training data to predict the future performance of the ALN, thereby eliminating the need for data base partitioning. Specifically, it is assumed that the model errors are zero-mean, pairwise uncorrelated, and have common variance σ^2. The errors need not be Gaussian random variables and need not have the same distribution. The model may be nonlinear in the input variables. The derivation assumes linearity in the coefficients. Individual elements of ALNs are linear in their coefficients, but networks of elements are not. Nevertheless, the criterion is believed to be a useful and realistic result for ALNs.

Under the assumptions noted above, expected squared error on future data is given by (see Chap. 4)

$$\sigma^2 + \frac{\sigma^2 \text{ trace } (\underline{R}_F \underline{R}_T^{-1})}{n} \tag{2}$$

where $\underline{R}_T$ is the "training covariance" matrix composed of average cross-products between the transformed input variables used in the ALN (the average being over the n training observations) and $\underline{R}_F$ is the corresponding "future covariance" matrix (the average being over any set of observations with errors uncorrelated with training errors).

The two terms in formula (2) correspond to two factors contributing to error on future data. The first term, σ^2, is the expected squared error of the "true" or optimum (but unknown) model. The second term is the expected squared difference between the trained model and the true model when evaluated on data not used for training. If the "covariance structure" of training observations is nearly the same as for future observations, then $\underline{R}_F \underline{R}_T^{-1} \cong \underline{I}$, the identity matrix of dimension k. In that case the expected squared error (2) reduces to

$$\sigma^2 + \frac{\sigma^2 k}{n} \tag{3}$$

where k is the number of estimated coefficients. Thus models of high complexity and many estimated coefficients are not expected to perform well unless there are enough training observations that the second term in (3) is negligible. Small training data bases necessitate simpler models.

Since the true model, its error variance (σ^2), and number of parameters k are regarded as unknown, formula (3) is not yet in the form of a usable criterion for selecting the structures of ALNs. We need a family of estimates of (3) which for each ALN indicates how well it will perform on new data. Then we adopt that ALN which we estimate will perform best. The suggested estimate of an ALN's performance, used in the most recent ALN synthesis algorithm, PNETTR IV, is the predicted squared error (PSE):

$$\text{PSE} = \text{TSE} + \frac{2\sigma_P^2 k}{n} \tag{4}$$

where TSE is the (empirical) average squared error of the ALN on the training data and σ_P^2 is a prior estimate of σ^2 (which does not depend on the ALN being examined). The fixed σ_P^2 in the penalty term is used because we do not want PSE to underestimate future squared error when the particular ALN considered is incorrect (e.g., an overly complex ALN with low error on training data). The factor 2 appears because the training squared error (TSE) is biased below σ^2 by a factor of $\sigma^2 k/n$. From analysis of the statistical properties of PSE, it is found that prior knowledge of σ^2 need not be accurate, although having $\sigma_P^2 \geq \sigma^2$ is helpful to avoid overfitting [54]. Typically, it is reasonable to assume that σ^2 is less than the variation in the dependent variable y, which is given by

$$\sigma_0^2 = \frac{1}{n} \sum (y_i - \bar{y})^2$$

where $\bar{y} = \Sigma y_i / n$. If no prior value is provided, PNETTR IV uses $\sigma_P^2 = \sigma_0^2/2$. Experience has verified that this choice gives acceptable results.

The PSE criterion resembles criteria proposed by Mallows [55], and Akaike [52, 53]. The differences and reasons for preferring the PSE criterion are discussed by A. R. Barron. Implicit in the derivations of Mallows and Akaike are quantities such as (3) [but not (2)]. Bibby and Toutenburg [56] have also examined expected performance on future observations [and (2) can be derived from their equations 1.5.5 and 1.5.13]; however, they do not derive a criterion for selecting among many models.

The result (2) can be very useful after an ALN has been selected and is being applied to new data for which one does not know the true value of the dependent variable y and wishes to know how accurate is the ALN estimate. Result (2) indicates that one should monitor the "covariance structure" R_F (the matrix of average cross-products) of the new input data. If trace $(R_F R_T^{-1})$ is less than or comparable to k, the ALN should be satisfactory. However, if trace $(R_F R_T^{-1})$ tends to be much greater than k, the ALN may not be suitable for the new data. In such a case the ALN should be adapted or retrained.

The work of A. R. Barron led him to the development of versions III and IV of PNETTR. Because PNETTR III and PNETTR IV do not use data groups (subsets of data for cross validation), they are not, strictly speaking, examples of the group method of data handling. Instead, polynomial ALNs are created via PNETTR III and IV using information-theoretic criteria to govern the selection of terms in the elements and which elements remain in the model. Table 1 summarizes versions II to IV of the ALN polynomial network training routine.

Numerous applications of the ALN synthesis methodology have been made. The preponderance of these applications have arisen in signal, image, and time-series estimation (forecasting, target detection, target discrimination, and the like), but uses have also appeared in discrete modeling solutions. Because of the emphasis on signal processing and analyses of time series, considerable attention has been devoted to the development of computer routines for extraction of signal (time-series) parameters ("features"). Practitioners of ALN modeling now have at their disposal a battery of related computational aids, including time-domain and frequency-domain feature extraction, clustering algorithms (hyperellipsoidal and hypercubical), multimodal Bayes decision-rule synthesis algorithms, and so on.

These same numerous applications have generally been conducted under the assumption that the model to be found,

$$y = f(x_1, \ldots, x_N)$$

expresses an analytic relationship between the dependent quantity y and the feature vectors x_1 through x_N. The Kolmogorov–Gabor (K–G) polynomial discussed at length elsewhere in this volume represents such an analytic relationship. The need to generalize to a class of nonanalytic polynomial representations has been recognized by Cook and Craig [57]. The generalization admits singularities into the analytic background represented by the K–G polynomial. Using this representation, basic building blocks of ALNs can be constructed by PNETTR IV in such a manner that the resulting networks can represent symmetric and asymmetric singularities, discontinuities, and jumps embedded in analytic backgrounds.

It is particularly noteworthy that elementary forms of the ALN synthesis algorithms now exist in new instruments that exploit advances in microprocessor (integrated circuit) technology to achieve completely self-contained ALN capabilities in portable devices. Although it will be some time before the power and generality of these compact devices rival the capabilities of, say, PNETTR IV installed in mainframe scientific computers, it is visualized that the gap in capabilities between portable and fixed-base ALN systems will be gradually closed over the next several years.

One other future trend is discernible at the time of this writing. Traditionally, artificial intelligence (AI) techniques have been oriented almost totally toward realization of AI via emulation of the behavior of human experts. GMDH and ALN techniques provide a useful augmentative or alternative approach for the realization of artificial intelligence; that is, some AI capabilities may be achieved by methods that are relatively alien to human thought processes but exploit the arithmetic and logical powers of computers.

Table 1 Adaptive Learning Network Synthesis Algorithms

Algorithm	Performance criterion	Element structure	Layer structure	Other attributes
PNETTR II (1971)— A. G. Ivakhnenko, A. N. Mucciardi	Coefficients adjusted and structure selected so as to minimize (empirical) average squared error on, respectively, fitting and selection subsets of training data	Two inputs: linear or quadratic (preassigned); all terms included	Possible inputs include 16 best outputs from preceding layer and key (preassigned) original input variables; complete elements for all pairs of possible inputs are computed	Clustering procedure is used to ensure representativeness of data subsets
PNETTR III (1979)— A. R. Barron	Coefficients and structure determined so as to minimize sum of error and model complexity terms; criterion evaluated using all training data	Two or three inputs: linear, quadratic, or cubic; various subsets of terms considered (best automatically selected); input pairs are screened prior to computing complete elements to predict which pairs will perform well	Possible inputs include variable number of outputs from immediately preceding layer (typically 64 from first layer, fewer from following layers) and all original input variables	Multiple networks may be trained for as many dependent variables; possible interactions (links) between networks are treated; faster execution and lower cost than PNETTR II while considering a greater variety of ALN structures

50

| PNETTR IV (1982)— A. R. Barron | Coefficients and structure determined so as to minimize error (PSE) criterion; criterion evaluated using all training data | Same as PNETTR III but with improved screening on input pairs; includes cube roots and exponentials of inputs (including some outputs of preceding layers); rates of the exponentials and critical points of cube roots are determined automatically from data | Possible inputs include variable number of outputs from all preceding layers (typically 30 from first, 25 from second, ..., 5 from sixth and following layers) and all original input variables | Same as PNETTR III, but includes extensive automatic analysis of classifier ALN (confusion matrices, ROC curves, nonsymmetric density estimations of ALN outputs, results of Bayes' classifiers); final networks and classifiers outputted as ready-to-use subroutines; user friendly; fast; inexpensive |

The challenge for ongoing research will be to meld an effective union between the traditional (rule-based) and ALN (inductive) approaches toward AI [58].

V. APPLICATIONS OF ADAPTIVE LEARNING NETWORKS

Table 2 summarizes recent U.S. applications of ALNs, and the following sections of this chapter detail a few of these applications with a view to illuminating the principles involved.

A. Scene Classification of LANDSAT Multispectral Scanner Data

The adaptive learning network (ALN) methodology was used to classify a LANDSAT scene into three terrain classes: water, forest, and nonforest. The U.S. Army Engineer Waterways Experimental Station provided training and evaluation sets of LANDSAT multispectral scanner data taken from the 13 October 1975 LANDSAT 2 scene located 40 km northwest of Vicksburg, Mississippi.

The training data consisted of 545 classified radiance vectors. Each radiance vector contained a mean radiance for each of four spectral bands which was obtained from 3×3 pixel arrays. The number of vectors in the water, forest, and nonforest classes of the training data was 8, 156, and 381, respectively. The geographical distribution of the 545 3×3 pixel arrays was not made available.

The ALN classifier synthesized from these training data was used to classify 52,000 pixels in an independent evaluation set. The classification results were essentially 100% accurate, based on a comparison with the actual terrain conditions. Considering the large evaluation set compared to the training data set, together with the speed and simplicity of the derived ALN, the ALN methodology was deemed to be ideally suited to rapid classification of large LANDSAT scenes [59, 60].

B. Target Recognition for Missile Guidance

The ALN synthesis methodology has been used to create a ground target image classification algorithm for infrared images representative of those obtained with seekers in tactical air-launched missiles. Using features extracted from transforms of the original image, the classification algorithm achieves range- and aspect-angle-independent separation of images that contain a specific target type from images that do not contain that type. A receiver operating characteristic (ROC) analysis of the algorithm, using 385 sample images, shows 95% detection rate, 5% false-alarm rate, and a small ($< 1\%$) false-dismissal rate. This study examined the potential for ALNs to recognize specific targets of interest in infrared images. Particular

Table 2 Representative U.S. Applications of Adaptive Learning Networks

---

Process control
  Hot strip steel mill runout table cooling sprays
  Crystallization processes
  Fermentation processes
Radar
  Reentry vehicle trajectory prediction
  Radar imagery target classification
  Detection and identification of tactical targets
  Radar pulse classification
Passive acoustic and seismic analyses
  Ocean platform detection and classification
  Seismic discrimination
Infrared
  Target acquisition and aim-point selection
  LANDSAT scene classification
X-ray
  X-ray image analysis for bomb detection in luggage
Ultrasonics and acoustic emission
  Ultrasonic imaging
  Feedwater nozzle inspection
  Turbine rotor inspection
  Ultrasonic pipe inspection
  Monitoring of crack-growth activity
Eddy currents
  Automatic bolthole inspection
  Recirculating steam generator tubing inspection
  Once-through steam generator tubing inspection
Missile guidance
  Air-to-air guidance law synthesis
Materials
  Radiation embrittlement modeling
  Modeling of single-particle erosion of heat shields
  Weld strength estimation
Multisensor signal processing
  Physical security systems
Microprocessor-based hardware
  "Smart" ultrasonic flaw discriminator
Biomedical modeling
  Sleep stage classification
  Crash injury modeling
Econometric forecasting
  Steel shipment forecasting
  Cost-estimating relationships

---

emphasis was placed on designing a target recognition algorithm that is independent of the target range and aspect angle [61, 62].

C. Missile Guidance Laws

The feasibility of using ALN techniques to provide <u>passive</u> implementation of modern optimal guidance laws has been demonstrated via simulations. A modified proportional navigation (MPN) optimal guidance law was used to establish the ALN training data base. (The details of MPN guidance laws are described in Ref. 63.) The resulting ALN guidance law was found to be superior to constant-gain proportional navigation and to have performance comparable to that of the ideal, but passively unrealizable MPN guidance law over the envelope of MPN launch conditions [57, 64, 65].

The MPN law was defined to be the one that provides desirable commands for missile acceleration, but for implementation the MPN law requires knowledge of passively unobservable range to target and range rate; and an estimate of the completely unobservable time to go, t_{go}, is also required. Thus the basic idea was to use MPN to produce a training data base of intercept engagements and have the ALN model the acceleration commands in this data base using, as ALN inputs, only passive-seeker observables. To perform as well as MPN, the ALN had to learn to infer the unobservable information (combinations of range, range rate, and t_{go} appearing as optimal gains in the MPN formulation) from passive observables in the missile-target engagements incorporated in the training data base. (A description of the training data base is given in Ref. 64.)

The ALN implementation of MPN showed excellent performance. In an independent six-degree-of-freedom simulation of 266 engagements of a maneuvering target in which the missile was under the control of the <u>passive</u> ALN guidance law, the missile "hit" the target in 212 engagements (79.7%). The corresponding figure for the <u>active</u> MPN guidance law was 200 engagements (75.2%), and for classical passive proportional navigation it was 159 engagements (59.8%). Because of the success of the ALN in modeling MPN guidance while requiring only passively observable inputs, this new approach is viewed as offering significant improvements over classical guidance techniques for passive systems and a promising approach for active and hybrid systems.

D. Development of a Distributed, Adaptive, Intrusion Detection System

A laboratory prototype of an adaptive, fixed-site, physical security system has been developed which incorporates a distributed microprocessor network, fiber-optic data links, and adaptive signal processing technologies [66-69]. The system achieves a high probability of detection and a low nuisance alarm rate by using ALN-based detection algorithms to perform multisensor signal processing. The decision logic simultaneously integrates the outputs of geophones and three other types of sensors (SPIR, RACON, and MILES). The detection algorithm is an "alerted classifier"; that is, energy in any of the detectors above a threshold initiates the classification of the combined waveforms as intruder or nonintruder induced. The ALN classifier is trained on representative data. The system features the ability to adapt to site-specific characteristics. The adaptation takes place by training/retraining the classifier on data collected by the ALN system at that site. This capability resides in the software resident on the system hardware. The system also has a capability of recognizing that its environment has changed and alerting the system operator accordingly.

E. Establishing Signal Processing and Pattern Recognition Techniques for In-Flight Discrimination Between Crack-Growth Acoustic Emissions and Other Acoustic Waveforms

Signal processing and pattern recognition algorithms have been developed to discriminate crack-growth acoustic emissions from other innocuous, extraneous acoustic sources. Laboratory experiments were performed to record thousands of crack-growth and noise waveforms on aircraft structural aluminum plates of different geometries and alloy compositions. The problem was separated into four stages, each solved in an automatic mode: detection of signal in background noise, windowing of various parts of the signal, feature extraction, and classification. The algorithms were designed keeping the limitations and requirements of real-time implementation in mind. The ALN methodology was used to select the most important features from the candidate feature list and to derive nonlinear classification functions. Results indicate that optimum combinations of temporal and spectral features result in significantly improved acoustic emission signal identification [70].

F. Quantitative Nondestructive Evaluation of Materials

Adaptive learning networks were first applied to quantitative nondestructive evaluation (NDE) in 1973. The combination of digital signal processing as a preprocessing step and ALNs to model time- and frequency-domain waveform features has led to a new level of performance of NDE systems. Prior to the introduction of ALNs, the detection and assessment of material defects (such as cracks in pipe welds) was based almost solely on the amplitude of a pulse-echo or through-transmission ultrasonic response.

Typical candidate features that are computed from the ultrasonic (i.e., "RF") signal for possible use as ALN inputs are given in Table 3. Usually, the time-domain features are derived from the analytic envelope of the signal, estimated via the Hilbert transform. The frequency-domain features are obtained from the Fourier transform.

Typical ALNs for determining defect type ("characterization") and size are shown in Fig. 4. These models were derived using the PNETTR IV

Table 3 Typical Ultrasonic RF Waveform Features

Rise time 25-90%
Pulse width at 60% amplitude   } Computed from the Hilbert transform
Pulse width at 25% amplitude   }   analytic envelope
Fall time 90-25%
Polarity at 25% on rising edge
Polarity at 60% on rising edge
Polarity at 90% on rising edge
Polarity at peak   } Computed from the Hilbert transform
Average phase between 25% points
Average phase between 60% points
Average phase between 90% points
Energy band, analytic power spectrum at 0.156 MHz
Energy band, analytic power spectrum at 0.313 MHz
Energy band, analytic power spectrum at 0.469 MHz
Energy band, analytic power spectrum at 0.625 MHz
Energy band, analytic power spectrum at 0.781 MHz
Energy band, analytic power spectrum at 0.938-1.094 MHz
Energy band, analytic power spectrum at 1.250-1.406 MHz
Energy band, analytic power spectrum at 0.156-0.625 MHz
Energy band, analytic power spectrum at 0.781-1.250 MHz
Energy band, analytic power spectrum at 1.406-2.500 MHz
Energy band, RF power spectrum at 3.44-3.75 MHz
Energy band, RF power spectrum at 3.75-4.38 MHz
Energy band, RF power spectrum at 4.38-5.00 MHz
Energy band, RF power spectrum at 5.00-5.63 MHz
Energy band, RF power spectrum at 5.63-6.25 MHz
Energy band, RF power spectrum at 3.44-4.06 MHz
Energy band, RF power spectrum at 4.06-4.84 MHz
Energy band, RF power spectrum at 4.84-6.25 MHz
Energy band, RF power spectrum at 3.44-4.53 MHz
Energy band, RF power spectrum at 4.53-6.25 MHz
Ratio of RF energy bands: feature 27/feature 28
Ratio of RF energy bands: feature 28/feature 29
Ratio of RF energy bands: feature 27/feature 29
Moment of the RF power spectrum, center of mass
Moment of the RF power spectrum, standard deviation
Moment of the RF power spectrum, skewness
Moment of the RF power spectrum, kurtosis
Moment of the analytic signal, center of mass
Moment of the analytic signal, standard deviation
Moment of the analytic signal, skewness
Moment of the analytic signal, kurtosis

(a) Sizing Model

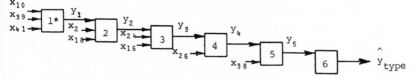

*Mathematical form of element blocks

Single-Feature Blocks:

$$y_i = a_o + a_1 x_1 + a_2 x_1^2 + a_3 x_1^3$$

Two-Feature Blocks:

$$y_i = a_o + a_1 x_1 + a_2 x_2 + a_3 x_1 x_2 + a_4 x_1^2 + a_5 x_2^2 + a_6 x_1^3 + a_7 x_2^3$$

Three-Feature Blocks:

$$y_i = a_o + a_1 x_1 + a_2 x_2 + a_3 x_3 + a_4 x_1 x_2 + a_5 x_1 x_3 + a_6 x_2 x_3$$

$$+ a_7 x_1^2 + a_8 x_2^2 + a_9 x_3^2 + a_{10} x_1^3 + a_{11} x_2^3 + a_{12} x_3^3$$

(b) Characterization Model

*Note: Input features are shown in Table 3. Model structure and coefficients
(a's) learned from the training data.

Fig. 4 Adaptive learning network models to (a) estimate the size of
surface-connected cracks, and (b) discriminate between crack and inclusion-
type defects in turbine rotor bores.

algorithm. The following are representative examples of ALN NDE applications.

1. Under-Fastener Cracks in Aircraft Components

The feasibility of adaptively synthesized nonlinear signal processing techniques for characterization of ultrasonic waveforms to detect and evaluate under-fastener fatigue cracks in aluminum was demonstrated in 1976. Reliable detections and accurate size measurements over the entire flaw size range of 10 to 270 mils represented by the tested specimens was achieved. Previous NDE techniques provided no detection capability below approximately 35 mils and no size measurement capability whatever [71].

2. Multilayered Adhesively Bonded Materials

The ultrasonic waveforms obtained from pulse-echo testing of multilayered, adhesively bonded materials is a complex function of the number and composition of the layers and the adhesive bonds, multipath reflections, the incident angle of the main ultrasonic beam with respect to the material surface, and the presence of numerous spurious reflectors. ALNs can successfully detect bond-line defects, classify the types of these defects (disbonds, delaminations, porosity, etc.), and report their sizes and locations [72].

3. Austenitic Pipe Welds

ALNs have been synthesized for detecting, locating, and classifying flaws produced by intergranular stress-assisted corrosion cracking in austenitic pipe welds. These cracks are a critical problem in the operation of reactors for nuclear power generation. The detection of the cracks is complicated by the presence of numerous geometrical reflectors and by the high attenuation of acoustic signals propagated through stainless steel. The geometrical reflectors consist mainly of the stainless steel grain structure and other surfaces induced in the machining and welding steps. True defects can be accurately discriminated by ALNs, and information regarding flaw sizes, locations, and orientations can be extracted from the ultrasonic waveforms. As a direct outgrowth of this project and related work, a nondestructive evaluation "smart" automatic pipe inspection system for use during in-service ultrasonic inspections of nuclear power plant piping has been developed [73].

4. Turbine Rotor Boresonic Inspection

Advanced signal processing techniques have been applied to the evaluation of ultrasonic data collected from nuclear power plant turbine rotors. The data have been analyzed and algorithms developed to increase the transducer output signal-to-noise ratio by both temporal and spatial beamforming techniques. A smart signal processing system based on ALNs for turbine rotor bore inspection has been developed [74].

5. Spot Weld Strength

An NDE system has been developed to measure and document the strengths of resistance spot welds in automobile bodies. Experimental instrumentation was used to measure and record a variety of welding variables, including electrical, mechanical, and acoustic emission signals. Sheet metal coupons, welded and monitored in laboratory and plant environments, were tested destructively to determine their tensile shear strengths. Using a data base consisting of the waveforms recorded during the welding process and the net strengths obtained from the destructive tests, ALN techniques were used to synthesize models to predict net strength from the instrumented signals. The feasibility of estimating the strengths of resistance spot welds from measurements made during the welding operation was established based on the results of modeling the strength of more than 600 welds [75].

6. Molten Metal Inspection

In the history of nondestructive evaluation of materials, attention has focused almost exclusively on detection (and, more recently, characterization) of flaws in materials inspected in their solid states. Procedures and a hardware system for detection of flaws in molten aluminum have been developed. With this system, flaws can be detected and removed before the metal solidifies, greatly reducing costs of production and significantly improving product reliability [76].

G. Modeling of Behavior of Materials

The behavior of materials has been characterized using the ALN method. The following applications are representative.

1. Single-Particle Heat Shield Erosion Analysis

The applicability of ALNs to modeling single-particle heat shield erosion test data has been investigated. The erosion mass loss from carbon-carbon composites was modeled successfully in terms of material properties, manufacturing process variables, and test parameters. In addition, crater core depth and radius were successfully modeled [77].

2. Radiation Embrittlement

Radiation embrittlement toughness curves and the variability in fracture properties of nuclear reactor steels have received research attention in the nuclear power industry. A large data base for unirradiated materials has been assembled to permit analysis of fracture-toughness characteristics of pressure vessel steels. A reduced irradiated data base has also been acquired. ALNs have been used to model successully a measure of toughness-test response which, in addition to being a function of temperature and neutron fluence, depends on irradiation time, neutron energy density, radiation temperature, material properties, reflecting the microstructure, and chemical impurity content [78].

REFERENCES

1. N. Wiener, A. Rosenblueth, and J. Bigelow, Behaviour, Purpose, and Teleology, 1943. [Appears in Modern Systems Research for Behavioral Scientists, W. Buckley (Ed.), Aldine, Chicago, 1968, pp. 221-226.]
2. N. Wiener, Cybernetics, Wiley, New York, 1948.
3. C. W. Gwinn and R. L. Barron, Recent Advances in Self-Organizing and Learning Controllers for Aeronautical Systems, Proc. AGARD Symp. Appl. Digital Computers to Guidance and Control, London, June 2-5, 1970.
4. W. R. Ashby, Design for a Brain, Wiley, New York, 1952.
5. W. R. Ashby, Introduction to Cybernetics, Wiley, New York, 1956.
6. R. J. Lee, Self-Programming Information and Control Equipment, Melpar, Inc., Falls Church, Va., 1959.
7. R. J. Lee, Generalization of Learning in a Machine, Proc. 14th Natl. Meet. ACM, Sept. 1-3, 1959.
8. L. O. Gilstrap, Jr., and R. J. Lee, Learning Machines, Proc. Bionics Symp., Air Force Wright Air Development Division, WADD TR 60-600, Sept. 1960, pp. 437-450.
9. R. J. Lee, L. O. Gilstrap, Jr., R. F. Synder, and M. J. Pedelty, Theory of Probability State Variable Systems, 6 vols., Adaptronics, Inc., Final Technical Report, Air Force Avionics Laboratory, ASD-TDR-63-664, Dec. 1963.
10. M. A. Ostgaard and L. M. Butsch, Adaptive and Self-Organizing Flight Control Systems, Aerosp. Eng., Sept. 1962.
11. R. J. Lee and R. F. Snyder, Functional Capability of Neuromime Networks for Use in Attitude Stabilization Systems, Adaptronics, Inc., Final Technical Report, Air Force Aeronautical Systems Division, ASD-TDR-63-549, 1963, AD 429 116.
12. R. L. Barron et al., Self-Organizing Spacecraft Attitude Control, Adaptronics, Inc., Final Technical Report, Air Force Flight Dynamics Laboratory, AFFDL-TR-65-141, 1965, AD 475 167.
13. Anonymous, Design, Fabrication, and Flight Testing of Self-Organizing Flight Control System, Adaptronics, Inc., Final Technical Report, Air Force Flight Dynamics Laboratory, AFFDL-TR-70-77, June 1970.
14. R. L. Barron, Adaptive Flight Control Systems, in Principles and Practice of Bionics, AGARD Conf. Proc. 44 (1968), Technivision Services, Slough, England, 1970, pp. 119-167.
15. R. L. Barron, Self-Organizing Control of Aircraft Pitch Rate and Normal Acceleration, Adaptronics, Inc., Final Technical Report, Air Force Flight Dynamics Laboratory, AFFDL-TR-66-41, 1966, AD 801 157.

16. R. L. Barron, Self-Organizing and Learning Control Systems, AD 811 244, in Cybernetic Problems in Bionics (1966 Bionics Symposium), Gordon and Breach, London, 1968, pp. 147-203.

17. R. L. Barron et al., Analysis and Synthesis of Advanced Self-Organizing Control Systems, Adaptronics, Inc., Final Technical Report, Air Force Avionics Laboratory, AFAL-TR-67-93, Apr. 1967, AD 813 918.

18. R. L. Barron and R. M. McKechnie III, Design Principles for Self-Organizing Control System Flight Hardware, Proc. 19th Ann. NAECON, May 1967, pp. 465-473.

19. R. L. Barron et al., Synthesis of a Spacecraft Probability State Variable Adaptive Control System, Adaptronics, Inc., Final Project Report, NASA Goddard Space Flight Center, June 1967.

20. R. L. Barron, Self-Organizing Controller (Mark V), U.S. Patent No. 3,519,998, Re. 671, 743, July 7, 1970.

21. R. L. Barron, Self-Organizing Control: The Next Generation of Controllers, Control Eng., Part I: The Elementary SOC, Feb. 1968, pp. 70-74; Part II: The General Purpose SOC, Mar. 1968, pp. 69-74.

22. R. L. Barron, Analysis and Synthesis of Advanced Self-Organizing Control Systems—II, Adaptronics, Inc., Final Technical Report, Air Force Avionics Laboratory, AFAL-TR-68-236, Sept. 1968, AD 840 295.

23. R. L. Barron, Self-Organizing Controller with Constrained Performance Assessment, U.S. Patent No. 3,591,778, Nov. 5, 1968.

24. R. L. Barron et al., Application of Self-Organizing Control Techniques to High-Performance Missiles, Adaptronics, Inc., Final Technical Report, Naval Air Systems Command, NASC, Apr. 1969, AD 852 869.

25. R. L. Barron et al., Self-Organizing Control of Advanced Turbine Engines, Adaptronics, Inc./Hamilton Standard Division of United Aircraft Corporation, Final Technical Report, Air Force Aeropropulsion Laboratory, AFAPL-TR-69-73, Aug. 1969, AD 857 616.

26. R. L. Barron, Self-Organizing Controller (Mark III), U.S. Patent No. 3,460,096, Aug. 5, 1969.

27. D. Cleveland, R. L. Barron, et al., Research and Development on Self-Organizing Control Systems for Air Launched Missiles, Adaptronics, Inc., Final Technical Report, Naval Air Systems Command, Apr. 1970, AD 867 918.

28. R. L. Barron and J. R. Gouge, Jr., Redundant Self-Checking, Self-Organizing Control System, U.S. Patent No. 3,593,307, July 13, 1971.

29. Anonymous, Application of Self-Organizing Control to Remotely Piloted Vehicles, Adaptronics, Inc., Final Technical Report, Air Force Aeronautical Systems Division, ASD XR-72-19, Apr. 1972.

30. R. L. Barron, K. S. Kelleher, and G. C. Vieth, Jr., Self-Programming Antenna Tracking System, U.S. Patent No. 3,680,126, July 25, 1972.

31. R. L. Barron and R. A. Gagnon (Major, USAF), Application of Self-Organizing Control to Remote Piloting of Vehicles, Remotely Manned Systems, Proc. NASA/Cal. Tech. First Natl. Conf. Remotely Manned

Systems, E. Heer (Ed.), Sept. 13-15, 1972, Pasadena, Calif.,
pp. 409-422. (Publ. 1973)

32. D. Cleveland, R. L. Barron, J. R. Binkley, Jr., and L. O. Gilstrap,
Jr., RPV/Self-Organizing Control Demonstration System, Vol. I: SOC
Equation Development, Logic Configurations, and Control Modes;
Vol. II: Hardware Description, System Operation and Maintenance,
and RPV Simulation, Adaptronics, Inc., Final Technical Report, Aero-
space Medical Research Laboratory, AMRL TR-73-66, June 1973.

33. R. L. Barron and D. Cleveland, Self-Organizing Control System, U.S.
Patent No. 3,794,271, Feb. 26, 1974.

34. R. F. Snyder, R. L. Barron, et al., Advanced Computer Concepts for
Intercept Prediction, Vol. I: Conditioning of Parallel Networks for
High-Speed Prediction of Re-entry Trajectories, Adaptronics, Inc.,
Final Technical Report, Army Nike-X Project Office, Redstone Arsenal,
Ala., Nov. 1964.

35. L. O. Gilstrap, Jr., An Adaptive Approach to Smoothing, Filtering,
and Prediction, Proc. 1969 NAECON, Dayton, Ohio, May 1969,
pp. 275-280.

36. L. O. Gilstrap, Jr., Keys to Developing Machines with High-Level
Artificial Intelligence, ASME Paper 71-DE-21. (Presented at ASME
Design Eng. Conf., New York, Apr. 19-22, 1971.)

37. R. L. Barron, Inference of Vehicle and Atmosphere Parameters from
Free-Flight Motions, AIAA J. Spacecraft Rockets $6(6):641-648$ (1969).
(Paper presented at AIAA Guidance, Control Flight Dyn. Conf.,
Huntsville, Ala., Aug. 1967.)

38. S. H. Brooks, A Discussion of Random Methods for Seeking Maxima,
Oper. Res. $6:244-251$ (1958).

39. J. Matyas, Random Optimization, Avtomatika Telemekhanika (Automation and Remote Control) $26:246-253$ (1965).

40. A. N. Mucciardi, Neuromime Nets as the Basis for the Predictive
Component of Robot Brains, Cybernetics, Artificial Intelligence, and
Ecology, H. W. Robinson and D. E. Knight (Eds.), Spartan Books,
Bensalem, Pa., 1972, pp. 159-193. (Presented at 4th Ann. Symp. Am.
Soc. Cybern., Washington, D.C., Oct. 1970.)

41. L. Devroye, A Bibliography on Random Search, Technical Report
SOCS-79.9, McGill University, Montreal, May 1979.

42. R. L. Barron, Adaptive Transformation Networks for Modeling, Prediction, and Control, Proc. IEEE/ORSA Joint Natl. Conf. Major Syst.,
Anaheim, Calif., Oct. 25-26, 1971.

43. R. L. Barron, Theory and Application of Cybernetic Systems: An
Overview, Proc. IEEE 1974 Natl. Aerosp. Electron. Conf., May
1974, pp. 107-118.

44. A. G. Ivakhnenko, Polynomial Theory of Complex Systems, IEEE
Trans. Syst. Man Cybern. $SMC-1(4):364-378$ (1971).

45. A. G. Ivakhnenko, G. J. Krotov, and V. N. Visotsky, Identification of
the Mathematical Model of a Complex System by the Self-Organization

Method, Theoretical Systems Ecology, Academic Press, New York, 1969, pp. 325-352.

46. R. L. Barron, Learning Networks Improve Computer-Aided Prediction and Control, Comput. Des., Aug. 1975, pp. 65-70.

47. A. N. Mucciardi and E. E. Gose, An Automatic Clustering Algorithm and Its Properties in High-Dimensional Spaces, IEEE Trans. Syst. Man Cybern. SMC-2(2):247-254 (1972).

48. A. N. Mucciardi, Information Filtering Using the CLUSTR Algorithm, Proc. Computer Image Processing and Recognition Conf., Univ. of Missouri, Columbia, 1972, Vol. 2, pp. 15-3-1 to 15-3-8.

49. R. L. Barron, Three Approaches for Extrapolation with Black Box Models in Process Control Systems, Proc. 11th Conf. on Use of Digital Computers in Process Control, Louisiana State University, Baton Rouge, La., Feb. 25-27, 1976.

50. A. N. Mucciardi, New Computational Techniques in the Evaluation of Drug-Induced EEG Changes, Psychotropic Drugs and the Human EEG: Modern Problems in Pharmacopsychiatry, Vol. 8, Turan M. Itil (Ed.), Karger, Basel, 1974, pp. 350-377.

51. A. N. Mucciardi, Self-Organizing Probability State Variable Parameter Search Algorithms for Systems That Must Avoid High-Penalty Operating Regions, IEEE Trans. Syst. Man Cybern. SMC-4(4):350-362 (1974).

52. H. Akaike, Statistical Predictor Identification, Ann. Inst. Stat. Math. 22:203-217 (1970).

53. H. Akaike, Information Theory and an Extension of the Maximum Likelihood Principle, in Proceedings of the Second International Symposium on Information Theory, B. N. Petrov and F. Csaki (Eds.), Akadémia Kiádo, Budapest, 1972, pp. 267-281.

54. A. R. Barron, Properties of the Predicted Square Error: A Criterion for Selecting Variables, Ranking Models, and Determining Order, Adaptronics, Inc., McLean, Va., 1981.

55. C. L. Mallows, Some Comments on C_p, Technometrics 15:661-675 (1973).

56. J. Bibby and H. Toutenburg, Prediction and Improved Estimation in Linear Models, Wiley, New York, 1970, pp. 13-15.

57. F. J. Cook and J. N. Craig, Adaptive Learning Networks and Image Processing for Missile Guidance, Proc. Soc. Photo-Optical Instrum. Eng., Vol. 238: Image Processing for Missile Guidance, 1980, pp. 293-301.

58. R. L. Barron, Adaptive Learning Network Algorithms: Bringing a New Innovation to Market, seminar presentation to Rensselaer Polytechnic Institute, Oct. 14, 1982.

59. P. Horvath and F. J. Cook, Scene Classification of Landsat Multispectral Scanner Data by Means of the Adaptive Learning Network Methodology, Proc. IEEE Comput. Soc. Conf. on Pattern Recognition and Image Processing, Aug. 1981, pp. 473-477.

60. F. J. Cook, Analysis of LANDSAT Radiance Variance Sets, Adaptronics, Inc., Final Technical Report, Army Engineer Waterways Experimental Station, Contract DACW39-78-M-4955, Dec. 1978.

61. J. N. Craig, F. J. Cook, and M. F. Whalen, A Priori Training of Guidance and Control Algorithms for Tactical Missiles, Task II—Target Acquisition and Aim-Point Selection, Adaptronics, Inc., Final Technical Report, Air Force Armament Technology Laboratory, AFATL-TR-80-95, Aug. 1980.

62. J. N. Craig, M. F. Whalen, and F. J. Cook, Target Recognition for Missile Guidance Using Adaptive Learning Networks, Proc. Soc. Photo-Optical Instrum. Eng., Vol. 238: Image Processing for Missile Guidance, 1980, pp. 309-315.

63. T. L. Riggs, Jr., Linear Optimal Guidance for Short Range Air-to-Air Missiles, Proc. IEEE 1979 Natl. Aerosp. Electron. Conf., p. 757.

64. J. N. Craig, R. L. Barron, and F. J. Cook, A Priori Training of Guidance and Control Algorithms for Tactical Missiles, Task I—Air-to-Air Guidance Law Implementation, Adaptronics, Inc., Final Technical Report, Air Force Armament Technology Laboratory, AFATL-TR-80-102, Sept. 1980.

65. J. N. Craig, R. L. Barron, and F. J. Cook, A New Class of Guidance Laws for Air-to-Air Missiles, 3rd Meet. Coordinating Group on Modern Control Theory, Redstone Arsenal, Ala., Oct. 1981.

66. F. A. Bick, M. Inbar, F. J. Cook, and F. J. Kline, The Utilization of Emerging Technologies in Physical Security Systems: Phase I Status Report—System Feasibility, Effects Technology, Inc./Adaptronics, Inc. Draft Final Technical Report for Defense Nuclear Agency, Contract DNA001-80-C-0271, Nov. 1980.

67. F. J. Kline and F. J. Cook, The Utilization of Emerging Technologies in Physical Security Systems: Adaptive Learning Network Performance Demonstration—Phase II, Adaptronics, Inc., Draft Final Technical Report for Effects Technology, Inc., Nov. 1980.

68. A. R. Hunt, F. J. Cook, G. S. Lapman, A. H. Sanders, and F. E. Lanham, Development of a Distributed, Adaptive, Intrusion Detection System: Hardware and Software Operation, Proc. Conf. on Crime Countermeasures and Security Technology, May 11-13, 1983.

69. F. J. Kline, F. J. Cook, A. R. Hunt, and F. E. Lanham, Development of a Distributed, Adaptive, Intrusion Detection System: Algorithm Operation and Field Test Results, Proc. Conf. on Crime Countermeasures and Security Technology, May 11-13, 1983.

70. P. Horvath and F. J. Cook, Establishing Signal Processing and Pattern Recognition Techniques for Inflight Discrimination Between Crack-Growth Acoustic Emission and Other Acoustic Waveforms, in Review of Progress in Quantitative Nondestructive Evaluation, D. O. Thompson and D. E. Chimenti (Eds.), Plenum Press, New York, 1982, pp. 463-473.

71. R. Shankar, A. N. Mucciardi, D. Cleveland, W. E. Lawrie, and H. L. Reeves, Adaptive Nonlinear Signal Processing for Characterization of Ultrasonic NDE Waveforms; Task 2: Measurement of Subsurface Fatigue Crack Size, Adaptronics, Inc., Final Technical Report, Air Force Materials Laboratory, AFML-TR-76-44, Apr. 1976.

72. M. H. Loew, J. M. Fitzgerald, A. N. Mucciardi, R. K. Elsley, and G. A. Alers, Exploratory Development of Adhesive Bond Flaw Detection, Adaptronics, Inc., Final Technical Report, Air Force Materials Laboratory, AFML-TR-78-206, Dec. 1978.

73. A. N. Mucciardi, ALN 4000 Ultrasonic Pipe Inspection System, in Nondestructive Evaluation Program: Progress in 1981, Electric Power Research Institute Report NP-2088-SR, Jan. 1982.

74. M. F. Whalen et al., Advanced Signal Processing of Turbine Rotor Bore Waveforms, Adaptronics, Inc., Final Technical Report, Electric Power Research Institute, Contract RP502-8, Aug. 1981.

75. D. Cleveland, B. Decina, J. M. Jamieson, and A. N. Mucciardi, Development of Adaptive Learning Network Models to Predict the Tensile Shear Strengths of Resistance Spot Welds, Adaptronics, Inc., Final Technical Report, General Motors Corp., Contract MD 874157, Sept. 1981.

76. Anonymous, ALN 5000 Final Software Documentation for the 4M System[T.M.], June 1982; 4M System[T.M.] Users Manual, June 1982; 4M System[T.M.] Hardware Documentation, June 1982; Adaptronics, Inc. for Reynolds Metals Company.

77. J. N. Craig and F. J. Cook, Single Particle Erosion Modeling Using Adaptive Learning Networks, Adaptronics, Inc., Final Technical Report, Defense Advanced Research Projects Agency, Contract MDA903-80-C-0291, 1981.

78. J. N. Craig and F. J. Cook, Application of ALN Modeling to Radiation Embrittlement, Adaptronics, Inc., Final Technical Report, Electric Power Research Institute, Contract RP1553-2, 1981.

3

Regression Without Models: Directions in the Search for Structure

JAMES M. MALONE II/Innovative Information Technology, Schenectady, New York

I. INTRODUCTION

One can safely say that regression analysis is the study of what a sample of values of a dependent variable y associated with various combinations of values of some independent variables x has to say about the relationship of y to the x's. (We shall later give a more general and more precise definition equally different from the one found in elementary texts.) If we have no other information, except perhaps some idea of the "smoothness" of the relationship, and do not in any way limit our study a priori, this is regression without models (or nonparametric regression).

We are interested in regression because it is useful. It is one of the "big guns" of scientific inquiry (Clough, 1978) and hence a revolution in regression analysis implies a revolution in science. We are interested in regression without models, or in not assuming specific functional forms, because our most extensive experience, which is with one independent variable, convinces us that a variety of noninterchangeable functional forms abounds in real-world relationships.

Regression without models is a neglected area in which many tools are immediately available for the practitioner's use. One such tool is Ivakhnenko's group method of data handling (GMDH) and it provides a convenient focus for a broad overview of the subject. We shall first briefly suggest why it is a neglected area (very little concerning it ever having appeared in the methodological literature despite important developments as long as 20 years ago—although the need for such developments has sometimes been cited!) and then survey the state of the art as accurately as we can. The scope of the subject and limits of space dictate that the survey must be painfully succinct.

II. REGRESSION WITHOUT MODELS:
DATA ANALYSIS AND STATISTICS

Attitude toward data analytic methodologies is shaped not by mathematics but by world view. The frequentist does not reject the Bayesian's deductions, but rather the latter's belief that prior knowledge is commonplace. The advocate of regression without models holds that total ignorance of structure is commonplace, and must base his or her belief on experiences and associations vastly different from those of the more arrogant Bayesian. We would claim that our age is not so enlightened, and that a majority of phenomena and processes, far from being understood, have never been seriously examined. To see that this is so even in "elementary" physics, one need but browse through Jearl Walker's "Amateur Scientist" column appearing for a number of years in Scientific American.

The frontiers of knowledge, like other frontiers, are unappealing to those used to the amenities of more civilized places, explored first because they were more accessible. These frontiers can only be reached now because the computer is the key that has unlocked a kit of new exploratory tools. Those with a vested interest in sharpening the old tools are not enthusiastic about the new ones. Much of statistical practice is less a consequence of logic than of history—a history of development dependent on hand calculation. When the computer appeared, statistics divided into two unfriendly camps: those happy at being able to do what they had been doing on a more ambitious scale, and those happy at being able to cease what they had been doing in favor of doing what they thought more appropriate. The colloquial names of these camps are statistics and data analysis. The former has the advantage of numbers.

Nonparametric regression has tended to be excluded from statistics because randomness plays a nontraditional role in it—in the sense that the nature of the pattern recognition problem is scarcely altered when all measurements are error-free. Yet application is statistical (in general to conditional distributions, as we shall see), terminology is statistical, and the ideas underlying tactics are largely statistical. There is also an experimental design problem (for what patterns of independent variable or factor levels do the methods work well?) and it is unexpectedly useful to view the design matrix as stochastic (Brillinger, 1977, 1982).

The quest by data analysts for more elaborate—and accurate—representations of reality requires tolerance of complexity but not, as sometimes alleged, disaffection with regard to simplicity. Simplicity is to be sought where it is to be found (but not by drawing blood with Occam's razor). In contemplating the techniques described below, one should be alert to observe their strong capabilities for eliciting parsimonious models compared with traditional methods.

Finally, nonparametric regression is long on heuristic and short on rigor (as exploratory data analysis perhaps has to be). Its motto might be the same as Picasso's: "If one knows where one is going, why go?"

III. APPROACHES TO THE SEARCH FOR STRUCTURE

A quite general definition of regression analysis is construction of a description of the distribution of a response variable conditional on the values of several predictor variables. This definition requires extension if a priori relationships are to be maintained between descriptions constructed for more than one response, such as when the responses are not conditionally independent, or if a composite criterion of joint fit of several such distributions is being optimized. The conditional distribution may be represented by a parametric form or by descriptors such as moments. Parameters and descriptors are assumed to be nonrandom functions of the predictors, so that search for the structural form of functional dependence for each is the essence of the analysis. We are particularly concerned with situations where nothing is known about the functional forms.

A. Regression from Nonparametric Density Estimation

If the predictors are random variables, the conditional distribution of the responses may be derived from the joint distribution of responses and predictors. One contemporary approach is to approximate the joint distribution by nonparametric density estimators. There are two notable variants of this approach.

Let z_1, z_2, $\ldots$, z_n be a size n random sample of the vector variable z, whose components include predictors and responses of interest. Let $d(z, z_i)$ be a measure of distance between an arbitrary z and an observed z_i. Let $H(t)$ be a monotonically decreasing function of t such that $H(d(z, z_0))$ is a density function where z_0 is arbitrary. Then a kernel function estimator of the density of z is

$$\hat{\delta}_H(z) = \frac{1}{n} \sum_{i=1}^{n} H(d(z, z_i))$$

Let k be an index vector belonging to an index set K, and $\{\phi_k(z), k \in K\}$ a set of functions complete in a suitable topology and orthogonal with respect to a suitable inner product. Then a Fourier series estimator of the density of z is

$$\hat{f}_\phi(z) = \sum_{k \in K} \hat{a}_k \phi_k(z)$$

where $\hat{a}_k$ is the Fourier coefficient estimated from the sample values z_1, z_2, $\ldots$, z_n. This estimator may be refined by dropping terms whose coefficients are deemed not significant.

Such nonparametric density estimators are compared and discussed at length and numerous references are given in the excellent paper by Scott and Factor (1981).

B. Identification of Nonlinearly Related
    Linear Combinations

1. The Method of Parallel Tangents

We now focus on the search for a single functional form relating y to predictors x_1, x_2, $\ldots$, x_p. We agree to refer to y as the response, although
it may be a parameter or descriptor of a conditional distribution in the
original regression problem, and we often ignore any lack of precision in
our knowledge of y.

An important strategy is to reduce the dimension of the predictor space
by identifying <u>predictive components</u> which are linear combinations

$$u_i = \sum_{j=1}^{p} a_{ij} x_j, \quad i = 1, 2, \ldots, m < p$$

such that $y = f(u_1, u_2, \ldots, u_m)$, where the form of the function f is supposed
entirely unknown. Assuming differentiability of f,

$$\frac{\partial y}{\partial x_j} = \sum_{i=1}^{m} a_{ij} \frac{\partial f}{\partial u_i}, \quad j = 1, 2, \ldots, p$$

Thus the $\partial y / \partial x_j$ have an m-dimensional basis in function space. For <u>any</u>
such basis,

$$\frac{\partial y}{\partial x_j} = \sum_{i=1}^{m} a'_{ij} v_i$$

we wish to show that y is a function of $u'_i = \Sigma_j a'_{ij} x_j$. But

$$dy = \sum_{j=1}^{p} \left(\sum_{i=1}^{m} a'_{ij} v_i \right) dx_j = \sum_{i=1}^{m} v_i \left(\sum_{j=1}^{p} a'_{ij} dx_j \right) = \sum_{i=1}^{m} v_i \, du'_i$$

which is the result required.

Now choose a set S of $m' \geq m$ points in predictor space and let y_{kj} be
the value of $\partial y / \partial x_j$ at the kth point of S. Suppose that the m' vectors $(y_{k1},
y_{k2}, \ldots, y_{kp})$ span a space of dimension m and have a representation

$$y_{kj} = \sum_{i=1}^{m} a'_{ij} v_{ki}$$

The v_{ki} are, of course, linear combinations of the y_{kj}. It is claimed that the <u>same</u> linearly independent combinations of the $\partial y/\partial x_j$ may be taken to be the v_i. That is, the representation in S may be extended to the function space, and hence the a'_{ij} may be taken to be the coefficients of the predictive components.

Next, perform classical multiple linear regressions of y on x_1, x_2, ..., x_p for m' subsets of data points (complete observations). The <u>parallel tangent heuristic</u> assumes that there exists for each subset a hyperplane tangent to the response surface $y = y(x_1, x_2, \ldots, x_p)$ and parallel to the regression hyperplane. At any point of tangency, $\partial y/\partial x_j = b_j$ the regression coefficient of x_j. By taking a point of tangency for each regression to be a point of S, we complete the outline of a method for computing the a_{ij} (Malone, 1965).

Now for details of a specific approach toward constructing the basis for the y_{kj}. Suppose by analogy with principal components that we wish to choose the first basis vector

$$v_{k1} = \sum_{j=1}^{p} c_{1j} y_{kj}, \quad \sum_{j=1}^{p} c_{1j}^2 = 1$$

to

$$\text{maximize} \sum_{k} w_k v_{k1}^2 = \sum_{k,j,j'} w_k y_{kj} y_{kj'} c_{1j} c_{1j'}$$

for suitable positive weights w_k. This leads immediately to the symmetric eigenvalue problem

$$\sum_{j'=1}^{p} \left(\sum_{k} w_k y_{kj} y_{kj'} \right) c_{1j'} = \lambda_1 c_{1j}, \quad j = 1, 2, \ldots, p$$

From this, considering all nonzero eigenvalues (by assumption there are exactly m), is derived a complete set of orthogonal basis (eigen)vectors. Then we must have

$$\sum_{i=1}^{m} c_{ij} v_{ki} = \sum_{i,j'} c_{ij} c_{ij'} y_{kj'} = \sum_{j'=1}^{p} \delta_{jj'} y_{kj'} = y_{kj}$$

so the c_{ij} are the required a_{ij}.

(It can be shown that this solution is invariant under orthogonal transformations in the p-dimensional predictor space. That is, if a unique solution

is desired, the original predictors should first be transformed to a normalized orthogonal basis before carrying out the analysis.)

Next, suppose that for each linear regression the response is measured
with independent homeoscedastic Gaussian errors. We might attempt to
choose the weights w_k to moderate errors in estimating the matrix elements
$\Sigma_k w_k y_{kj} y_{kj'}$. Ideally for this purpose, as is well known, we would wish the
w_k inversely proportional to the variance of $y_{kj} y_{kj'}$. The latter may be seen
to be proportional to the square of the error variance provided that differences in the m' regression design matrices can be ignored. Thus if homeo-
scedasticity subsists within but not between regressions, estimates of error
variance may be utilized in a simple albeit crude "rule of thumb" for assigning weights. Among such estimates is one assuming the error variance to
be a constant fraction of the residual variance.

2. Brillinger's Result for a Single Component

The principal issue left unresolved in the preceding discussion is when the
parallel tangent heuristic is valid. Brillinger (1977, 1982) has in essence
shown for the case m = m' = 1 when predictor values are sampled from a
multivariate normal distribution that identification of the predictive component is consistent, that is, becomes exact almost surely as the sample
size increases without limit. The slightly stronger parallel tangent property
then follows from Rolle's theorem.

This limited and rather curious result leaves open difficult questions,
some theoretical, others practical, concerning the effectiveness of the
heuristic. It has, however, been observed to perform well. [Actually, the
essential feature of the predictor (design) distribution in Brillinger's (1977)
argument is ability to deduce independence from null correlation. Brillinger
(1982) ignores this point.]

3. The Method of Optimizing Nonlinear Association

An alternative approach to identification of predictive components is estimation of the coefficients by optimizing a measure of nonlinear association
between y and u_i. (By nonlinear association between y and u, we mean in
the most general sense the ability to predict one by some function of the
other.) We therefore digress momentarily for the purpose of examining
some nonlinear association measures.

Suppose that $y = y(\underline{x})$ is a smooth function of vector $\underline{x}$. Let $d_x(\underline{x}_k, \underline{x}_\ell)$ be
a measure of the distance between points $\underline{x}_k$ and $\underline{x}_\ell$ in x space. We shall
always assume that $\underline{x}_k$ and $\underline{x}_\ell$ are distinct for $k \neq \ell$. Let $\bar{y}$ be the mean and
s_y^2 be the variance of n values of $y_k = y(\underline{x}_k)$, $k = 1, 2, \ldots, n$. Two interesting measures of association of y with $\underline{x}$ are

$$I_{y:x} = \sum_{k \neq \ell} \frac{(y_k - \bar{y})(y_\ell - \bar{y})}{d_x(\underline{x}_k, \underline{x}_\ell)}$$

$$C_{y:x} = \frac{1}{2} \sum_{k \neq \ell} \frac{(y_k - y_\ell)^2}{d_x(\underline{x}_k, \underline{x}_\ell)}$$

With appropriate normalization factors involving n, s_y^2, and $\Sigma_{k \neq \ell}[1/d_x(\underline{x}_k, \underline{x}_\ell)]$, these are called generalized Moran's I and generalized Geary's C. Informally, one expects I to become large and C to become small as smoothness of the functional relationship increases, becasue the sums are dominated by terms involving y_k, y_ℓ for nearby $\underline{x}_k$, $\underline{x}_\ell$. It is easy to see that we may write either I or C [and also the "correlation ratio" for y grouped by $\underline{x}$] in the form

$$K_{y:x} = \sum_{k, \ell} b_{k\ell} y_k y_\ell$$

where $b_{k\ell} = b_{\ell k}$ depends only on the values of the $\underline{x}_k$ (and $\Sigma_\ell b_{k\ell} = 0$).

Here we are interested in smoothness of relationship of a response y to a function $u(\underline{x}, \underline{\theta})$ where $\underline{x}$ is a predictor vector and $\underline{\theta}$ is a vector of parameters to be estimated. We could proceed by estimating $\underline{\theta}$ to optimize $K_{y:u}$. However, if y is monotonically related to u, we might alternatively optimize $K_{u:y}$, which in practice is computationally simpler. (We might hope that optimizing $K_{u:y}$ will still be effective even when y is not monotonically related to u.)

Write $u_k(\theta) = u(\underline{x}_k, \underline{\theta})$. We shall without loss of generality assume that the parameter vector is subject to a constraint $h(\underline{\theta}) = 0$.

To proceed, we shall iteratively replace a current estimate of $\underline{\theta}$ by an improved estimate $\underline{\theta} + \delta\underline{\theta}$ using the Taylor series approximations

$$u_k(\underline{\theta} + \delta\underline{\theta}) = u_k(\underline{\theta}) + \sum_j u_{kj}(\underline{\theta}) \delta\theta_j$$

$$h(\underline{\theta} + \delta\underline{\theta}) = h(\underline{\theta}) + \sum_j h_j(\underline{\theta}) \delta\theta_j$$

where $u_{kj}(\theta) = \partial u_k(\underline{\theta})/\partial\theta_j$ and $h_j(\theta) = \partial h(\underline{\theta})/\partial\theta_j$. Then we obtain first-order conditions for an optimum K which are linear in $\delta\underline{\theta}$ and a Lagrange multiplier λ:

$$\sum_j \left(\sum_{k, \ell} b_{k\ell} u_{ki} u_{\ell j} \right) \delta\theta_j - \lambda h_i = - \sum_{k, \ell} b_{k\ell} u_{ki} u_\ell$$

$$\sum_j h_j \delta\theta_j = -h$$

This might be called a nonparametric form of nonlinear estimation. (In the iteration we are essentially seeking predictive components where the predictors are u and the $\partial u / \partial \theta_j$.)

For ordinary predictive component analysis $\underline{\theta} = (c_{11}, c_{12}, \cdots, c_{1p})$,
$u(\underline{x}, \underline{\theta}) = \Sigma_{j=1}^{p} c_{1j} x_j$. Let $h(\underline{\theta}) = \frac{1}{2}(\Sigma_j c_{1j}^2 - 1)$ and $\underline{x}_k = (x_{k1}, x_{k2}, \cdots, x_{kp})$.
Then it is easy to see that the c_{1j} satisfy the symmetric eigenvalue problem

$$\sum_j \left(\sum_{k, \ell} b_{k\ell} x_{ki} x_{\ell j} \right) c_{1j} = \lambda_1 c_{1i}$$

(This is, as before, invariant under orthogonal transformations in the predictor space.)

Use of $K_{y:x}$ to test the hypothesis that y is independent of $\underline{x}$ is considered at length in Cliff and Ord (1973). $C_{y:x}$ was used to obtain predictive components in Carroll and Chang (1964), and for "parametric mapping" in Shepard and Carroll (1966).

Next write

$$y_k^x = \frac{\Sigma_{\ell \neq k} b_{k\ell} y_\ell}{\Sigma_{\ell \neq k} b_{k\ell}}$$

and observe that y_k^x may be regarded as an estimate of y at $\underline{x}_k$ smoothed or filtered by $\underline{x}$ (i.e., a weighted average dominated by y values for $\underline{x}$'s near $\underline{x}_k$). Also notice that

$$K_{y:x} = \sum_k \left(\sum_{\ell \neq k} b_{k\ell} \right) y_k y_k^x + \sum_k b_{kk} y_k^2$$

so that if the final term is constant, $K_{y:u}$ is optimized by y_k^u's, which maximize a weighted product moment correlation between y_k and y_k^u.

Some alternative procedures where u is a predictive component are immediately apparent. One may use any (in general nonlinear) filter to obtain the y_k^u and ordinary (unweighted) Galton-Pearson correlation. This is projection pursuit regression (Friedman and Stuetzle, 1981). It is possible to obtain several predictive components one at a time by requiring each to be orthogonal to those preceding or by working with "residuals" from the previous fit, either $y_k - y_k^u$ or $y_k - f(u)$. It is not obvious that such a stepwise approach will prove effective, but results reported by Friedman and Stuetzle are encouraging.

Finally, what of ordinary correlation (perhaps with differential weighting of observations) as a measure of nonlinear association between y and u? This is, in fact, the simplest case $m = m' = 1$ of the method of parallel tangents, investigated by Brillinger.

To summarize the state of predictive components analysis, successful procedures have been discovered independently by Malone, Carroll, Brillinger, and Friedman, with Cliff and Ord shedding further light on the ideas underlying Carroll's work. This survey of the subject is believed to be the first to appear in print.

4. Kolmogorov Representation Theorem

In passing we record the

Theorem (Kolmogorov-Sprecher-Kahane). Every continuous function $f(x_1, x_2, \ldots, x_p)$ has a representation

$$f(x_1, x_2, \ldots, x_p) = \sum_{i=1}^{2p+1} g\left(\sum_{j=1}^{p} a_j x_j^{(i)} \right)$$

where g is continuous and $x_j^{(i)}$ denotes a continuous increasing function of x_j. "Most selections of $(2p + 1)$-tuples $x_j^{(i)}$ will do."

This remarkable theorem is an offspring of the "thirteenth problem of Hilbert," and is discussed at length by Lorentz (1976).

Lorentz says: "I do not know of any applications of Kolmogorov's theorem." Nor do we. However, it has sufficient kinship with predictive components analysis to suggest directions for future development.

C. Further Applications of Parallel Tangent Methods

Consider the model

$$y = f(x_1) + g(x_2, x_3, \ldots, x_p)$$

where the forms of the functions f and g are again entirely unknown. Then

$$\frac{\partial y}{\partial x_1} = f'(x_1)$$

We again perform m' multiple linear regressions on data subsets and assume that in each case there is a hyperplane parallel to the regression hyperplane and tangent to the response surface. At the point of tangency for the kth regression $\partial y/\partial x_1 = b_{1k}$, the regression coefficient of x_1. We now further assume that the point of tangency is approximately at the centroid of the data subset, so that $b_{1k} \approx f'(\bar{x}_{1k})$, where $\bar{x}_{1k}$ is the mean of x_1 in the kth subset. Therefore, plotting the points $(\bar{x}_{1k}, b_{1k})$, we determine the form of f', hence of f, and finally $g(x_2, x_3, \ldots, x_p) = y - f(x_1)$, so that we have a residual problem in one less predictor.

Notice that if we apply this approach to

$$y = \sum_{j=1}^{p} f_j(x_j)$$

we get the points $(\bar{x}_{jk}, b_{jk})$ for determining f'_j simultaneously for all $j = 1, 2, \ldots, p$.

Another model among several which yield to similar treatment is

$$y = x_1 f(x_2, x_3, \ldots, x_p) + g(x_2, x_3, \ldots, x_p)$$

$$\frac{\partial y}{\partial x_1} = f(x_2, x_3, \ldots, x_p)$$

A rather special form is

$$y = f_1(x_1) + f_2(x_2) + f_3(x_3) + f_{12}(x_1 x_2) + f_{13}(x_1 x_3) + f_{23}(x_2 x_3)$$

Here

$$b_1 \bar{x}_1 - b_2 \bar{x}_2 - b_3 \bar{x}_3 \approx \bar{x}_1 f'_1(\bar{x}_1) + \text{terms in } \bar{x}_2 \text{ and } \bar{x}_3$$

This is of the first form above in the "barred" predictors with $f(\bar{x}_1) = \bar{x}_1 f'_1(\bar{x}_1)$, so it is clear how f_1, and similarly f_2 and f_3, might be determined. This leaves a residual problem of the form

$$y = f_{12}(x_1 x_2) + f_{13}(x_1 x_3) + f_{23}(x_2 x_3)$$

where

$$b_1 \bar{x}_1 + b_2 \bar{x}_2 - b_3 \bar{x}_3 \approx 2 f_{12}(\bar{x}_1 \bar{x}_2)$$

but for this form it is better to make a change of variables to $x_1 x_2$, $x_1 x_3$, $x_2 x_3$.

D. The Matrix Factor Method

Let a function of two predictors be described by a complete two-way table or matrix of observed responses

$$y_{i_1 i_2} = y(x_{1 i_1}, x_{2 i_2}), \quad i_j = 1, 2, \ldots, p_j, \quad j = 1, 2$$

It is well known that this matrix may be canonically decomposed or factored:

$$y_{i_1 i_2} = \sum_{k=1}^{m} u_{1ki_1} u_{2ki_2} = \sum_{k=1}^{m} u_{1k}(x_{1i_1}) u_{2k}(x_{2i_2}), \quad m \leq \min{(p_1, p_2)}$$

where each u_{jk} is a function of a single predictor. This idea perhaps first proposed in this context by Tucker (1958) is developed extensively in Mandel (1971). It is recommended that the reader consult the latter reference. Where there are missing values in the table, it is often possible to determine the u_{jki_j} by "alternating least squares" (Carroll and Chang, 1970;

Frisen, 1979; Gabriel and Zamir, 1979). This approach is trivially extended to the case where the predictors are a pair of vector variables, and can be iteratively applied to a multiway table to decompose the functional relation into a sum of products of functions of single (scalar) variables (Carroll and Chang, 1970).

The approach is generally useful only if m is relatively small and the functions u are "smooth." The problem of identifying a felicitous decomposition (noticed at the outset by Tucker) is aggravated by the "rotational indeterminacy" of factor analysis. In matrix notation if $Y = U_1 U_2$, then also $Y = V_1 V_2$, where $V_1 = U_1 M$, $V_2 = M^{-1} U_2$, and M is any nonsingular m × m matrix. The present writer has examined this issue in an unpublished working paper, and it has been attacked in a rather different manner by Cornelius et al. (1979). However, more work is needed.

E. Iterative Partitioning Methods: POP and PEP

Another class of approaches to the search for structure, useful for response prediction and for evaluation of explanatory power of predictors, is the iterative partitioning or prediction-oriented partitioning (POP) methods (Fielding, 1977), the first of which was the automatic interaction detector (Morgan and Sonquist, 1963; Sonquist et al., 1971). Extensions have been reported by Gillo (1972), Messenger and Mandell (1972), Morgan and Messenger (1973), Gillo and Shelly (1974), Kass (1980), and Perreault and Barksdale (1980). A more general paradigm of which all these are special cases will now be described. It accommodates multiple responses.

Initially, one has a single group of observations of responses and predictors. At any stage one chooses an as yet "unsplit" group whose responses are most heterogeneous, and identifies that predictor by dichotomizing the levels of which two new groups with most dissimilar responses are created. Thus at each stage an unsplit group is split according to a single predictor. A dichotomy is constructed by beginning with "clusters" associated with individual levels of the relevant (suitably discretized) predictor and continuing to combine ("agglomerate") the pair of clusters with most similar responses until exactly two clusters remain.

Splitting continues until there are "enough" groups, no group can be split, all unsplit groups have homogeneous responses, or splitting would produce too small a group. The user of the procedure is permitted to define the stopping rules and measures of "similarity" of responses and response groups to themselves and each other.

The paradigm, which of course constructs a binary tree of groups, is a predictor-directed response clustering so that many ideas of "cluster analysis" apply. A useful reference is Anderberg (1973). It is possible to use random subsamples of the data to obtain approximations to required similarity measures as the procedure is executed. This, together with the agglomerative method for constructing dichotomies, makes it possible to handle large data sets as well as multiple responses, and thus to overcome difficulties and limitations encountered with earlier approaches. Of course, combining predictor levels can, where desired, be restricted to adjacent levels when levels are ordered.

The structure of each final unsplit group is revealed by using a multivariate data display technique to exhibit typical responses. Every time an unsplit group is examined for possible splitting, a "local" measure of ability to explain response differences is obtained for each predictor (provided that it appears at two or more levels). The several such measures for a predictor can be aggregated to provide a global measure of potential explanatory power (PEP) of that predictor when interacting with all the others. While such PEP statistics originally conceived by Finifter (1971a, b) have obvious shortcomings, they also have virtues not possessed by other association measures.

The binary tree does not identify a response pattern for predictor-level combinations not present among the observations. In this case, having the PEP measures permits one to proceed as follows.

Begin with all observations. For the predictor with most PEP not yet used, retain observations with the same level of that predictor if any; else, adjacent levels if levels are on an ordinal or interval scale; else, all levels. (Discard observations not retained.) Continue until all predictors (or all "salient" predictors) have been used. Then the response pattern is predicted to be similar (or if meaningful "most similar") to the observations remaining.

Perhaps it should be remarked that the generalized PEP statistics described here and their utilization for a prediction scheme go well beyond Finifter's original proposals.

F. Data Graphics

The analyst's own pattern recognition capabilities can be brought to bear in the model identification process with the assistance of data graphics. Several new developments in this area will be described here.

Responses depending on three primary predictors can be displayed, possibly together with associated values of other variables (including cell

frequencies of "grouped" variables), using perspective drawings or stereograms which display points and objects in three-dimensional space.

A stereogram is a "binocular pair" of plane images viewed through a
stereoscope such as the inexpensive Stereopticon 707 sold by Taylor Mer-
chant Corporation (25 W. 45th St., New York, N.Y.). A stereogram may
be constructed by mapping each point (x, y, z) in the unit cube into the pair
of points (X_1, Y), (X_2, Y) in the plane according to the formulas

$$X_1 = 0.74 + r(x - 0.32) \quad Y = 0.84 + r(y - 0.50)$$

$$X_2 = 3.26 + r(x - 0.68) \quad r = \frac{2.74}{2.64 - z}$$

(Here the positive z axis is assumed to point upward.)

At the point with coordinates (x, y, z) may be placed either a symbol
encoding the level of a variable or one or a set of glyphs (metroglyphs)
showing levels of several variables. The glyph is a set of rays emanating
from a point, the orientation of a ray identifying a variable and its length
encoding the level of that variable observed at the levels of the coordinates.
These code or glyph stereograms have been found well suited to many data
sets and are readily understood by nonspecialists. Should one not have
access to a plotter, matrices of level codes (called "shadow boxes") provide
useful visual displays of multivariate observations.

An interesting aid to identification of functions of a single variable is
"rubber-sheet curve fitting." One draws the curve $y = y(x)$ and the straight
line $y = x$ on an elastic membrane (or pleated graph paper or a CRT screen).
Then the membrane is irregularly stretched parallel to the x-axis, giving
new curves, say $y = f(u)$, $x = g(u)$, until the latter are of recognizable form.
The parameter u is then eliminated to give the relationship between x and y.
This idea can be extended to contour plots.

G. Multiphase Predictive Modeling

As often seen in the foregoing, model identification may include multiple
phases of analysis which combine a wide variety of methodological tools.
Much more can be said on this subject than space permits.

A traditional approach is to create subproblems for fixed levels of some
predictors, model dependence of the response on the remaining predictors
using a single parametric form, and consider the parameter values as
responses to the predictors which were fixed. Other traditional tactics include the transformation of predictors (e.g., introduce squares and cross
products, or "dummy" variables for occurring levels) and responses (e.g.,
take logarithms).

When a tentative model is obtained, it should be tested for adequacy.
For example, one may investigate whether $y - \hat{y}$ or $y/\hat{y}$ is independent of
the predictors where $\hat{y}$ is the predicted value of y. For this purpose we

notice that the (in general, vector valued) variables x and y are independent if an only if the distribution consisting of the product of the marginal distributions of x and y is the same as the joint distribution of x and y. A subsample of x, y pairs may be converted into a sample from the product distribution by randomly permuting the x values, so we are left with the problem of testing homogeneity of distributions underlying two multivariate samples.

To attack the latter, one can randomly select pairs of points without replacement to construct univariate samples of between and (pooled) within sample interpoint distances. Testing these resulting univariate samples for homogeneity may be carried out by standard methods. This randomized distribution-free strategy works surprisingly well even with very modest sample sizes.

If $y - \hat{y}$ or $y/\hat{y}$ is not independent of the predictors, analysis of the residual dependence appears to be a problem conceptually identical to the one we began with. However, when $\hat{y}$ was developed by certain methods, it often happens that there exists a function of $\hat{y}$ which is an adequate predictor. Indeed, if $\hat{y}$ is the classical multiple regression predictor, this is another way of expressing Brillinger's result! Hence in that case (and others) one should always plot y against $\hat{y}$. It is astonishing that this simple tactic is virtually unknown, and never suggested in existing discussions of so-called "regression diagnostics."

H. Ivakhnenko's Approach

All the above is viewed as necessary background material. Unfortunately, this had to be presented with little elaboration and at appreciable length, much of it not having previously appeared in print. Having explored the context in which it arises, some elements of which it is composed, and ideas that will suggest variants, we may now turn to Ivakhnenko's approach.

Ivakhnenko's approach is a formalized paradigm for multiphase predictive modeling which iterates subject to a stopping rule between selection of predictors based on explanatory power and the creation of new predictors by representation of explanatory power. The three elements of representation, selection, and stopping rule will be discussed in turn.

1. Representation of Predictive Power

Each representation step produces derived predictors designed to embody the explanatory power of existing predictors. Ivakhnenko's first proposal is that since model development will be carried out in a sequence of steps, each step may be kept relatively simple (naive, myopic), taking due account of the limitations imposed thereby on the final result. For instance, if each derived predictor is a low-degree polynomial in a few existing predictors (iterated polynomial regression), the final model may be a high-degree polynomial in many predictors.

a. <u>Iterated Polynomial Regression</u>   The usual representation step is indeed construction of a (not necessarily unique) polynomial in the "carriers" x_i, x_j, x_i^2, x_j^2, and $x_i x_j$, where x_i and x_j are a pair of existing predictors. [If there is but a single predictor x, we introduce x^2 as a predictor before the representation step.]

For iterated polynomial regression in general, there are several matters to notice. First, as is well known, for integer n sin nt and cos nt can be expressed as polynomials of degree n in sin t and cos t. Thus if sin t_j and cos t_j are included among the initial predictors to represent independent variable t_j, one may to advantage use iterated polynomial regression to perform harmonic analysis. Similarly, if the initial predictors are binary ("dummy") variables associated with factor levels, one might use iterated polynomial regression as a substitute for analysis of variance. These observations suggest that Ivakhnenko's paradigm leads to computing algorithms of unexpectedly wide applicability.

Construction of the polynomial $y_{ij}(x_i, x_j)$ consists of assigning values to its (in this case six) coefficients. Here it is desired that the polynomial should be of maximum usefulness in the final model for prediction of y. Again for expediency we may adopt a myopic view and choose the coefficients to optimize a measure of association between y_{ij} and y. We described a number of association measures in the preliminary discussion but are constrained to selecting one that imposes acceptable computational overhead.

Product-moment Galton-Pearson correlation has been nominated by convenience. Why has it been more successful than might be anticipated? While the five "carriers" are in no sense sampled from a multivariate normal distribution, results and conclusions of Brillinger (1977) are strongly suggestive.

The final predictive model (of whatever form) after iteration ceases is required not merely to explain y but to provide an estimator of y, so that goodness of fit rather than explanatory power becomes the criterion to be optimized. Some popular measures of badness of fit to the observations are of the types $\Sigma_k c_k |y_k - \hat{y}_k|^\alpha$ and max $|y_k - \hat{y}_k|$, where $\hat{y}_k$ is the estimate of y_k. (These are, of course, so-called L^α and L^∞ norms.)

It is not so well known that minimization of these norms may be accomplished by iteratively minimizing $\Sigma_k w_{k\ell}(y_k - \hat{y}_{k\ell})^2$, where $w_{k\ell}$ is the weight attached to the kth observation and $\hat{y}_{k\ell}$ is the estimate of y_k at iteration ℓ (i.e., by weighted least squares). The $w_{k\ell}$ weights are given by

$$w_{k\ell} = c_k |y_k - \hat{y}_{k\ell-1}|^{\alpha-2}, \quad \alpha \neq \infty, \ \ell \neq 1$$

$$w_{k1} = c_k, \quad\quad\quad\quad\quad \alpha \neq \infty$$

$$w_{k\ell} = \frac{w_{k\ell-1}|y_k - \hat{y}_{k\ell-1}|}{\Sigma_{k'} w_{k'\ell-1}|y_{k'} - \hat{y}_{k'\ell-1}|}, \quad \alpha = \infty, \quad \ell \neq 1$$

$$w_{k1} = 1, \qquad\qquad\qquad \alpha = \infty$$

b. <u>Alternatives to Polynomial Regression</u>   As a first generalization of
polynomials, one might consider rational functions (quotients of polynomials).
In fact, one can in principle build up quotients of high-degree polynomials
by iteratively forming quotients of first-degree polynomials (linear functions).

Suppose that a rational function P/Q is to be fitted by weighted least
squares where P and Q are polynomials taking the values $P_{k\ell}$ and $Q_{k\ell}$ for
observation k at stage ℓ. A possible approximate procedure (Turner et al.,
1961) is: For $\ell = 1, 2, \ldots$, determine parameters of P and Q to

$$\text{minimize} \sum_k \frac{c_k(y_k Q_{k\ell} - P_{k\ell})^2}{Q_{k\ell-1}^2}, \quad Q_{k0} = 1$$

Of particular interest in curve fitting are the rational functions of a
single variable x which are polynomials in $1/(x + c)$. Hastings et al. (1955,
p. 106) give a simple yet very effective method for choosing c, as follows.
Suppose that the curve is $y(x)$, where

$$y_0 = y(x_0) \leq y(x) \leq y_1 = y(x_1)$$

and

$$y_2 = y(x_2) \approx \frac{y_0 + y_1}{2}$$

If $x_2 = (x_0 + x_1)/2$, fit the polynomial directly in x. Otherwise, take

$$c = \frac{x_2(x_0 + x_1) - 2x_0 x_1}{x_0 + x_1 - 2x_2}$$

At the representation stage it is possible to provide a "menu" of functional forms in one or several predictors. An obvious, often used method
for selecting from a limited menu is to try all the choices and compare
results. A formalized method based on quite different principles which
attracted some attention when first proposed is described by Simonsen and
Anketell (1966). An "unlimited" menu is possible for the analyst working
interactively with the computer using state-of-the-art data graphics.

To sum up our discussion of the representation step, however, it
should be said that the $y_{ij}(x_i, x_j)$ models may be expected to work well unless
the response surface is "angular" (Hastings et al., 1955, p. 105).

2. Identification of Predictive Power—Selection of Predictors

Each selection step of Ivakhnenko's approach sifts the available predictors, discarding some to prevent exponential growth in the computational burden of the representation steps, and perhaps to limit the complexity of the final model. This has to be done by measuring nonlinear association between a predictor or sets of predictors and the response.

In the selection step we may consider more elaborate measures than in the representation step since here they need be evaluated only once, rather than possibly repeatedly as required in iterative parameter estimation algorithms. Some candidates that now appear are statistics used in testing independence (including the randomized distribution-free strategy described earlier) and Finifter-type PEP statistics. Since we wish to save predictors with joint explanatory power, PEP-like statistics are clearly what is desired; however, their computing cost will often be prohibitive. Once again the correlation coefficient is the poor man's choice.

We may wish in some manner to "penalize" the measure of association between a predictor and response for the predictor's complexity or lack of smoothness. One example of this idea is Akaike's "information criterion" (Amemiya, 1980). Also, there are sometimes disutilities such as measurement costs associated with use of certain original predictors.

Golub et al. (1979; see also Wahba, 1977) employ the sum of squared magnitudes of response gradients at observation points as a measure of lack of smoothness, and mean-squared residual as the criterion. They show that this idea then leads to ridge regression, and propose "generalized cross-validation" as a method for choosing the penalty coefficient (ridge parameter).

In the case of iterated polynomial regression it is far from obvious how complexity ought to be measured. Although it is believed that some GMDH users are employing penalty terms, the merits of this strategy are not likely to be entirely clear for some time to come.

One strategy that is not good is simply to retain a fixed number of predictors having high association with the response, for these may include large numbers of equivalent predictors (i.e., predictors which explain each other). For instance, if both $y_{ij}(x_i, x_j)$ and $y_{ik}(x_i, x_k)$, $j \neq k$, are "dominated" by the x_i and x_i^2 terms, these derived predictors will be practically indistinguishable. A better approach is a stepwise one, retaining a predictor only if it is highly associated with the response but not highly associated with any previously selected predictor.

Notice that the predictors examined in the selection step ought to include not only those derived in the preceding representation step, but also those retained in the preceding selection step, whenever the association measures used in the two types of step are different! This is the case with GMDH data grouping, where a "training set" of observations is used in the representation step and a different "testing set" is used in the selection step. [Another case would be when the y_{ij} polynomial is fitted by least squares but evaluated by PRESS (Cady and Allen, 1972).]

Commentators on Ivakhnenko's approach have tended to concentrate on general principles rather than algorithmic details. They are especially vague about the selection step, which is not altogether unlike deciding how many factors to retain in factor analysis. However, it is primarily a matter of establishing a trade-off between computing cost and quality of the final model.

Our recommendation is stepwise selection, modified as above, in descending order of association with the response, stopping prematurely as soon as all original independent variables occur in one mor or more of the retained predictors.

3. Stopping Rules

The final element of Ivakhnenko's approach is a stopping rule for terminating iterations after a selection step. One might decide to stop when a fixed number r of successive selection steps fails to find a predictor better by ϵ than the best predictor previously found. Usually, $r = 1$, $\epsilon = 0$; and there is evidence that this is a "safe" choice (Farlow, 1981).

IV. CONCLUDING REMARKS

We have sought to suggest with regard to Ivakhnenko's approach underlying principles, possible competitors, and unsettled issues. Concerning the latter, much remains to be explored. What is important is that, as said at the outset, regression without models is a neglected area in which, however, many tools are immediately available for the practitioner's use.

REFERENCES

Amemiya, T. (1980). Selection of Regressors. Int. Econ. Rev. 21:331-354.
Anderberg, M. R. (1973). Cluster Analysis for Applications. Academic Press, New York.
Brillinger, D. R. (1977). The Identification of a Particular Nonlinear Time Series System. Biometrika 64:509-515.
Brillinger, D. R. (1982). A Generalized Linear Model with "Gaussian" Regressor Variables. Annu. Meet., Stat. Soc. Can., Ottawa, June 3-5.
Cady, F. B., and Allen, D. M. (1972). Combining Experiments to Predict Future Yield Data. Agron. J. 64:211-214.
Carroll, J. D., and Chang, J. J. (1964). A General Index of Nonlinear Association and Its Application to the Interpretation of Multidimensional Scaling Solutions. Am. Psychol. 19:540.
Carroll, J. D., and Chang, J. J. (1970). Analysis of Individual Differences in Multidimensional Scaling via an N-Way Generalization of "Eckart-Young" Decomposition. Psychometrika 35:283-319.

Cliff, A. D., and Ord, J. K. (1973). Spatial Autocorrelation. Pion, London.

Clough, D. J. (1978). Biased Reflections on MS/OR Consulting, Part 2—
How and What to Use for $ucce$$ful Con$ulting. Interfaces 8:21-33.

Cornelius, P. L., Templeton, W. C., Jr., and Taylor, T. H. (1979).
Curve Fitting by Regression on Smoothed Singular Vectors. Biometrics
35:849-859.

Farlow, S. J. (1981). The GMDH Algorithm of Ivakhnenko. Am. Stat. 35:
210-215.

Fielding A. (1977). Binary Segmentation, the Automatic Interaction Detector
and Related Techniques for Exploring Data Structure. In The Analysis
of Survey Data I, C. A. O'Muircheartaigh and C. Payne (Eds.). Wiley,
New York.

Finifter, B. M. (1971a). Program ERIV and the PEP Statistic for Measuring Relative Importance. Unpublished memorandum, Dept. of Sociology,
Michigan State University, East Lansing.

Finifter, B. M. (1971b). ERIV: A Computer Program for Evaluating Relative Importance of Variables in the Analysis of Interaction Effects.
Behav. Sci. 16:511-512.

Friedman, J. H., and Stuetzle, W. (1981). Projection Pursuit Regression.
J. Am. Stat. Assoc. 76:817-823.

Frisen, M. (1979). A Note on Alternating Estimation in Non-linear Regression. J. Stat. Comput. Simul. 9:19-23.

Gabriel, K. R., and Zamir, S. (1979). Lower Rank Approximation of
Matrices by Least Squares with Any Choice of Weights. Technometrics
21:489-498.

Gillo, M. W. (1972). MAID, a Honeywell 600 Program for Automatized
Survey Analysis. Behav. Sci. 17:251-252.

Gillo, M. W., and Shelly, M. W. (1974). Predictive Modeling of Multi-
variable and Multivariate Data. J. Am. Stat. Assoc. 69:646-653.

Golub, G. H., Heath, M., and Wahba, G. (1979). Generalized Cross Validation as a Method for Choosing a Good Ridge Parameter. Technometrics 21:215-223.

Hastings, C., Jr., Hayward, J. T., and Wong, J. P., Jr. (1955). Approximations for Digital Computers. Princeton University Press, Princeton, N.J.

Kass, G. V. (1980). An Exploratory Technique for Investigating Large
Quantities of Categorical Data. Appl. Stat. 29:119-127.

Lorentz, G. G. (1976). The 13th Problem of Hilbert. In Mathematical
Developments Arising from Hilbert Problems, F. E. Browder (Ed.).
American Mathematical Society, Providence, R.I.

Malone, J. M., II (1965). Dimension Reduction—A Non-linear Regression
Exploration Technique. Technical Memorandum 06.02.004, Engineering
and Scientific Computer Services, U.S. Steel Corporation, Pittsburgh,
Pa.

Mandel, J. (1971). A New Analysis of Variance Model for Non-additive
Data. Technometrics 13:1-18.

Messenger, R. C., and Mandell, L. M. (1972). A Modal Search Technique for Predictive Nominal Scale Multivariate Analysis. J. Am. Stat. Assoc. 67:768-772.

Morgan, J. N., and Messenger, R. C. (1973). THAID: A Sequential Analysis Program for the Analysis of Nominal Scale Dependent Variables. University of Michigan, Ann Arbor.

Morgan, J. N., and Sonquist, J. A. (1963). Problems in the Analysis of Survey Data and a Proposal. J. Am. Stat. Assoc. 58:415-434.

Perreault, W. D., and Barksdale, H. C., Jr. (1980). A Model-Free Approach for Analysis of Complex Contingency Data in Survey Research. J. Mark. Res. 17:503-515.

Scott, D. W., and Factor, L. E. (1981). Monte Carlo Study of Three Data-Based Nonparametric Probability Density Estimates. J. Am. Stat. Assoc. 76:9-15.

Shepard, R. N., and Carroll, J. D. (1966). Parametric Representation of Nonlinear Data Structures. In Multivariate Analysis, P. R. Krishnaiah (Ed.). Academic Press, New York.

Simonsen, R. H., and Anketell, D. L. (1966). Mechanization of the Curve Fitting Process: DATAN. Commun. ACM 9:299-304.

Sonquist, J. A., Baker, E. L., and Morgan, J. N. (1971). Searching for Structure. University of Michigan, Ann Arbor.

Tucker, L. R. (1958). Determination of Parameters of a Functional Relation by Factor Analysis. Psychometrika 23:19-23.

Turner, M. E., Monroe, R. J., and Henry, L. L., Jr. (1961). Generalized Asymptotic Regression and Non-linear Path Analysis. Biometrics 17:120-143.

Wahba, G. (1977). A Survey of Some Smoothing Problems and the Method of Generalized Cross Validation for Solving Them. In Applications of Statistics, P. R. Krishnaiah (Ed.). North-Holland, Amsterdam.

4

Predicted Squared Error: A Criterion for Automatic Model Selection

ANDREW R. BARRON* / Adaptronics, Inc., Subsidiary of Flow General, Inc., McLean, Virginia

I. INTRODUCTION

Whereas conventional empirical modeling techniques require an assumed model structure, new procedures have been developed which generate the model structure as well as the model coefficients from a data base. These procedures include the GMDH and PNETTR algorithms for creating polynomial networks. Key to any automatic procedure for generating models is the criterion for ranking different model structures and selecting the best.

The objective of empirical modeling is to identify and train a model that will perform with low error on as yet unseen data. Experience has shown that this objective is met by selecting that model which minimizes an estimate of future performance that we call the predicted squared error (PSE). This criterion is incorporated in the PNETTR 4 algorithm developed by the author at Adaptronics, Inc. (see Chap. 2). This chapter presents a statistical analysis of PSE that explains why it is a good estimate of future performance.

First an intuitive understanding of PSE is helpful. PSE is the sum of two terms: the training squared error and overfit penalty. The training squared error (TSE) is given by the (empirical) average squared error of a model on n training observations. Let k be the number of coefficients in the model that are estimated so as to minimize TSE. The overfit penalty is given by $2\sigma_p^2(k/n)$, where σ_p^2 is a prior estimate of the true error variance that does not depend on the particular model being considered. Thus the predicted squared error is given by

---

*Present affiliation: Information Systems Laboratory, Stanford University, Stanford, California

$$PSE = TSE + 2\sigma_p^2 \frac{k}{n} \qquad (1)$$

The PSE is used at all stages of network construction to rank and select the better model structures. The network that achieves the least PSE is the final product of network synthesis. A minimum will always be attained because TSE decreases with each additional coefficient but always remains nonnegative, whereas the overfit penalty linearly increases in the number of coefficients.

The TSE term favors models that perform well on the training data; however, by itself it can be a poor estimate of future performance. Factors that make TSE underestimate future error (this condition is often called "overfitting" the training data) include overly complicated model structure and many coefficients each adjusted to lower TSE. The overfit penalty term penalizes complex models. It will be shown that this penalty term accounts for the expected squared difference between the estimated model and the true model on future data and accounts for the bias of TSE below the true error variance. The presence of both TSE and penalty terms ensures that PSE favors simple models with low error.

Central to the derivation of PSE and also of independent interest is an understanding of the expected squared error on unseen data (data not yet available to train the model). Section II delves into this topic with some interesting and useful results. Section III discusses PSE as an estimate of this expected squared (future) error. Section IV relates PSE to hypothesis-testing procedures. Section V compares the PSE to other criteria for model selection: including those proposed by Akaike (1970, 1972), Mallows (1973), and Schwarz (1977).

II. EXPECTED PERFORMANCE ON FUTURE DATA

In general, data used to train a model will differ from data the model will encounter in the future. If future data are vastly different, we expect that the model will not perform as well. This is especially true if the model is required to extrapolate far beyond the range of the training data. On the other hand, if the training data are representative of future observations, we expect reasonable performance. This intuitive reasoning is substantiated by the results of this section. A simple expression is derived which provides an exact description of expected squared error when the model is linear in its coefficients and an approximate description when the model is nonlinear in its coefficients.

Consider for now models that are linear in their coefficients. For example, elements (the building blocks of polynomial networks) are linear in the coefficients even if quadratic or cubic terms in the inputs are included. Furthermore, polynomial networks are equivalent to models that are linear in the coefficients if they are composed of elements with nonlinear terms in original input variables only (i.e., without nonlinear terms in intermediate inputs). Let z denote a row vector of transformed input variables that correspond to the terms in the model. For n training observations (input vectors x_i and dependent variables y_i, $i = 1, 2, \ldots, n$) consider the k by k symmetric matrix R_T composed of normalized (by $1/n$) sums of cross-products of the transformed inputs. In matrix notation $R_T = T'T/n$, where $T = (z_1, z_2, \ldots, z_n)'$ is the training data matrix consisting of the n transformed vectors. Now suppose that after training, the model will be applied to n_F "future" observations (x_{iF} and y_{iF}, $i = 1, 2, \ldots, n_F$; here y_{iF} need not be observable) with data matrix $F = (z_{1F}, z_{2F}, \ldots, z_{n_F F})'$ and $R_F = F'F/n_F$. Note that R_T and R_F can be thought of as the "covariance" structure of the training and future data, respectively (R_T and R_F would be sample covariance matrices if the input data were regarded as random; however, here the input data are regarded as fixed). Now suppose that for some unknown value of the coefficient vector, the difference (errors) between the model output and the dependent variable are independent random variables with mean zero, and common variance σ^2 (no further assumptions regarding the error distribution are necessary; in particular, the errors need not be Gaussian). The expected squared error on the F data (of the model trained on the T data) is given by

$$\sigma^2 + \sigma^2 \frac{\text{trace}(R_F R_T^{-1})}{n} \tag{2}$$

This result is derived, using standard matrix manipulation, in the Appendix. [See also (Bibby and Toutenburg, 1977); the result above follows from their equations 1.5.5 and 1.5.13.]

The two terms in formula (2) correspond to two factors contributing to error on future data. The first term, σ^2, is the expected squared error of the ideal (but unknown) model. The second term is the expected squared difference (on the F data) between the trained model and the ideal model. This term shows that the expected performance depends on the degree of similarity between training and future data.

Although (2) is derived to motivate the PSE criterion for model selection, the result may be equally important for the areas of experimental design and model adaptation. Experimental design is concerned with locating training input data (in the space of possible input variable values), so that a model can be efficiently trained to perform well in the future. If we have some notion of how future input data will be dispersed, in particular, if we know (or can approximate) R_F, then the training input data can be chosen*

---

*Cluster analysis (see Chap. 2) is an invaluable aid in understanding how input data are dispersed and in choosing the training data.

or generated to be representative of future data such that $\underline{R}_T \cong \underline{R}_F$. In this case the expected squared error (2) simplifies to

$$\sigma^2 + \sigma^2 \frac{k}{n} \tag{3}$$

It is tempting to try to reduce the future squared error by designing training data with $\underline{R}_T$ much "larger" than $\underline{R}_F$, so that trace $(\underline{R}_F \underline{R}_T^{-1}) \ll k$. Then we expect the future squared error to be less than (3). For fixed n, $\underline{R}_T$ larger means that the training data are more spread out. But unless we are confident that the model considered has the right structure, increasing $\underline{R}_T$ can cause a decrease in model accuracy due to interpolating between more distant training observations. Thus if we know that the model structure is correct, we should choose $\underline{R}_T$ such that trace $(\underline{R}_F \underline{R}_T^{-1}) \ll k$; however, if the structure is one of many possible (as with polynomial network training), it is better to choose $\underline{R}_T \cong \underline{R}_F$ such that trace $(\underline{R}_F \underline{R}_T^{-1}) \cong k$.

The result (2) can be very useful after a model has been trained (on data with "covariance" $\underline{R}_T$) and is being applied to new data for which the true values of the dependent variable y are unknown or unavailable. We wonder how accurate is the model's estimate of y. Result (2) indicates that the "covariance" $\underline{R}_F$ (of the new input data) should be monitored. If $\underline{R}_F$ is such that trace $(\underline{R}_F \underline{R}_T^{-1})$ is less than or comparable to k, the model should be satisfactory. However, if $\underline{R}_F$ is such that trace $(\underline{R}_F \underline{R}_T^{-1}) \gg k$, the model is no longer suited for the data. In this case, we are attempting to extrapolate the model to data points consistently outside the range of training data. The model should be adapted or retrained.

Thus both before and after training models the skilled analyst or engineer works to ensure $\underline{R}_T \cong \underline{R}_F$ so that trace $(\underline{R}_F \underline{R}_T^{-1}) \cong k$. Then the expected squared error on future data is given by $\sigma^2 + \sigma^2 (k/n)$. Note that models with large k (many estimated coefficients and hence high complexity) are not expected to perform well, unless there are enough training observations that k/n is adequately small.

The results given above have two shortcomings that must be addressed. One deficiency is the assumption that the model considered is of the correct form (i.e., for some unknown coefficient values the errors in the output of this model are independently distributed with zero mean and common variance). During the synthesis of polynomial networks, many different model structures are considered, not all of which can approximate the actual dependencies in the process. For "wrong" structures, there is an additional term in the expected squared (future) error, which is the average squared difference between the unknown correct model and the wrong model (where the coefficients of the wrong model are such that this average squared difference is minimum). When the correct structure is not known in advance, it is difficult to account for this term.

If a model selection procedure is successful in weeding out those model structures that cannot approximate the true relationships in the data, then results (2) and (3) accurately assess the expected squared (future) error for the remaining models. The PSE criterion is designed (in Sec. III) to be effective in both weeding out the clearly incorrect models and in estimating the future error for the remaining models.

The other deficiency is the assumption that the model considered is linear in its coefficients. This is not true for general polynomial networks. However, the results remain valid to the extent that polynomial networks can be approximated by some first-order Taylor expansion in the coefficients. Not all the coefficients of a general network are needed for this expansion. In particular, the coefficients that correspond to linear combinations of past element outputs can be regarded as fixed (since by varying the other coefficients we can obtain any linear combination of these element outputs). Similarly, not all of the coefficients corresponding to constant terms need to be counted or included in the expansion. Let the model function be denoted by $f(\underline{x}, \underline{\beta})$, where $\underline{x}$ is the vector of inputs and $\underline{\beta}$ is the column vector of k free coefficients, Let $\underline{z}_\beta$ be the row vector composed of the partial derivatives of the model $f(\underline{x}, \underline{\beta})$ with respect to the coefficients. A model is nonlinear in its coefficients whenever $\underline{z}_\beta$ depends on $\underline{\beta}$. Let $\hat{\underline{\beta}}$ be the vector of estimated coefficients. The first-order Taylor expansion is given by

$$f(\underline{x}, \underline{\beta}) \cong \underline{z}_{\hat{\beta}}(\underline{\beta} - \hat{\underline{\beta}}) + f(\underline{x}, \hat{\underline{\beta}}) \tag{4}$$

Note that this expansion may be highly nonlinear in the input variables $\underline{x}$ even though it is linear in the coefficients $\underline{\beta}$. The results summarized in formulas (2) and (3) will remain valid if the first-order expansion is an accurate approximation of $f(\underline{x}, \underline{\beta})$ for $\underline{\beta}$ equal to the ideal coefficient values.

The partial derivatives $\underline{z}_\beta$ correspond to $\underline{z}$ the vector of transformed inputs and these coincide when $\underline{z}_\beta$ does not depend on $\underline{\beta}$. For models that are nonlinear in coefficients, approximate $\underline{R}_T$ and $\underline{R}_F$ matrices can be constructed by evaluating (for each observation) these partial derivatives using the estimated coefficient values [e.g., $\underline{R}_T = \underline{T}'\underline{T}/n$, where $\underline{T} = (\underline{z}_{1\hat{\beta}}, \underline{z}_{2\hat{\beta}},$

$\ldots, \underline{z}_{n\hat{\beta}})'$]. The partial derivatives can be computed analytically by applying the chain rule of calculus to the layers of polynomial elements (since a network is a composition of element functions). In practice, it is usually best to leave the derivatives in network form and reapply the chain rule whenever a value is desired.

Further research may resolve the issue of expected performance of models which are nonlinear in the coefficients. It may be possible to obtain a more general expression for the expected squared (future) error. No such results are known, but it is conjectured that the dominant terms are the same as in formulas (2) and (3). It is interesting to note that some of the

viewpoints (other than expected squared error) advocated for deriving model
selection criteria use no linearity assumptions, yet result in criteria that
are similar to PSE. These other viewpoints are mentioned in Sec. V.

III.   ESTIMATING EXPECTED PERFORMANCE

This section demonstrates that PSE is a good estimate of the expected
squared (future) error. When the training data are representative of future
data and when the model considered has a structure that can approximate
the "correct" model, then (from Sec. II) the expected squared error on
future data is given by Eq. (3)

$$\sigma^2 + \sigma^2 \frac{k}{n}$$

To estimate (3) from the training data, it is natural to consider using an
unbiased estimator of σ^2 which is given by

$$\hat{\sigma}_u^2 = \frac{n}{n-k} \text{TSE} = \frac{1}{n-k} \sum_{i=1}^{n} [y_i - f(\underline{x}_i, \hat{\underline{\beta}})]^2 \tag{5}$$

This yields the final prediction error FPE estimate of (3) proposed by
Akaike [1970]:

$$\text{FPE} = \frac{n+k}{n-k} \text{TSE} \tag{6}$$

If, indeed, the model considered has the correct structure, then FPE is an
unbiased estimate of $\sigma^2 + \sigma^2(k/n)$. Furthermore, if the errors are Gaussian,
FPE has minimum variance among unbiased estimates. Akaike proposed
FPE as a model selection criterion, and it has proved quite valuable in
selecting subsets models from "complete" models in classical linear regression. This good performance is not surprising, because the classical linear
regression setup restricts consideration to models that are no more complex than an assumed correct model. However, when training polynomial
networks, many wrong (and typically complicated) models are considered
that need to be rejected (the bonus, of course, is that we have more chance
of finding an accurate and simple nonlinear model). Our experience is that
FPE tends to favor some of the complicated overfit models. Why this is so
will be evident from some of the analysis of this section.
    The PSE estimator of $\sigma^2 + \sigma^2(k/n)$ is given by Eq. (1),

$$\text{PSE} = \text{TSE} + 2\sigma_p^2 \frac{k}{n}$$

where σ_p^2 is a prior estimate of σ^2 that does not depend on the model considered. A simple motivation for the PSE estimate is that the two sources of future error should be identified and estimated separately. The squared error of the ideal model can be estimated by the average squared error on the training data TSE. However, the expected value of TSE is $\sigma^2 - \sigma^2(k/n)$, where the subtracted term is the expected squared difference between the estimated model and ideal model on the training data. This expected squared difference on the training data plus the expected squared difference on future data [total of $2\sigma^2(k/n)$] is estimated using the penalty term [given by $2\sigma_p^2(k/n)$]. The fixed σ_p^2 in the penalty term is used because we do not want PSE to underestimate future squared error when the particular model considered is incorrect (e.g., an overly complex network with low TSE).

PSE is a biased estimator of (3): that is, the expected value of PSE exceeds (3) by the amount

$$\text{bias (PSE)} = \frac{2k}{n} (\sigma_p^2 - \sigma^2) \tag{7}$$

For simple models with few coefficients (small k relative to n), this bias is negligible. Only for overly large k (when typically we want to reject the model) is this bias significant. It is important that our prior σ_p^2 be at least as large as σ^2, so that these models will be rejected [some criteria correspond to using $\sigma_p^2 \cong \frac{1}{2}\sigma^2 \log n$; see Sec. V]. The bias of PSE is not bad; it is an important contribution to the ability of PSE to reject overly complicated (and usually wrong) models. In fact, the bias can help account for the additional term in the expected squared error when the model considered has the wrong structure. High TSE (error in the training set) usually rejects the overly simple models; the $2\sigma_p^2(k/n)$ penalty term (which includes the bias) is needed to reject the overly complex models.

In addition to rejecting wrong models, we want PSE to be an accurate estimate of the expected squared (future) error when the model has correct structure. A natural estimate of the accuracy of PSE is its mean-squared error: that is, the expected value of the squared difference between PSE and $\sigma^2 + \sigma^2(k/n)$. The mean-squared error is the sum of variance and squared bias of PSE.

$$\text{mse (PSE)} = \text{var (PSE)} + \text{bias}^2(\text{PSE}) \tag{8}$$

In order to compute the variance of PSE, we need an additional assumption on the distribution of the errors. If the errors were Gaussian, the sum of squared errors (SSE = nTSE) would be chi-squared on n - k degrees of freedom, which has variance $2(n - k)\sigma^4$. Therefore, as a benchmark for comparison, suppose that

$$\text{var (TSE)} = \frac{\text{var (nTSE)}}{n^2} = \frac{2(n - k)\sigma^4}{n^2} \tag{9}$$

Then the mean-squared error of PSE is given by

$$\text{mse (PSE)} = \frac{2(n-k)\sigma^4}{n^2} + 4\left(\frac{k}{n}\right)^2 (\sigma_p^2 - \sigma^2)^2 \tag{10}$$

Note that the variance term is decreasing with increasing k. The decreasing variance and increasing bias implies that the probability that PSE underestimates $\sigma^2 + \sigma^2(k/n)$ decreases with increasing k. Similarly, the probability that PSE is less than $\sigma^2 + \sigma^2(k/n) - \epsilon$ (for any fixed threshold $\epsilon > 0$) decreases. (From the frequentist's point of view, the proportion of times PSE overfits the data decreases as we increase the number of coefficients considered.)

To see how accurate PSE is, we compare the mean-squared error of PSE to the mean-squared error of the unbiased estimator FPE. The mean-squared error of FPE has a variance term only:

$$\text{mse (FPE)} = \text{var}\left(\frac{n+k}{n-k}\text{TSE}\right) = \left(\frac{n+k}{n-k}\right)^2 \text{var (TSE)} = \frac{2(n+k)^2}{n^2(n-k)}\sigma^4 \tag{11}$$

Note that the variance of FPE is greater than the variance of PSE by a factor of $(n+k)^2/(n-k)^2$. The variance of FPE is increasing in k. Thus it is more likely that FPE is less than $\sigma^2 + \sigma^2(k/n) - \epsilon$ (for any threshold $\epsilon > 0$) as k increases. Loosely speaking, PSE has less probability of selecting an overfit model than does FPE.

From Eqs. (10) and (11), it can be shown that PSE has less mean-squared error than FPE if and only if

$$|\sigma_p^2 - \sigma^2| < \sigma^2\left[\frac{2n}{k(n-k)}\right]^{\frac{1}{2}} \tag{12}$$

Thus from the point of view of mean-squared error, PSE is superior whenever our prior σ_p^2 is reasonably close to the correct σ^2. For example, if $n = 32$ and $k = 8$, then PSE is better for $0.42\sigma^2 < \sigma_p^2 < 1.58\sigma^2$.

The mean-squared errors of PSE and FPE depend on the unknown error variance σ^2. We can compare the estimators further by computing weighted averages of the mean-squared error. Suppose that the weighted average of σ^2 is σ_p^2 and that the weighted average of $(\sigma^2 - \sigma_p^2)^2$ is $\gamma^2\sigma_p^4$. This analysis is equivalent to the Bayesian point of view that the parameter σ^2 has an a priori distribution with mean σ_p^2 and standard deviation $\gamma\sigma_p^2$. The Bayes risk is the average mean-squared error. To compute the risks for PSE and FPE, note that the average of σ^4 is the variance of σ^2 plus the squared mean of σ^2. Thus

$$\text{risk (FPE)} = E(\text{mse FPE}) = \frac{2(n+k)^2}{n^2(n-k)}E(\sigma^4) = \frac{2(n+k)^2}{n^2(n-k)}(\gamma^2+1)\sigma_p^4 \tag{13}$$

and similarly,

$$\text{risk (PSE)} = \left[\frac{2(n-k)}{n^2}(\gamma^2 + 1) + 4\left(\frac{k}{n}\right)^2\gamma^2\right]\sigma_p^4 \tag{14}$$

Under what conditions is the risk of PSE less than the risk of FPE? It is straightforward to show that risk (PSE) < risk (FPE) if and only if

$$\frac{\gamma^2}{\gamma^2 + 1} < \frac{2n}{k(n-k)} \tag{15}$$

For large n, the inequality can be simplified to risk (PSE) < risk (FPE) provided that the number of coefficients $k \leq 2(1 + \gamma^2)/\gamma^2$. The right-hand side of inequality (15) is minimized for $k = n/2$ and then equals $8/n$. Thus the risk of PSE is less for all k when $n < 8(1 + \gamma^2)/\gamma^2$. For example, suppose that it is vaguely known that σ^2 is about σ_p^2 with uncertainty (standard deviation) $\pm\frac{1}{2}\sigma_p^2$. Then PSE has less risk for all n if $k \leq 10$ and for all k if $n < 40$.

Clearly, PSE is the better estimator (in terms of mean-squared error) when the number k of estimated coefficients is small (e.g., $k \leq 10$). What can be said about the performance of PSE for larger k? The risk of PSE may be greater than the risk of FPE, but the percent difference remains small for $k^2 < n$:

$$\frac{\text{risk (PSE)} - \text{risk (FPE)}}{\text{risk (FPE)}} = \frac{k^2(n-k)}{(n+k)^2}\frac{2\gamma^2}{\gamma^2 + 1} - \frac{4kn}{(n+k)^2} \tag{16}$$

It should be remembered that these risks have been computed assuming a correctly specified model. Polynomial network training algorithms build the more complex models from the simpler models with smaller k. If incorrect decisions are made on the fewer coefficients, the large k model is incorrectly specified and the risk comparison is not valid. The importance of making correct decisions on small models, even if k will be large, suggests that PSE may be preferred over FPE. The definition of risk as average mean-squared error is misleading. It fails to account for the benefits of positive bias $\sigma_p^2 > \sigma^2$ for discouraging overfit and for helping to account for error due to incorrect model structure. Instead, mean-squared error treats both positive and negative bias as equally bad.

An objection to the PSE estimate is that it requires σ_p^2, a prior upper bound to the error variance σ^2 which might be hard to determine. Fortunately, there is a simple estimate which usually upper-bounds σ^2 and that does not depend on the model considered: specifically, the variation in the dependent variable y, given by

$$\sigma_0^2 = \frac{\sum(y_i - \bar{y})^2}{n} \quad \text{where} \quad \bar{y} = \frac{\sum y_i}{n} \tag{17}$$

The variation σ_0^2 will be greater than the TSE of every model considered (with the sole exception of the constant model). Thus if σ^2 is greater than σ_0^2, there is little hope of identifying a model. For reasonable data, σ^2 will be less than σ_0^2. If no prior value is provided, PNETTR 4 uses $\sigma_p^2 = \sigma_0^2/2$. [This choice corresponds to asserting that σ^ι is uniformly distributed or completely unknown in $0 < \sigma^2 < \sigma_0^2$ and hence that σ^2 has "mean" $\sigma_0^2/2$ and "standard deviation" $\pm(1/2\sqrt{3})\sigma_0^2$.] Since σ_0^2 depends indirectly on the random errors (of the true model), there is an additional contribution to the variance of PSE. However, this additional variance can be shown to be negligible (Barron, 1981).

IV. HYPOTHESIS TESTING

If the PSE criterion is viewed as a sequential hypothesis-testing procedure, we gain additional understanding of its behavior. Suppose that we have a nested sequence of linear hypotheses. For example, within a particular element of a polynomial network we may have up to eight terms and we wish to sequentially test whether to include individual terms (given that preceding terms have or have not been included). Let PSE(k) be the predicted squared error when a term (corresponding to the kth free coefficient in the entire network) is included and PSE(k - 1) when only preceding terms are included. The term is included if and only if

$$PSE(k - 1) - PSE(k) > 0 \tag{18}$$

Multiplying by n and then adding $2\sigma_p^2$ to both sides, Eq. (18) can be rewritten as

$$\Delta SSE(k) > 2\sigma_p^2 \tag{19}$$

where $\Delta SSE(k) = SSE(k - 1) - SSE(k)$ is the reduction of the sum of squared errors (SSE = nTSE) if the new term is included. The test (19) is recognized as the sequential chi-squared test assuming Gaussian errors. However, the test is robust in that regardless of the shape of the error distribution, the expected reduction in residual error is σ^2 (under the null hypothesis that the term should be excluded) or σ^2 plus the increase in variation of the model (under the hypothesis that the term should be included).

What would be the corresponding test if the final prediction error FPE were used as the criterion? The FPE criterion would include the term if and only if

$$\frac{n + k - 1}{n - k + 1} SSE(k - 1) - \frac{n + k}{n - k} SSE(k) > 0 \tag{20}$$

Assuming that SSE(k) > 0 and multiplying (20) by the positive factor
(n - k)(n - k + 1)/[(n + k - 1)SSE(k)] and then adding 2n/(n + k - 1) to both
sides, inequality (2) reduces to

$$(n - k) \frac{\Delta SSE(k)}{SSE(k)} > \frac{2n}{n + k - 1} \tag{21}$$

This test is recognized as a sequential F-test assuming Gaussian errors.
Note that the threshold on this test is automatically set (and is usually
near 2). The F-test is not as robust as the chi-square-type test. The disadvantage of the F-test for the kth term is the sensitivity to incorrect decisions on the other k - 1 terms. If other terms have been included when they
should have been omitted, then SSE(k) will be smaller—biasing the F-test
high—so that this term has a greater chance of being included. Thus it is
possible with the F-test to have a "snowballing" inclusion of terms and
hence large overfit models.

This section has shown that in a restricted framework the PSE criterion
can be viewed as a robust hypothesis-testing procedure. However, it is important to note that traditional hypothesis-testing procedures are not able
to compare and rank models of entirely different structure as is essential
in synthesizing polynomial networks. A criterion, such as PSE, that can
assess the performance of a model irrespective of the other candidate
models is necessary.

V. OTHER CRITERIA

A natural way to estimate future squared error is to withhold a subset of
observations from the training data and to evaluate the (empirical) average
squared error on this subset. If this evaluation subset is kept independent
of the training process, and if the set is representative of the range of
potential observations, it then provides a reasonable estimate of the performance of the model. When there are ample data for both training and
evaluating, the practice above is strongly recommended. If the error on the
evaluation set is comparable to the PSE, it gives the analyst additional confidence in the model selected. If the evaluation set error is much larger
than PSE, it suggests that one or both of the subsets has not been designed
to be representative of potential data.

Cross-validation, a common criterion for GMDH model selection, involves "withheld" data actively in the synthesis of the model. One subset
of data is used to fit the coefficients of each model structure considered
and a second subset is used to select the better structures. However, the
selection subset does not provide an independent measure of the expected
performance. Both the fitting and selection subsets are involved in training
the model. If enough different model structures are considered, one can
often be found that has low error on the fitting and selection sets, but will

not generalize well to new data. GMDH algorithms such as PNETTR 2 use
fitting and selection sets but often need additional checks in network growth
to avoid overfit. Additionally, careful attention to the partitioning of the
observations into representative data groups (using a cluster algorithm—
see Chap. 2) is required. If the number of observations is not large, the
accuracy of the trained models is curtailed by the splitting into subsets.
These difficulties led the author to develop PNETTR 3 and PNETTR 4 and
the PSE criterion for model selection. The PSE criterion does not require
data base partitioning. If desired, all the data may be used for training.
Furthermore, PSE automatically restricts the network growth.

The PSE criterion resembles and was partially motivated by model
selection criteria proposed by Mallows and Akaike. The criterion proposed
by Mallows (1973) is to select that model which has minimum C_p, where
he defines

$$C_p = \frac{SSE(k)}{\sigma_c^2} + 2k - n \tag{22}$$

If σ_c^2 is a prior estimate or upper bound to σ^2, then (setting $\sigma_p^2 = \sigma_c^2$) it is
simple to show that minimizing C_p is equivalent to minimizing PSE. How-
ever, Mallows suggests using $\sigma_c^2 = SSE(p)/(n - p)$, where p is the order of
a completely specified model and k < p. In that way, C_p is not a tool for
creating models but rather a tool by which insignificant terms are removed
from a known model. Even when a "completely specified" model is known,
the C_p statistic may encourage overfit, since $SSE(p)/(n - p)$ is frequently
less than σ^2. In the context of polynomial network synthesis, no correct
structure is assumed known a priori (even in traditional linear regression,
the assumption that the linear model in all inputs is "complete" seems
doubtful). Using a fixed σ_p^2, which we believe to exceed σ^2, is preferred.

Akaike has proposed two criteria for model selection. The first (1970)
is the final prediction error FPE criterion which has been analyzed in
Secs. III and IV. The other is the Akaike information criterion AIC (1972).
The AIC is based on a distribution assumed for the vector $\underline{y}$ of dependent
variables and on the number of parameters adjusted to maximize the likelihood of $\underline{y}$.

$$AIC = -2 \log \ell(\underline{y}, \hat{\sigma}^2, \hat{\underline{\beta}}_k) + 2k \tag{23}$$

where $\ell(\underline{y}, \sigma^2, \beta)$ is the likelihood function which is maximized for $\sigma^2 = \hat{\sigma}^2$
and $\beta = \hat{\beta}_k$ (vector with k estimated coefficients). If the distribution is
assumed to be independent Gaussian errors $e_i = y_i - f(\underline{x}_i, \underline{\beta})$, then minimizing AIC is equivalent to minimizing

$$\log (TSE) + \frac{2k}{n} \tag{24}$$

However, minimizing (24) has a serious flaw. The "minimum" is attained
by having k sufficiently large that TSE = 0 (e.g., k = n linearly independent
terms in a linear model). A more realistic criterion is obtained if σ^2 is
assumed known. Then minimizing AIC is equivalent to minimizing

$$\text{TSE} + 2\sigma^2 \frac{k}{n} \tag{25}$$

Clearly, this is equivalent to the PSE criterion with σ^2 known ($\sigma_p^2 = \sigma^2$).
The analysis of this chapter has shown that assuming σ^2 known is a stronger
restriction than necessary. Possibly, a generalized AIC could be derived
which incorporates vague knowledge of some parameters.

Akaike did not require linear models in his derivation of the AIC, but
derived that asymptotically the log-likelihood is quadratic in the unknown
parameters (i.e., it behaves like a Gaussian log-likelihood for a linear
model with known error variance.

The AIC is one of several proposed criteria that depend explicitly on
the assumed family of distribution. In principle such criteria are applicable
to a wide range of problems. However, for a particular problem it is difficult to know what is the "true" family of distributions. PSE is a criterion
that does not depend on the particular shape of distributions (e.g., Gaussian).
Instead, PSE is derived from a specific "loss" function and is applicable
whenever minimizing squared error on independent data is a realistic goal.

Schwarz (1977) proposed that if a parametric family of distribution is
assumed, the model should be selected that is a posteriori most probable.
He showed that if the log-likelihood function is of a common form (specifically, Koopman-Darmois, which includes the Gaussian), then for almost any
prior distribution on the parameters, minimizing

$$-\log \ell(\underline{y}, \hat{\underline{\beta}}_k) + \frac{k}{2} \log n \tag{26}$$

is asymptotically equivalent to maximizing the a posteriori probability of
the model. This procedure guarantees consistency (which means that asymptotically, i.e., as $n \to \infty$, the correct model will be selected). For models
with independent Gaussian errors of known variance σ^2, minimizing (26) is
equivalent to minimizing

$$\text{TSE} + \sigma^2 \frac{k}{n} \log_e n \tag{27}$$

This criterion corresponds to the PSE but with $\frac{1}{2}\sigma^2 \log_e n$ in the penalty
term in place of σ_p^2. Thus for large n, Schwarz's criterion restricts model
dimensionality (e.g., size of a polynomial network) more than does PSE
(with σ_p^2 near σ^2). The quantity (27) is biased above the expected squared
error on new data by a factor of

$$\frac{k}{n}\sigma^2(\log_e n - 2) \tag{28}$$

which remains negligible provided that the number of estimated coefficients k remains much less than $n/(\log_e n - 2)$.

Recently, a new philosophy for model selection has been proposed, first by Rissanen (1978, 1983) and then independently by this author (Barron, 1982). The goal proposed is to find that model which induces the shortest description for the data available. If a parametric family of distributions is assumed, then for each candidate model there is a description of the data that corresponds to a concatenation of a description of the model (including the estimated parameters) and a Shannon code for the data (given the parameters and input variables). Rissanen and Barron have each shown that minimizing

$$\frac{k}{2}\log n - \log \ell(\underline{y},\hat{\underline{\beta}}_k) \tag{29}$$

is asymptotically equivalent to finding the shortest description. The first term amounts to using $(1/2)\log n$ bits for each of the coefficients and the second term corresponds to the length of the Shannon code. Note that this criterion is equivalent to (26), the criterion proposed by Schwarz. Moreover, the description length criterion (29) does not require the model to be linear in the coefficients. Furthermore, the notion of minimum description length permits improvements in (29) for finite n (Rissanen, 1983; Barron, 1982). If the Gaussian distribution is used to Shannon-code the data [i.e., $\underline{y} = (y_1, y_2, \ldots, y_n)'$ is described by describing the errors $\hat{e}_i = y_i - f(\underline{x}_i, \hat{\underline{\beta}}_k)$, $i = 1, 2, \ldots, n$, according to a zero-mean, covariance $\sigma^2 I$ Gaussian distribution], then minimizing (29) is equivalent to minimizing (27) given above.

Does the shortest description of data available now provide a good explanation of statistically similar data in the future? This may be a philosophical question. But the similarity of criteria based on minimum description length [such as (27)] and the predicted squared error (1) seems to be a first step toward a quantitative answer. The goals of good data description and good prediction are not incompatible; however, there are intriguing differences [e.g., σ_p^2 versus $(1/2)\sigma^2 \log n$ in the penalty term]. Whenever the primary objective of empirical modeling is to identify a model that will perform with low error on as yet unseen data, the predicted squared error criterion is strongly recommended.

APPENDIX

This appendix presents a derivation of the expected squared error of a model (trained on one set of data) when the model is applied to a new set of data. The result will be Eq. (2) discussed in the body of this chapter.

Let two sets of observation be denoted by

$$\{(\underline{x}_i, y_i),\ i = 1, 2, \ldots, n\} \quad \text{and} \quad \{(\underline{x}_{iF}, y_{iF}),\ i = 1, 2, \ldots, n_F\}$$

The first set is the training data, and the second can be thought out as future data. Suppose that $y_i = f(\underline{x}_i, \underline{\beta}) + e_i$, where f denotes a candidate model with k unknown coefficients represented by the column vector $\underline{\beta}$. Similarly, $y_{iF} = f(\underline{x}_{iF}, \underline{\beta}) + e_i$. Let $\hat{\underline{\beta}}$ be the coefficients estimated from the training data. We want to compute the expected squared error on the new data when using $\hat{\underline{\beta}}$.

$$E\left[\frac{1}{n_F} \sum_{i=1}^{n_F} [y_{iF} - f(\underline{x}_{iF}, \hat{\underline{\beta}})]^2\right] \tag{A-1}$$

Adding and subtracting the unknown coefficient values, (A-1) becomes

$$E\left[\frac{1}{n_F} \sum_{i=1}^{n_F} [y_{iF} - f(\underline{x}_{iF}, \underline{\beta}) + f(\underline{x}_{iF}, \underline{\beta}) - f(\underline{x}_{iF}, \hat{\underline{\beta}})]^2\right] \tag{A-2}$$

This expression can be expanded into three important terms:

$$E\left[\frac{1}{n_F} \sum_{i=1}^{n_F} [y_{iF} - f(\underline{x}_{iF}, \underline{\beta})]^2\right] + E\left[\frac{1}{n_F} \sum_{i=1}^{n_F} [f(\underline{x}_{iF}, \underline{\beta}) - f(\underline{x}_{iF}, \hat{\underline{\beta}})]^2\right]$$

$$+ 2E\left[\frac{1}{n_F} \sum_{i=1}^{n_F} [y_{iF} - f(\underline{x}_{iF}, \underline{\beta})][f(\underline{x}_{iF}, \underline{\beta}) - f(\underline{x}_{iF}, \hat{\underline{\beta}})]\right] \tag{A-3}$$

Substituting the error $e_i = y_{iF} - f(\underline{x}_{iF}, \underline{\beta})$, expression (A-3) simplifies to

$$E\left[\frac{1}{n_F} \sum_{i=1}^{n_F} e_{iF}^2\right] + E\left[\frac{1}{n_F} \sum_{i=1}^{n_F} [f(\underline{x}_{iF}, \underline{\beta}) - f(\underline{x}_{iF}, \hat{\underline{\beta}})]^2\right]$$

$$+ 2E\left[\frac{1}{n_F} \sum_{i=1}^{n_F} e_{iF}[f(\underline{x}_{iF}, \underline{\beta}) - f(\underline{x}_{iF}, \hat{\underline{\beta}})]\right] \tag{A-4}$$

The first term of (A-4) is the expected squared error of the ideal model on future data. Under the assumption of zero mean and common variance $[E(e_{iF}^2) = \sigma^2$ for each i] this term is just σ^2. Since the input vectors $\underline{x}_{iF}$ are regarded as fixed, the third term represents interaction between random errors in training and future data. If we assume independence, this term is zero. (If the model is linear in the coefficients, then uncorrelated errors is sufficient for this term to be zero.) The Eq. (A-4) for expected squared error has now been reduced to

$$\sigma^2 + E\left[\frac{1}{n_F} \sum_{i=1}^{n_F} [f(\underline{x}_{iF}, \underline{\beta}) - f(\underline{x}_{iF}, \hat{\underline{\beta}})]^2\right] \qquad (A\text{-}5)$$

Now assume that the model can be approximated as linear in the coefficients $f(\underline{x}_i, \beta) = \underline{z}_i\beta$, where $\underline{z}_i$ is a (row) vector of k possibly nonlinear transformations of the input variables. Similarly, $f(\underline{x}_{iF}, \beta) = \underline{z}_{iF}\beta$. Define n by k training data matrix $\underline{T} = (\underline{z}_1, \underline{z}_2, \cdots, \underline{z}_n)'$ and n_F by k future data matrix $\underline{F} = (\underline{z}_{1F}, \underline{z}_{2F}, \cdots, \underline{z}_{n_F F})'$. Similarly, define column vectors for the dependent variables $\underline{y}$ and $\underline{y}_F$ and for the errors $\underline{e}$ and $\underline{e}_F$. Using the notation of matrix algebra, (A-5) becomes

$$\sigma^2 + \frac{1}{n_F} E[\|\underline{F}\beta - \underline{F}\hat{\beta}\|^2] \qquad (A\text{-}6)$$

From traditional regression analysis, the coefficients that minimize the (empirical) average squared error on the training set are given by

$$\hat{\underline{\beta}} = (\underline{T}'\underline{T})^{-1}\underline{T}'\underline{y} \qquad (A\text{-}7)$$

Also, since $\underline{y} = \underline{T}\beta + \underline{e}$ we can write the difference $\underline{F}\beta - \underline{F}\hat{\beta}$ in terms of the data matrices and the error $\underline{e}$.

$$\underline{F}\beta - \underline{F}\hat{\beta} = \underline{F}\beta - \underline{F}(\underline{T}'\underline{T})^{-1}\underline{T}'(\underline{T}\beta + \underline{e}) = \underline{F}\beta - \underline{F}\beta - \underline{F}(\underline{T}'\underline{T})^{-1}\underline{T}'\underline{e} = -\underline{F}(\underline{T}'\underline{T})^{-1}\underline{T}'\underline{e}$$

$$(A\text{-}8)$$

Now substituting (A-8) into the expected squared error (A-6) yields

$$\sigma^2 + \frac{1}{n_F} E[\|\underline{F}(\underline{T}'\underline{T})^{-1}\underline{T}'\underline{e}\|^2] = \sigma^2 + \frac{1}{n_F} E[\underline{e}'\underline{T}(\underline{T}'\underline{T})^{-1}\underline{F}'\underline{F}(\underline{T}'\underline{T})^{-1}\underline{T}'\underline{e}] \qquad (A\text{-}9)$$

The quantity in the brackets in (A-9) is a scalar. The trace of a scalar leaves the scalar untouched. Furthermore, within a trace operation, matrices commute. Thus (A-9) becomes

$$\sigma^2 + \frac{1}{n_F} E[\text{trace } (\underline{e}'\underline{T}(\underline{T}'\underline{T})^{-1}\underline{F}'\underline{F}(\underline{T}'\underline{T})^{-1}\underline{T}'\underline{e}]$$

$$= \sigma^2 + \frac{1}{n_F} E[\text{trace } (\underline{T}'\underline{e}\underline{e}'\underline{T}(\underline{T}'\underline{T})^{-1}\underline{F}'\underline{F}(\underline{T}'\underline{T})^{-1})] \qquad (A\text{-}10)$$

Now if the errors are uncorrelated and have common variance σ^2, then $E(\underline{e}\,\underline{e}') = \sigma^2 I$. So the formula reduces to

$$\sigma^2 + \frac{1}{n_F}\sigma^2 \text{ trace } ((\underline{T}'\underline{T})(\underline{T}'\underline{T})^{-1}(\underline{F}'\underline{F})(\underline{T}'\underline{T})^{-1})$$

$$= \sigma^2 + \frac{1}{n_F}\sigma^2 \text{ trace } ((\underline{F}'\underline{F})(\underline{T}'\underline{T})^{-1}) \qquad\qquad \text{(A-11)}$$

Defining $\underline{R}_T = \underline{T}'\underline{T}/n$ and $\underline{R}_F = \underline{F}'\underline{F}/n_F$, the notation is simplified. The result for the expected average squared error of the trained model when applied to new data is

$$\sigma^2 + \sigma^2 \frac{\text{trace }(\underline{R}_F\underline{R}_T^{-1})}{n} \qquad\qquad \text{(A-12)}$$

REFERENCES

Akaike, H. (1970). Statistical Predictor Identification. Ann. Inst. Stat. Math. 22:203-217.

Akaike, H. (1972). Information Theory and an Extension of the Maximum Likelihood Principle. In Proceedings of the Second International Symposium on Information Theory, B. N. Petrov and F. Csaki (Eds.), Akadémiai Kiadó, Budapest, pp. 267-281.

Barron, A. R. (1981). Properties of the Predicted Squared Error: A Criterion for Selecting Variables, Ranking Models, and Determining Order. Adaptronics, Inc., McLean, Va.

Barron, A. R. (1982). Complexity Approach to Estimating the Order of a Model. Electrical Engineering 378B Final Report, Information Systems Laboratory, Stanford University.

Bibby, J., and Toutenburg, H. (1977). Prediction and Improved Estimation in Linear Models. Wiley, New York.

Mallows, C. L. (1973). Some Comments on C_p. Technometrics 15:661-675.

Rissanen, J. (1978). Modeling by Shortest Data Description. Automatica 14: 465-471.

Rissanen, J. (1983). A Universal Prior for Integers and Estimation by Minimum Description Length. Ann. Stat. 11(2):416-431.

Schwarz, G. (1977). Estimating the Dimension of a Model. Ann. Stat. 6(2): 461-464.

5
Past, Present, and Future of GMDH

A. G. IVAKHNENKO / Institute of Cybernetics, Ukrainian Academy of
Sciences, Kiev, Ukraine, Union of Soviet Socialist Republics

I. CYBERNETICS—THE SCIENCE OF MODELING

A. Introduction

Cybernetics was originally conceived by Norbert Wiener as the science of
"control and communication in animals and machine" [1]. Rapid growth of
cybernetics resulted in the development of many allied sciences. But in
recent years, beginning in the late sixties, its outward development slowed
and turned inward. Mathematical modeling was the new focus necessary for
the solution of a great variety of society's problems, including the optimal
design of computers and computer networks. Cybernetics, indeed, had become the general and unifying science of the modeling of communications
and control of complex systems.

B. The Basic Problem of Modeling

The following four problems are the basic goals of modeling:

1. Systems analysis or interaction between variables in complex systems
2. Structural and parametric identification of the system
3. Long-range qualitative and quantitative prediction of a system
4. Decision making and planning

This review is written in response to an invitation made by Dr. Stanley
Farlow of the University of Maine and is devoted to the theory and application of GMDH. It is written in conjunction with the planned international
conference on GMDH.

The system analysis of the interaction between variables precedes the identification problem. It requires that we not only define the set of important variables, but also the subdividing of them into two smaller sets—the output and the state variables. In identification, the output variables are given, and it is necessary to find the structure of the system (equations, coefficients, ...). Identification results in the so-called physical model of the system and therefore can be called the discovery of the physical law. When the data are noisy, the model can only be used for studying short-term predictions (long-range quantitative predictions are impossible). Nevertheless, we can often organize fuzzy or qualitative long-range predictions with the aid of a scenario game on the principle, if-then. In other words, for noisy data, one is only able to make short-term quantitative predictions and long-range qualitative ones. The problem of quantitative prediction for noisy data can only be solved by means of self-organizing models.

Another philosophy of modeling suggested by Gabor [2], based on the principle of consequent decisions (not final), is less developed.

C. Modeling Developed from Information and Control Theory

The problem of long-range quantitative prediction illustrates the distinction between classical modeling and self-organizing models. One should only use one's experience gained in observing the system or by taking observations to serve as the basis for the design of a mathematical model. The experience can be used in the form of already completed ideas of the researcher or in a table of data (the results of an active or a passive experiment). The first form of experience leads to simulation modeling, while the second one leads to the experimental method of the theory of self-organization. Well-known models of world dynamics developed by the American scientist J. Forrester [3] is a classical example of simulation modeling. The author of this model calls his method dynamic and this confirms its links with automatic control theory. This theory is characterized by the somewhat exaggerated attention to the problems of dynamic and process stability and by insufficient investigations of the quasi-static problem (trends). Information (or communication) theory pays more attention to the quasi-static component of the process (which is called the optimal input signal or signal trend). This component is considered to be white noise and is called the fluctuation disturbance. The ideas of information theory give rise to a number of directions in modeling. The first is the modeling of the input/output matrix, connected with the name of the English scientist Richard Stone and the author of the model on the British economy [4]. The information theory approach is also represented in the theory of model self-organization demonstrated by the GMDH algorithms. We can also call the self-organizing models information-type models [30, 40].

D. The Future of Cybernetics: A Combined Method

The two control theory approaches mentioned above (information and automatic control) are very different in their philosophy, but that is not to say that they cannot both be used in the development of cybernetics. Beginning with Wiener's initial work, nobody considered automatic control theory as a competitor of information theory in spite of differences in their approaches. A combined method using both methods is often the most effective.

One can imagine the cybernetics of the future as some metascience of modeling which uses the applications of all possible methods jointly: simulation, probability, and self-organization. The main facet of this combined method is: everything known is to be given to the computer. The weak point of simulation nowadays is that the researcher must know the laws of the system, including those which he does not know at all. By means of this combined method, complex systems are modeled with the aid of experimental methods of self-organization. Only the most obvious system components are to be chosen by the researcher himself.

E. Peculiarities of the Combined Method

Let us list the main stages of construction of physical models by the combined method:

1. The collecting of the data (it is desirable to take measurements in equal time intervals and equal steps in space)—the maximum values of the intervals and steps can be found using Shannon-Kotolnikov's theorem [40]
2. The choice of the physical law to be used as the reference function (this is a component of the simulation modeling method)
3. The replacement of the continuous mathematical model by its discrete analog (finite-difference equations)
4. The estimation and adaptation of the difference equations (this is the component of the self-organization method)

F. The Inductive Method of Self-Organization [36]

For those systems that are difficult to model, the inductive approach can be used to determine the structure of the model of optimal complexity. This optimal model is selected by sifting or searching through many different models according to some selection criterion. For long-range prediction, the prediction balance criterion is used as the criterion [15, 28, 31].

II. HISTORY OF THE GMDH ALGORITHM

A. An Overview

The method of mathematical modeling of complex systems, known as the group method of data handling (GMDH), is the essential part of the combined

method [37] of cybernetics and is an example of an inductive method [36].
It will be 16 years old soon—reaching its blossom of youthful strength.

Parallel to the development of the GMDH algorithm is the development
of the mathematical theory of computer-based model self-organization. The
applications and theory of these two areas are now being developed simultaneously, each benefiting the other. It is remarkable that there is no end
in sight. New horizons open continuously, turning the GMDH algorithm into
an apparatus unifying many engineering sciences. The GMDH algorithm has
become the center of cybernetics and of a universal metascience for studying
the modeling of communication and control of complex systems.

The history of the GMDH development is easy to follow. It is sufficient
to look through the volumes of the journal Avtomatika, published by the
Institute of Cybernetics of the Academy of Sciences of the Ukrainian SSR,
beginning from 1968. Almost every issue contains papers related to GMDH
problems. In 1969, a book [8] was published in Russian with an appendix
devoted to the new method. It was the first publication on GMDH. In this
treatise, it was shown that there existed an analogy between the perceptron
of Rosenblatt [16] and multilayered GMDH algorithms.

The complete Kolmogorov-Gabor polynomial (discrete analog to the
Volterra series)

$$y = a + \sum_{i=1}^{m} b_i x_i + \sum_{i=1}^{m} \sum_{j=1}^{m} c_{ij} x_i x_j + \sum_{i=1}^{m} \sum_{j=1}^{m} \sum_{k=1}^{m} d_{ijk} x_i x_j x_k + \cdots$$

can be shown to be replaced by a composition of lower-order polynomials
of the form

$$y = A + B x_i + C x_j + D x_i^2 + E x_j^2 + F x_i x_j$$

To find these polynomials (the coefficients A, B, C, D, E, and F) it is
sufficient to have only six data points at our disposal. Repeated solution of
the quadratic polynomials above enables us to construct the complete polynomial of any complexity using only these six points.*

The system of partial descriptions (the quadratic polynomials) is
allowed to grow as long as the root-mean-square error, computed from
two or three additional data points, is decreasing (external regularity
criterion).

A cover-to-cover translation of Avtomatika was published in the United
States from January 1968 until May 1971 under the title Soviet Automatic

---

*Editor's note: The details of how the Kolmogorov-Gabor polynomial (or
what some people working in GMDH call the Ivakhnenko polynomial) is constructed can be found in Chap. 1.

Control—Ukrainian Journal of Automation by Scripta Electronica, Inc.
Since then it has been published under the title Soviet Automatic Control
by the Scripta Publishing Corporation. Additional papers [9, 10, 12] have
been published in English in other publications.

One of the earliest studies of GMDH abroad was Shankar's dissertation
[13]. He compared GMDH to regression analysis for accuracy of modeling
with limited data. Later, other GMDH-type algorithms were published by
Japanese authors [18, 21, 23, 29] and by the Polish scientist Maciejowsky [27].
The conclusion of [27] was that GMDH is the best attack for the solution of
identification and short-range prediction problems. Unfortunately, the
study was not complete because the author did not consider the possibilities
of the harmonic and multilevel GMDH algorithms with the balance-of-
variables criterion. Had he done this, the author would have concluded that
for long-range prediction, the GMDH algorithm is the best method (probably
the only one) for ensuring the accuracy of prediction of an almost unlimited
lead time.

The mathematical theory of GMDH has shown that statistical regression
is a particular case of GMDH algorithms [34, 42–44]. Hence GMDH, being
more general, cannot be worse. For this reason, it is of no value to compare GMDH to regression.

Now in 1982, tens of young scientists have joined the development of
GMDH theory and applications. GMDH belongs to many authors, not to one
of them. Nevertheless, it is here in Kiev where we can see more clearly
what has been done, what is to be done, and what we can expect in the future.

The period from 1968 to 1971 was characterized by the use of only one
selection criterion and by the complete neglect of noise (this disregard of
noise continues in papers published abroad). In the period from 1972 to
1975, the first attack was made on noisy data. The multicriteria choice of
models appeared at that time, as well as special forms of criteria and the
GMDH harmonic algorithm. In the period from 1976 to 1979, important
results were obtained in the mathematical theory of GMDH (for the identification problem). During 1980 to 1982, the noise-resistant, multilevel,
long-range, detailed prediction algorithms appeared. In addition, new
objective system analysis of complex systems was developed. In the near
future we must undertake the problem of increased noise resistance of
GMDH algorithms (designed models insensitive to noise) with the help of
ideas of communication and information theory [40]. In addition, we must
continue the development of the GMDH theory for the prediction problem,
including applications to multicriteria control [25]. Let us discuss review
the history in detail now.

B. The Period 1968-1971

In the early stages, the GMDH algorithm was applied to the solution of
pattern recognition, identification, and short-range prediction problems.

These problems were solved by sifting (looking at them one at a time) the models by use of the single external criterion of regularity

$$\Delta(nc) = \frac{\Sigma_1^{nc} (q_f - q_m)^2}{\Sigma_1^{nc} q_j^2} \rightarrow min$$

where

q_f = observed dependent variable

q_m = computed estimate of dependent variable from the model

n = nc + nt = number of observations

nt = number of observations in the training set (about 70% of data)

nc = number of observations in the checking set (about 30% of data)

A more thorough description of nt and nc can be found in Chap. 1. Many proposals were suggested for dividing the input data set into the two parts nt and nc [20]. One suggestion was to order the data according to the variation of each observation. It was found experimentally that this method provides the minimum number of selection layers in multilayered GMDH algorithms. Only multilayered algorithms were considered during this period.

Another suggestion for data division was suggested in 1980 [33]. Sometimes the regularity criterion was used in the form of a correlation coefficient between the q_f and q_m variables or in the form of a correlation index (for nonlinear models).

For reference functions, polynomials, logical nets, and Zadeh fuzzy sets were used. In 1971, GMDH algorithms were designed with Bayes' formulas being used as reference functions (probabilistic GMDH algorithms were used successfully for solving ecological problems. The authors of the early papers were impressed by the high accuracy of the short-range predictions. The only models studied in this period were those with a low level of noise. The major weakness of the GMDH algorithm during this period was its sensitivity to incomplete information. We now progress to the next period.

C. The Period 1972-1975

The challenge of noisy data and incomplete information was met in the period 1972-1975. Two ways to handle noisy data in the GMDH algorithm were proposed:

1. Multicriteria choice of the model with a special form of the selection criteria [17, 24]
2. The use of additional a priori information (combined method mentioned earlier)

The basic ideas of multicriteria choice are similar to those of pattern recognition theory [30]. This theory can be used to find the set of most effective criteria. Suggested were the minimum-of-bias criterion and the criterion of two-variables balance [20].

The first criterion is recommended for those cases where one tries to obtain a physical model [14]. The second is used for long-range predictions [15]. Many other criteria were proposed in this period, criterion of short-range prediction accuracy, criterion of step-by-step prediction, several criteria of probabilistic stability for long-range prediction and criteria for the stability of distributions. We are convinced that the wide use of the minimum-of-bias and the balance-of-variables criteria, together with the solution of the noise resistance problem, are the major ways of improving present-day models.

The use of additional information can also be used to make the sifting shorter and, what is more important, it increases the accuracy and lead time of prediction. For example, if we know the general form of the characteristic roots, and if the number of hypotheses is small, there is surely a right model among them. The properties of the solution for definite forms of roots might be known, so many of them can be set aside and not considered. Thus we must only organize the sifting about the root properties and find (using Vieta formulas) the restrictions for the polynomial coefficients. Often we arrive at the GMDH harmonic algorithm [17].

Another example of the general method of searching among a priori hypotheses (where the number of feasible models is small and one of them is significant) is the following. We speak of the generalized GMDH algorithm. The sifting of hypotheses among cross-correlations of variables is the major facet in this method. For example, with four variables we could hypothesize:

Hypothesis 1: The variables are not correlated:

$$q = a_0 + f_1(x_1) + f_2(x_2) + f_3(x_3) + f_4(x_4)$$

Hypothesis 2: The first variable is not correlated, but the others are:

$$q = a_0 + f_5(x_1) + f_6(x_2, x_3, x_4)$$

Hypothesis 3: The first and second variables are correlated, and the third and four are correlated:

$$q = a_0 + f_7(x_1, x_2) + f_8(x_3, x_4)$$

and so on. Here f_1, f_2, f_3, f_4, and f_5 are all first-degree polynomials, while f_7 and f_8 are second degree. The function f_6 is third degree. We can present more examples, and it is clear that one of these hypotheses is optimal when the number of hypotheses is small.

Additional a priori information sharply increases the noise resistance of GMDH algorithms (the harmonic and generalized algorithms demonstrate the best noise resistance), but for practical purposes it is necessary to increase it by a factor of 10. To study noise resistance properties of GMDH algorithms, computational experiments have been developed [24, 26, 39]. It was found that multicriteria models are much better than the choice of the model by a single criterion. However, the best experiments have not yet given a noise-to-signal ratio a ratio better than 10 (the noise can be 10 times higher than the signal). In other words, it is not sufficient to handle practical problems in weather or ecological forecasting. Another way to improve noise resistance in GMDH algorithms is by the use of communication theory.

The originally proposed multilayered GMDH algorithms were based on the selection hypothesis used in agricultural practice [17]. These algorithms do not require many points of input data, and in fact the number of data points can be several times less than the number of parameters in the complete polynomial.

However, in 1972 a one-layered or combinatorial GMDH algorithm was suggested. This algorithm does not have the multilayerness error and will not miss the optimal model since it searches all possibilities. From a practical point of view it can be used for those problems where the number of the complete polynomial components is less than 20. This corresponds to 3 hours of computation on the BESM-6 computer. Each additional variable increases the calculation time twofold. There are three ways to combat this excessive computational time:

1. Preliminary statistical treatment to decrease the number of variables (the range correlation method)
2. The multilayered GMDH algorithms
3. The method of cutting the sifting triangle

We shall explain the last method. Polynomials used in combinatorial GMDH algorithms can be written in the following triangular form. In the first line we write the complete polynomial, in the second we write the next shorter one, and so on until the elementary quadratic polynomials are written. The sifting of models is applied only to those polynomials that have fewer than 20 terms. The explanation of the method is very simple; the laws of nature are always simple and visible. We shall not consider overly complicated multiple-component models, as they are always impossible. If we recall that nonphysical optimal models (which we obtain when noise is increasing) are always simpler than the physical ones, it becomes clear that the cutting of the sifting triangle can be applied for both physical and nonphysical models.

Multilayered GMDH algorithms are now rarely used when we use models having only one equation. They are generally used for choosing system equations in system analysis as well as in those cases where we have a very small number of data points (when modeling quickly changing processes).

D. The Period 1976-1979

The development of self-organization theory during the preceding period is described in two books [17, 19]. Here it was shown that many multilayered GMDH algorithms have a multilayerness error, which is small but sometimes essential. This error is similar to the statistical error in control systems. It was shown that multilayered algorithms, like statistical control systems, converge to an equilibrium point (concentration point).

Types of convergence were studied more thoroughly later. As a result, multilayered algorithms were proposed which have no multilayerness error [17, 34, 42]. In those cases where the input data change in time, additional data points are added and the ideas of the control system invariance are suitable for analysis [45].

There exists an area of application of control theory to self-organization theory, but so far only statistical problems with static input data have been investigated. The theory of servomechanism stability with its dynamic approach should be applied for studying long-range prediction problems when the input data change. Here, instead of algorithm convergence to the point of concentration, the convergence of trends in the processes should be studied.

Multilayered GMDH algorithms were proposed in 1977. The system analysis can be carried out on the level of trends (using algebraic equations) or on the level of differential equations or their finite-difference analogs [22, 41]. The most general analysis is based on equations of the form

$$q = f_1(t, u, T) + f_2(q_{-1}, q_{-2}, \ldots, q_{-T})$$

where f is called the source function in mathematical physics [36]. To simplify the analysis, it is divided into two parts:

1. Trend analysis: $q_1 = f_1(t, u, T)$
2. Dynamic analysis: $q_2 = f_2(q_{-1}, q_{-2}, \ldots, q_{-T})$

Although $q \neq q_1 + q_2$, to obtain $q = q_1 + q_2$ one can use a table of deviations from the trends in the second equation.

There was no objective system analysis before this work; only personal subjective analysis was used. It turns out that sifting, using a collection of criteria (including the minimum-of-bias criterion), enables us to select the unique optimal system of equations (algebraic or finite-difference) and therefore to find the components of complex systems.

We now have at our disposal a powerful method of modeling in an objective way many important questions facing humankind in ecology, economics, social, and other sciences. Computers become the objective judge in solving scientific discussions when the ensemble of criteria is agreed upon among experts. The promises of cybernetics given in its early years can now be fulfilled. The new form of system analysis must completely change the existing global economic models (such as the LINK model [5, 6] and others) and increase their performance. For example, in the BRITAIN block of the LINK model, 226 polynomial regression equations are used. But what is the value of all these equations from the point of view of long-range prediction if they strongly depend on the choice of the interval of the input-data table from which they are received? Simply, we cannot use them if sifting with a collection of criteria is not applied. The GMDH algorithm, using objective system analysis, has chosen for this model only four equations or four output variables [41]. They are the heart of the system and only they satisfy the minimum-of-bias criterion (they remain constant over time). All other equations change continuously both in their structure and in their coefficient estimates. Therefore, it is necessary to predict only the four essential variables. Predictions for the remaining 222 can be found later as functions of the leading four, using the combinatorial GMDH algorithm and sifting with the prediction balance criterion.

E. The Period 1980–1982

Many interesting results were found during this most recent period. First, it became clear that physical models (with noise) cannot be used for long-range predictions. The stronger the noise, the simpler the models chosen by the GMDH algorithms. So physical models are obtained only for use in identification and short-range prediction.

The only thing true of long-range research is that the prediction balance criterion will be useful. For long-range prediction, we must organize additional sifting using this balance criterion. We say additional because we want to avoid nonuniqueness. Only models already with the aid of some additional criteria may be compared using the balance criterion. The balance criterion becomes the main tool in choosing the model of optimal complexity.

Two-level GMDH algorithms for long-range prediction use the balance criterion to compare several predictions obtained with two different reference functions. For example, the values of the annual and seasonal predictions are compared and the best pair of predictions is chosen. The simultaneous use of two reference functions is a new development in modeling. The two-level, long-range prediction was illustrated on an example of extrapolation problems [28] and of the prediction of the Baikal lace ecological system [31]. The predictions were given for 20 years and they have already been fulfilled satisfactorily for 4 years.

It is important to find the leading variable in a set of variables of the given system. The leading one is the one that can predict more accurately than the others. We recommend the generalized or harmonic (or exponential-harmonic) GMDH algorithms for prediction of the leading variable since they have the highest noise-resistance properties [17]. When we know the leading variable, it is easy to predict the others using the balance criterion.

Since the paper [32], polynomials with variable coefficients have been used widely as reference functions. For example, if we have three control variables, u_1, u_2, and u_3, and four other variables, q_1, q_2, q_3, and q_4, we can write the complete polynomials for the GMDH combinatorial algorithm as

$$q_1 = (a_0 + a_1 u_1 + a_2 u_2 + a_3 u_3) + (b_0 + b_1 u_1 + b_2 u_2 + b_3 u_3)q_2$$
$$+ (c_0 + c_1 u_1 + c_2 u_2 + c_3 u_3)q_3 + (d_0 + d_1 u_1 + d_2 u_2 + d_3 u_3)q_4$$

and for the finite-difference equations we have

$$q_{+1(i)} = (a_0 + a_1 u_1 + a_2 u_2 + a_3 u_3) + (b_0 + b_1 u_1 + b_2 u_2 + b_3 u_3)q_{0(i)}$$
$$+ (c_0 + c_1 u_1 + c_2 u_2 + c_3 u_3)q_{-1(i)} + (d_0 + d_1 u_1 + d_2 u_2 + d_3 u_3)q_{-2(i)}$$

for $i = 1$, 2, 3, 4 and q_0, q_{-1}, and q_{-2} are delayed arguments. These equations have helped solve the problem of studying inflation stability and are described in [41]. By means of the computer, both equations can be united into a more general form.

If J. Forrester would use equations with variable coefficients, he could demonstrate an elastic variety of possible controls for world dynamics and not the stilted control that was obtained with the equations of constant coefficients.

The problem of multilayered algorithm stability has been studied in [42] and [43]. All results received before 1980 are summarized in two books [36, 37].

III. THE FUTURE OF GMDH

The main task is to handle problems with noise 10 to 100 times current levels. This is necessary to solve problems in the prediction of complex processes in ecology, meteorology, economics, and other areas. Such an increase can be achieved by carrying over the results of communication theory (including coding theory and pattern recognition) into the theory of self-organization [42]. Having noise-proof GMDH algorithms, we will be able to solve problems of long-range prediction even if the data are

incomplete [36]. To lower the sensitivity to noise with the aid of the correlation principle of signal reception, we have begun investigations into the
problem of inverse transforms [38]. The inverse transform promises many
wonderful applications, such as accurate prediction of earthquakes, prediction of pollution on the surface of lakes and seas, and the solution of many
problems stated in Thom's theory of catastrophes. What is the density of
the Baikal Lake? What about other lakes and seas? Is the water level in the
oceans rising? Is the climate of our planet changing? What are the forecasts for agricultural crops? What is the inflation rate? Questions like
these can be answered objectively only by the use of the GMDH algorithms
for the self-organization of models. All processes of the past, present, or
future correspond to definite points in Gilbert's space, widely used in the
theory of communications. The vector projection theory (including the
mean-squared error method) works in such a space, the dimension of which
increases with every new measurement. The necessity of comparison of
both theories is obvious. The ideal receiver in the Kotelnikov-Shannon
theory calculates the distance between the input signal and the reference
points of the signal in Gilbert's space and finds its minimum.

Pattern recognition proposes a more satisfactory way to find optimum
decisions in self-organizing models for identification and for short- and
long-range predictions. This is particularly useful with noisy data. Communication theory and pattern recognition differ from one another by the number
of dimensions used in model self-organization. On the other hand, communication theory and pattern recognition can take from the theory of self-
organization the concept of external criterion. The internal criteria currently
used in these theories does not allow one to distinguish the model of optimal
complexity from the more complex overfitted ones.

When the above-mentioned measures are taken into account, the existing
GMDH algorithms are often insensitive to noise. Hence the work before us
is to bring GMDH to a wider audience so that we will gain the support of
public opinion. A famous Ukrainian scientist, V. I. Vernadskiy, has said
that to solve problems facing humankind, it is necessary first to organize
public opinion around them.

For example, the system analysis of physical models and the subsequent
self-organization of long-range prediction by means of nonphysical ones
were used to clarify the mechanism of inflation in Britain [41]. It was shown
that the introduction of comparators (known in automatic control theory) can
stop inflation. Of course, nobody thinks comparators will be realized immediately, but an objective accurate result no doubt will influence public
understanding in the necessity of structural changes in economic systems.
This can possibly lead to important changes in the minds of people.

The last problem is this. There are no sessions at world or national
congresses or conferences devoted to GMDH. Publications in major journals
on control or communication theory in the United States and the USSR are
not devoted to GMDH papers. This situation must be changed for the benefit
of humankind.

REFERENCES

General References

1. N. Wiener, Cybernetics, Control and Communication in the Animal and the Machine, Technology Press—Wiley, New York, 1948.
2. D. Gabor, Freedom of Choice at Subsequent Time Instants as an Important Optimality Criterion for Control of Complex Random Processes, Sov. Autom. Control 5:58 (1972).
3. J. W. Forrester, World Dynamics, Wright Allen Press, 1971.
4. R. Stone, Computer Model of Economic Growth, Avtomatika, 39 (1963).
5. V. S. Dadayan, Global Economic Models, Nauka, (1981).
6. Interaction of Nature and the Economy in the Baikal Region, A. K. Alamazyan (Ed.), Nauka, 25 (1981).

Period 1968-1971

7. A. G. Ivakhenko, The Group Method of Data Handling—A rival of Stochastic Approximation, Sov. Autom. Control 3:43 (1968).
8. A. G. Ivakhnenko, Self-Teaching Systems of Recognition and Automatic Control, Tekhnika, 392 (1969).
9. A. G. Ivakhnenko, Heuristic Self-Organization in Problems of Engineering Cybernetics, Avtomatika 6:207 (1970).
10. A. G. Ivakhnenko, Lösung von Interpolationsproblemen der technischen Kybernetik nach der Methode der gruppenweise Argumentenberechnung, Mess. Steuern Regeln 13:134 (1970).
11. A. G. Ivakhnenko, Systems of Heuristic Self-Organization in Technical Cybernetics, Tekhnika, 372 (1971).
12. A. G. Ivakhnenko, Polynomial Theory of Complex Systems, IEEE Trans. Syst. Man Cybern. 4:364 (1971).

Period 1972-1975

13. Ramesh Shankar, The GMDH, Master's thesis, University of Delaware, 1972.
14. A. G. Ivakhnenko, V. V. Shervashidze, O. I. Sheludko, S. G. Paterern, and N. A. Ivakhnenko, Discovery of Physical Laws by GMDH Method with the Absence of Bias Criterion, Sov. Autom. Control 6:32 (1973).
15. A. G. Ivakhnenko and N. A. Ivakhnenko, Long-Term Prediction of Random Processes by GMDH Algorithms Using the Unbiasedness Criterion and the Balance-of-Variables Criterion, Sov. Autom. Control 7: 40 (1974).
16. A. G. Ivakhnenko (Ed.), Perception-System Pattern Recognition, Naukova Dumka, 430 (1975).
17. A. G. Ivakhnenko, Long-Term Prognosis and Control for Complex Systems, Tekhnika, 372 (1975).
18. J. Kondo, Air Pollution, Korona Company, Tokyo, 1975.

Period 1976-1979

19. A. G. Ivakhnenko, H. P. Zaichenko, and V. D. Dimitrov, Decision Making on the Principle of Self-Organization, Soviet Radio, 275 (1976).
20. A. G. Ivakhnenko, N. A. Ivakhnenko, and V. M. Vysotsky, Long-Term Prediction of Random Processes by GMDH Algorithms Using the Unbiasedness and Balance-of-Variables Criteria, Sov. Autom. Control 9:28 (1976).
21. S. Ikeda, M. Ochiai, and Y. Sawaragi, Sequential GMDH Algorithm and Its Application to River Flow Prediction, IEEE Trans. Syst. Man Cybern. SMC-6, 473 (1976).
22. R. Stone, A. G. Ivakhnenko, V. M. Vysotskyy, and L. P. Somina, Discovery of Laws Governing the Behavior of Complex Objects Whose Variables Are Not Indicated, Sov. Autom. Control 10:21 (1977).
23. V. Tamura and T. Kondo, Large-Spatial Pattern Identification of Air Pollution by a Combined Source Receptor Matrix and Revised GMDH, Proc. IFAC Symp. Environ. Syst. Planning, Design Control, 378 (1977).
24. A. G. Ivakhnenko, V. N. Vysotsky, and N. A. Ivakhnenko, Principal Versions of the Minimum Bias Criterion for a Model and an Investigation of Their Noise Immunity, Sov. Autom. Control 11: 27 (1978).
25. A. G. Ivakhnenko and Yu. Yurachkovskiy, Use of Long-Term Prediction in Selection of Optimum Control from the Domain of Pareto-Efficient Decisions, Sov. Autom. Control 11:31 (1978).
26. Vysotskiy and J. Ihara, Improvement of Noise Immunity of GMDH Selection Criteria by Using Vector Representation and Minimax Forms, Sov. Autom. Control 11:1 (1978).
27. J. M. Maciejowsky, The Modelling of Systems with Small Observation Sets, Lect. Notes Control Inf. Sci. 10:242 (1978).
28. A. G. Ivakhnenko and M. G. Rao, Prediction and Extrapolation of Meteorological Fields by Model Self-Organization, Sov. Autom. Control 12:13 (1979).
29. Y. Sawaragi, T. Soeda, and H. Tamura, Statistical Prediction of Air Pollution Levels Using Non-Physical Models, Avtomatika 15:453 (1979).

Period 1980-1982

30. A. G. Ivakhnenko, Yu. V. Koppa, D. N. Lantayeva, and N. A. Ivakhnenko, The Relationship Between Computer Self-Organization of Mathematical Models and Pattern Recognition, Sov. Autom. Control 13:1 (1980).
31. A. G. Ivakhnenko, G. I. Krotov, and V. I. Cheberkus, Multilayer Algorithm for Self-Organization of Long-Term Predictions, Sov. Autom. Control 13:22 (1980).
32. A. G. Ivakhnenko and G. I. Krotov, Self-Organization of Models with Variable Coefficients for Optimizing of the Planning of Water Reservoirs, Sov. Autom. Control 13:9 (1980).

33. Yu. P. Yurachkovskiy and A. N. Groshkov, Optimal Partition of Initial Data into a Training Sequence and a Checking Sequence by Analysis of the Distribution Function of the Criterion, Sov. Autom. Control 13: 1 (1980).

34. Yu. P. Yarachkovskiy, Convergence of Multilayer GMDH Algorithms, Sov. Autom. Control 14:36 (1981).

35. G. I. Krotov, Studies on Stability of Certain Two-Level Algorithms of Long-Range Prediction, Sov. Autom. Control 14:72 (1981).

36. A. G. Ivakhnenko, Inductive Method of Self-Organized Complex Models System, Naukova Dumka, (1982).

37. A. G. Ivakhnenko, N. N. Vostrov, and P. H. Pecka, Pairing Method of Selection and Management of Hydroelectric Plants and Oil Fields, Naukova Dumka, 850 (1983).

38. A. G. Ivakhnenko and V. V. Osipenko, An Algorithm of Inverse Transform from Probability Characteristics to Deterministic Prediction, Sov. Autom. Control 15:7 (1982).

39. A. G. Ivakhnenko, On the Golden Book of Long-Range Quantitative Predictions Which Came True, Sov. Autom. Control 15:87 (1982).

40. A. G. Ivakhnenko and A. M. Karpinsky, Model Self-Organization Using Computers in Terms of Information Theory, Sov. Autom. Control 15: (1982).

41. A. G. Ivakhnenko and Yu. V. Kostenko, System Based on Model Self-Organization, Sov. Autom. Control 15: (1982).

42. Yu. P. Yurachkovskiy, Structural Modelling by Observation Sets, Sov. Autom. Control 15: (1982).

43. Yu. P. Yurachkovskiy, The GMDH as an Effective Means for the Solution of the Structural Modelling Problem, Sov. Autom. Control 15: (1982).

44. A. G. Ivakhnenko and H. P. Hurachkovski, Self-Organization at a System of Complex Models, Radio and Signal (Moscow), (1981).

45. A. G. Ivakhnenko, E. A. Belozerskiy, and Yu. P. Yurachkovskiy, Invariance Principles in Self-Organization Theory, Sov. Autom. Control 15: (1982).

6

Forecasting Applications of GMDH in Agricultural and Meteorological Time Series

WILLIAM M. LEBOW, RAMAN K. MEHRA, and PAUL M. TOLDALAGI*/
Scientific Systems, Inc., Cambridge, Massachusetts

HARBERT RICE*/Strategic Information, Burlington, Massachusetts

I. INTRODUCTION

The group method of data handling (GMDH) can be considered as a statistical pattern recognition technique. This paper discusses an implementation of a modification of the algorithm, including the DUPLEX algorithm for data splitting and the PRESS criterion for model selection. We also provide numerical results in applications to climate/crop yield forecasts and to forecasts of meteorological time series.

II. GMDH ALGORITHM

The basic GMDH algorithm was proposed by Ivakhnenko. GMDH identifies polynomial models which are discrete-time analogs of Volterra series introduced into nonlinear analysis by Wiener [1]. The basic building block of GMDH is a quadratic polynomial in two variables.

Each "layer" of GMDH consists of a bank of quadratic polynomial functions with inputs from the previous layer having been passed through a selection layer. The initial input variables $(x_1, x_2, \ldots, x_N)$ are used in all pairwise combinations to construct $\binom{N}{2}$ quadratic polynomials. These outputs are filtered through a selection criterion that is heuristic. The better results are chosen and used as inputs to the next layer. Again, a bank of quadratic polynomials are formed and the outputs filtered. This process is continued until only one output passes the selection layer or until a predetermined stopping criterion is satisfied.

---

*Present affiliations: Analog Devices, Inc., Norwood, Massachusetts

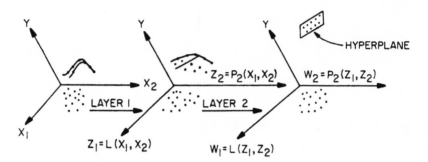

Fig. 1 GMDH transformation of independent variables.

Each layer of GMDH can be thought of as a transformation of the independent variables. The transformation is continued until the dependent variable becomes a linear function of the transformed independent variables. A simple example with two input variables x_1 and x_2 is shown in Fig. 1. Notice that the Y-axis (output variable) remains unchanged, but that (x_1, x_2) are successively transformed into (z_1, z_2) and (w_1, w_2) such that the nonlinear relationship $y = f(x_1, x_2)$ is well approximated by the linear relationship $y = L(w_1, w_2)$. In this sense, GMDH inverts the nonlinear relationship $f(\cdot, \cdot)$. Alternatively, one may think of GMDH as generating new "feature vectors" (w_1, w_2) such that distinct classes or clusters can be separated by hyperplanes rather than curved surfaces in the original input data space.

Various modifications of the original algorithm have been proposed. Duffy and Franklin [2] have suggested the use of stepwise regression in parameter estimation to control multicollinearity problems in the basic algorithm. Ikeda et al. [3] have proposed a sequential form of the algorithm for real-time forecasting. We have implemented both of these suggestions. In addition, we have implemented the DUPLEX algorithm for data splitting and the PRESS criterion for model selection.

A. DUPLEX—A Data-Splitting Algorithm [4]

The notion of data splitting is common in pattern recognition procedures. Where data are split, one set is used to train the model by estimating parameters ("estimation set," "training set," or "fit set"). The second set is used to evaluate the model based on its "prediction capability" ("prediction set," "testing set," or "check set"). Use of the predicting set can help to avoid the problem of overfitting. Relatively few methods are available for data splitting. Random selection and jackknifing techniques are two. Random selection works well where a large number of data points are available. For small data sets GMDH models will be sensitive to the data-splitting procedure.

An objective procedure is the DUPLEX procedure. According to Snee [4]: "The objective of DUPLEX is to divide the data into two sets which

cover approximately the same region and have similar statistical properties."
DUPLEX is akin to the CADEX algorithm of Kennard and Stone [5]. We
begin with a set of candidate points, which have been standardized and
orthonormalized as described in Ref. 5. The pair of candidate points
farthest apart in Euclidean distance is removed from the candidate set and
placed in the estimation set. Of the remaining candidate points, the pair
farthest apart in Euclidean distance is removed from the candidate set and
placed in the prediction set. Next, the candidate point farthest from the
existing estimation set is moved to the estimation set and the candidate
point farthest from the existing prediction set is moved to the prediction
set. This last step is repeated until the candidate set is empty.

It is helpful to have a measure of the statistical properties of the estimation and prediction sets. The volume of the region covered by a set
of points is related to $|X'X|$, where the k columns of X are the coordi-
nates of the points. Two determinants are compared by computing the
kth root of the ratio of the determinants. Hence we use $(|X'X|_{EST} /$
$|X'X|_{PRED})^{1/k}$ as a measure of the relative statistical properties of
the estimation and prediction sets.

To illustrate the operation of DUPLEX, 16 points in a 4 × 4 array
were arbitrarily specify to simulate a situation involving correlated
predictor variables. The convex hull of the 16 points form a parallelo-
gram, as shown in Fig. 2. The figure also shows how DUPLEX divides

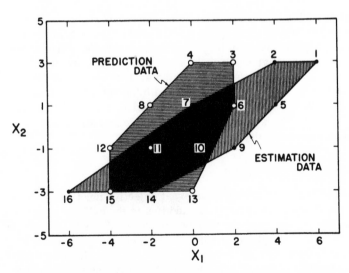

Fig. 2 DUPLEX two-factor example—convex hull of estimation and prediction data. (From Ref. 4.)

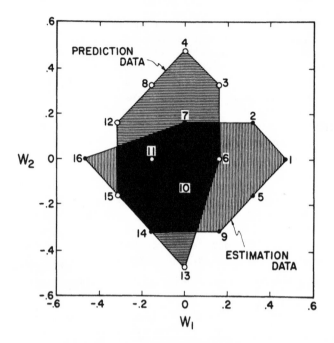

Fig. 3 DUPLEX two-factor example—orthonormalized units. (From
Ref. 4.)

the set of points. The points are first standardized by subtracting the
mean from each x and dividing by the square root of the sum of the x_i
squares to put both variables in the same units. Next, the standardized
variables are orthonormalized to make the region more spherical
(Fig. 3). Here $|X'X| = 1024$ for both data sets. The centroid of the
estimation set (1, 0) is in close agreement with the centroid of the prediction data (-1, 0) when one considers the x_1 and x_2 ranges. The equality
of the two determinants is specific to this example and should not be
interpreted as a general result. [4]

B.   PRESS—A Selection Criterion [6]

GMDH model identification is guided by external objective functions. These
functions are termed selection criteria. In addition to the criteria suggested
by Ivakhnenko in the original papers, we have found PRESS to be quite useful.
     According to Allen [6], PRESS (predicted residual sum of squares) is
defined by using the whole data set both as a fit set and as a check set. This
is achieved by changing the data points assigned to the two sets in a systematic fashion. To evaluate PRESS, each data point is used as a testing set of

size 1 and is predicted from the least-square coefficients developed from
the remaining N - 1 data points. One definition of PRESS is

$$\text{PRESS} = \sum_{i=1}^{N} [Y_i - \hat{Y}_i(i)]^2$$

where Y_i is the output for the ith observation and $\hat{Y}_i(i)$ is the estimated
value of the output evaluated from a model using the remaining N - 1 data
points $(x_1, x_2, \ldots, x_{i-1}, x_{i+1}, \ldots, x_N)$. Using the theory of recursive estimation (Kalman filtering) it can be shown that PRESS can be computed from
the results of one least-squares regression using all the data points. Heu-
ristically, we can write PRESS = fit variance + variance due to errors in
the model and the parameter estimates. It is the second term that penalizes
overfitting, since the variance of parameter estimates increases as the
ratio of p/N decreases (p is the number of parameters).

PRESS can also be used as an indication of outliers in the data set.
Consistently large values of PRESS indicate that one or more points are
estimated very badly by the remainder of the data set. In such cases it is
helpful to identify such points, check their correctness, and possibly discard these points from the data set.

III. APPLICATIONS

Three examples of GMDH applications to time-series forecasting are presented here. The first is the forecasting of Canadian crop yields using local
climatic data. The second is the forecasting of precipitation over a localized
region. The third is the forecasting of 700-mbar geopotential height averages over the northern hemisphere.

A. Climate/Crop Forecasting Models [7]

The following two examples use GMDH to forecast local climatic data and
crop yield data. Both examples are drawn from a series of GMDH model
identifications performed during 1977 using the Canadian prairie provinces
as a test area. These provinces comprise the majority of the Canadian
spring-wheat-growing region.

Figure 4 shows the Canadian crop district boundaries, and the spatial
distribution of the Thiessen polygon weighted climatic stations from which
the data set was drawn. For purposes of analysis, the spring wheat region
(Fig. 4) was divided into four regions:

1. Alberta
2. Northern Saskatchewan

Fig. 4 Map of Canadian provinces with MET stations.

Table 1 Input Variables Tested for Climate/Crop Models
in Canadian Prairie Provinces, 1977

Model type	Input variable[a]	Averaging time	Temporal coverage
Season	Total precipitation (TP)	Seasonal	January–April
	Average temperature (AT)	Seasonal	and
			May–September[b]
	Solar index (R)	Monthly	October–May[c]
Early season	Total precipitation (TP)	Monthly	October–May
	Average temperature (AT)	Monthly	October–May
	Solar index (R)	Monthly	October–May
Late season	Total precipitation (TP)	Monthly	October–September
	Average temperature (AT)	Monthly	October–September
	Solar index (R)	Monthly	October–August

[a]Climatic data from NOAA-CCEA (monthly TP and AT), Solar index from
NOAA.
[b]Seasonal May–September coverage for current season (lag 0) and preceding
season (lag 1 year).
[c]October of the preceding year.

3. Southern Saskatchewan
4. Manitoba

Northern and southern Saskatchewan were divided approximately equally in spatial coverage and were nearly equal in seeded acreage.

The range and source of input variables tested in forming climate/yield models is given in Table 1. The basic climatic data set consisted of monthly surface average temperature and total precipitation (9). In addition, formation of yield models was tested using a monthly solar index, R.

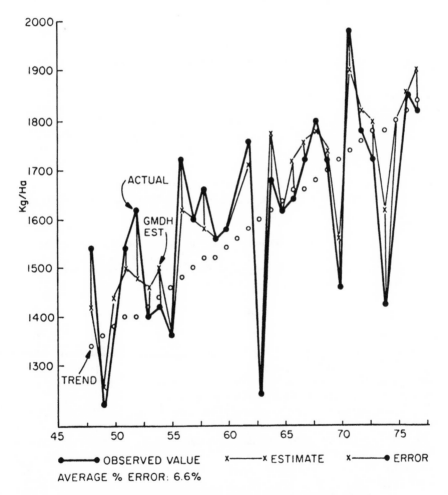

Fig. 5 GMDH Manitoba early season wheat model (1948-1976).

Table 2   Comparison of Early Season GMHD Wheat Models
and Regression Models for Manitoba Yields

Multiple regression[a]: input variables through June, August		GMDH identified model: input variables through May
Trend (1933-1974)[b]	Trend (1933-1974)[b]	Trend (1948-1976)[b]
April minimum temperature	April minimum temperature	February precipitation
June precipitation as potential evaporation	June precipitation	March precipitation
		April precipitation
	July precipitation	May precipitation
	August precipitation as potential evaporation	March average temperature
		May R
$r^2 = 0.29$-0.49	$r^2 = 0.48$-0.65	$r^2 = 0.73$
SEE = 232-252	SEE = 193-216	SEE = 132
		Average error 6.6%

[a]Lacie, NOAA-CCEA technical note 76-2. Lacie-00433. JSC-11658.
October 1976. Lacie models are for Manitoba regions 1 and 4.
[b]Standard deviation for Lacie series is 290-326; for the GMDH series, 227.

For purposes of model building, the growing season was defined as
October through September. Early season GMDH models were estimated
using information from October through May. Late season models were
estimated using information from October through September (Table 1).

1.   Early Season GMDH What Model

GMDH models were identified by forming a data matrix of lagged climatic
variables and solar index variables. These variables were selected without
regard to physiological assumptions. The data matrix was input to GMDH
and a check MSE was minimized. Twenty-one data points were used in the
training set and eight points in the check (testing) set.

Figure 5 shows the performance of an early season GMDH model for
Manitoba. This model was minimized at level 3. The average percent error
over the time history of the identification data is 6.6%. Models were identified both by detrending before input to GMDH, and also by allowing GMDH
to identify the trend. The model in Fig. 5 has a linear trend identified prior
to the GMDH run.

If the results of the GMDH model are compared to the regression models
identified for the same region, it is evident that the GMDH model is showing

performance characteristics equivalent to later season regression models. Table 2 compares the input variable structure and tracking characteristics of the Manitoba GMDH model to multiple regression models for the same region [8]. The regression model is identified on the basis of physical arguments for input variables.

The input structure for the GMDH model has five climatic variables plus a solar index variable. GMDH shows a higher R-squared value and a lower SEE value than those found using the regression model. With the addition of terms for July through September, the regression model improves considerably. From Table 2 it is apparent that GMDH is identifying a model that appears to extract more prior "information" from the data.

When the Manitoba early season model and related GMDH models were carried through to give a true step-ahead estimate of wheat yields in 1977, the average percent errors were similar to those encountered in identifying the models. Table 3 gives the 1977 estimates for the GMDH models and the corresponding estimates from Statistics Canada. The early season GMDH models estimated the relative wheat yields in the four test areas and the correct direction in yield from the preceding year. The combined Saskatchewan yield was underestimated.

B. Climate/Climate Models

Since we are treating the climate/crop system as an input/output system, it is feasible to examine the climatic component directly in the GMDH program. The same data may be used to form data matrices where the output variable has been changed to a climatic variable.

Table 3 GMDH Identified Models for Canadian Wheat Test, 1977

Region/model	Linear trend	1976 value	GMDH early	GMDH late	Statistics Canada[a]	Statistics Canada[b]	Average error (%)
Alberta	1939[c]	2186	1588	1529	1580	1639	6.7
Manitoba	1840	1823	1994	1974	2037	2131	6.4
Southern Saskatchewan	—	—	1416	1574	—	—	—
Northern Saskatchewan	—	—	1784	2033	—	—	—
Saskatchewan	1824	2119	1600	1803	1782	1938	7.0

[a]Statistics Canada 9/77 estimate.
[b]Statistics Canada 1/78 estimate.
[c]Yield as kilograms per hectare.

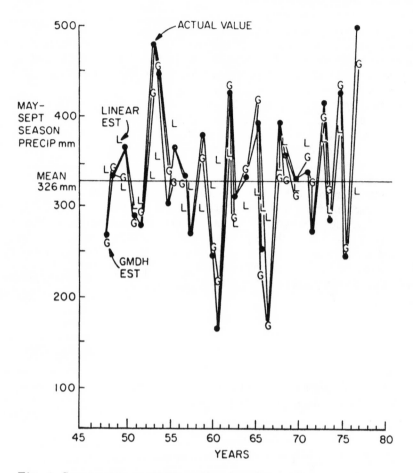

Fig. 6 Comparison of GMDH seasonal model and linear seasonal model for Manitoba precipitation (1948–1977).

A series of models were formed with seasonal precipitation as the output variable. Figure 6 shows the performance of the GMDH Manitoba growing-season model. This model gives a 5-months-ahead estimate of growing-season precipitation (May through September). The model was identified by searching the data through June of the preceding year over all four test regions. The GMDH model shown in Fig. 6 was identified by minimizing PRESS over 29 data points. The last point (1977) is a true step-ahead estimate.

The nonlinear model identified in GMDH shows a better tracking performance than that of a linear regression model. The linear model is a seasonal persistence model (10) which tends to track about the mean (Fig. 6).

Table 4 GMDH Canadian Prairie Provinces Seasonal Climate Tests, 1977

Region and model[a]		Output variable	Mean (mm)	Observed (mm)	Estimate (mm)	Error (%)
Manitoba	Mean	Precipitation	326	490	326	36
	Linear	—	—	—	314	37
	GMDH	—	—	—	448	9
Southern	Mean	Precipitation	248	257	248	4
Saskatchewan	Linear	—	—	—	196	23
	GMDH	—	—	—	277	8
Northern	Mean	Precipitation	251	298	251	16
Saskatchewan	Linear	—	—	—	262	12
	GMDH	—	—	—	274	8
Alberta	Mean	Precipitation	248	284	248	13
	Linear	—	—	—	289	1
	GMDH	—	—	—	270	4
				Average error:	Mean	17%
					Linear	18%
					GMDH	7%

[a]Climatic surface monthly input data + monthly geophysical input data.

It is evident that GMDH is identifying some structure which is different from seasonal persistence.

When carried over to make seasonal precipitation estimates for 1977, the average percent errors were similar to those expected from the tracking performance. Table 4 gives the seasonal estimates for four GMDH models for the test regions. The average percent error was 7%. The linear regression model showed no improvement in error (18%) over the climatic mean (7%).

As is the case of crop yields, the GMDH growing-season models are able to indicate the correct direction of the growing-season precipitation over the prairie provinces (Table 4).

1. Sources of "Information"

It is possible to examine the structure of the GMDH models in more detail to determine the source of "information" gain in the model determination. In Tables 5 and 6, the GMDH program was run to create models at level 3,

Table 5 Information Matrix for GMDH Southern Saskatchewan Seasonal Precipitation Models

Dependent variable	Information class	GMDH version	Years[a]	Run set statistics					
				Press	FMSE	r^2	Average error (%)	T^b (%)	T_2^c (%)
Southern Saskatchewan seasonal precipitation (May–September)	Regional time series	Press 3:1	29 + 1	6.10E4	2.26E3	0.35	21	77	53
	Add space-time lags	Press 3:1	29 + 1	3.38E4	1.25E3	0.77	12	86	70
	Add geophysical variable	Press 3:1	29 + 1	1.98E4	0.69E3	0.86	9	93	73
Climate mean model (May–September)			29 + 1	—	4.87E3	—	25	50	33

[a] 1948–1977. Track 29 with 1977 as true step-ahead estimate.
[b] T as direction from mean.
[c] T_2 as tercile of series.

Table 6 Information Matrix for GMDH Southern Saskatchewan Seasonal Precipitation Model

Dependent variable	Information class	Model input variables			Average error for information class (%)
Southern Saskatchewan seasonal precipitation (May–September)	Regional time series	1. January TP Southern Saskatchewan	1	1[a]	21
		2. October TP Southern Saskatchewan	0	0	
		3. November AT Southern Saskatchewan	1	1	
		4. August AT Southern Saskatchewan	0	0	
	Add space–time lag series	5. February TP Alberta	1	1	12
		6. August TP Alberta	0	0	
		7. March AT Alberta	0	0	
		8. December AT Alberta	1	1	
		9. April TP Manitoba	1	0	
		10. October TP Manitoba	1	1	
		11. March AT Manitoba	0	0	
		12. January AT Manitoba	1	0	
		13. November TP Northern Saskatchewan	0	0	
		14. February TP Northern Saskatchewan	1	1	
		15. October AT Northern Saskatchewan	1	0	
		16. March AT Northern Saskatchewan	1	1	
	Add geophysical variable	17. August R	0	0	9
		18. October R	0	0	
		19. January R	1	1	
		20. March R	0	0	
Climatic mean					25

[a]Code 1 indicates that variable entered selected model. 0 indicates that variable did not enter.

using PRESS as the selection criterion. Sequentially, three classes of information were added to the model determination:

1. Regional time series (constrained to southern Saskatchewan)
2. Time series from adjacent spatial regions
3. Solar index series

The average percent error in tracking is reduced from 21% to 9% relative to the climatic mean of 25% (Table 5). The major "information" gain arises from the model structure identified with the added space-time lagged climatic data (Table 6). When the solar index, R, is added to this structure, the model is simplified as GMDH deletes two temperature variables and a precipitation variable (Table 6). The solar index enters the model as a nonlinear term.

We interpret these results as indicating that GMDH is forming a nonlinear approximation to the long-range behavior in the climatic process. The resulting model is, in effect, a space-time transformation of the climatic process. The resulting model is, in effect, a space-time transformation of the climatic data for the southern Saskatchewan output variable. The resulting transform is detecting, or showing an interaction with the solar index without resorting to long-term cyclic models.

Although these results are attractive, we must consider the following: While the four climate models (Table 4) were derived independently, the seasonal output variables (precipitation) are correlated in any step-ahead estimate   From a forecasting standpoint, the estimates (1977) are not independent. Also, the models are regional models in that they do not describe global processes. We may safely anticipate that the GMDH-identified correlation structures are local ones which will not hold globally.

Nevertheless, both the GMDH climate/crop models and the GMDH growing- and precipitation-season models show tracking performances which exceed linear regression models. The GMDH wheat yield models show an improved early season estimate over regression models which primarily contain a trend estimate. The GMDH precipitation models show the identification of nonlinearities and an improved tracking performance over a simple linear seasonal persistence model.

C. Forecasting of 700-mbar Geopotential Heights [9]

If simple statistics which as the means and variances gathered from daily numerical predictions using powerful computers can be shown to have predictive skill above zero, then the averages of computerized forecasts for a month or a season in advance could contain economically valuable information. Unfortunately, this is not so at the present time, although the hope exists that better and more observations combined with more knowledge of atmospheric modeling will result in this advance. A tremendous effort in this direction is currently proceeding at numerous meteorological centers around the world. On a more pragmatic basis, many countries support operational and research units to issue monthly and seasonal forecasts, which despite their imperfections, are of primary interest to many industries.

In recent years several researchers have proposed the use of multivariate autoregressive models for long-range weather prediction. The problem of order determination, and identification of correct lags are the biggest problems that arise in linear ARIMA modeling and are difficult to handle [10, 11].

We now describe a study in which GMDH was used to forecast monthly and seasonal (3-month) averages of 700-mbar geopotential height fields over the northern hemisphere. The 700-mbar geopotential height is the altitude (above sea level) at which the atmospheric pressure is 700 mbar. This is a more global forecasting problem than the climate/climate forecasting reported previously. We will discuss the data reduction and representation process, and the effectiveness of GMDH models compared with other forecasting strategies.

1. Reduction and Analysis of Hemispheric 700 mbar Data

The data obtained from the National Weather Services (NWS) consisted of monthly average geopotential heights for the period January 1947 to December 1980. Actually, two 10×10 degree grids were interlaced to cover the northern hemispheric cap above 15 degrees north latitude. Due to the large amount of data received (541 values per grid), a preliminary data compression scheme was necessary to reduce the number of inputs to the GMDH algorithm.

After computing average maps for each month of the year as approximating the true (i.e., climatic) monthly means, the original data were transformed into a series of anomaly maps by subtracting all monthly averages. A weighted principal component analysis was then performed on the data, where if $X(t, \ell, L)$ represents the 700-mbar geopotential height anomaly at time t, latitude ℓ, and longitude L, truncated linear expressions were obtained of the form

$$X(t, \ell, L) = \sum_{i=1}^{N} a_i(t) E_i(\ell, L)$$

where the $E(\ell, L)$ are the first eigenvectors of the weighted sample covariance of $X(t, \ell, L)$ with weights proportional to the area surrounding each grid point. Such eigenvectors are also called empirical orthogonal functions (EOFs) in the meteorological literature and have been used by researchers since Lorenz's [12] and Gilman's [13] early work at MIT. The degree of truncation, N, was determined by studying the speed with which the eigenvalues decayed. N = 15 turned out to be an adequate choice for all months of the year, with approximately 90 to 93% of the total variance explained.

In a similar fashion, monthly anomalies were transformed into seasonal anomalies, using EOFs based on 3-month seasons. For example, the seasonal EOFs for January-February-March were obtained by computing the covariance of averaged anomalies in all January-February-March periods. Again, N = 15 turned out to be an adequate level of truncation for seasonal representations. Figures 7 to 9 show a typical average field, some of the EOF shapes, and a reconstruction of a field using different numbers of EOFs.

The result of this task was the compression of the original data maps into monthly and seasonal time series of dimension 15 each. Each series consisted of the projections of consecutive months or seasons onto the EOFs corresponding to that month or season. Each series contained 252 observations ranging from 1960 through 1980 (months prior to 1960 were discarded due to large numbers of missing data values). This is not an ideal procedure because successive elements in each series are describing different quantities (e.g., the grids are projected onto different EOFs), but the alternative of using the same EOFs for each month or season is unappealing since it tends to wash out strong seasonal phenomena. We are planning to study alternative representations in the near future.

2. Results

The long-range prediction problem was posed in the following terms: Given a specific month (or season) and EOF order from 1 to 15, what is the best linear or nonlinear polynomial model capable of predicting its corresponding component $a_i(t)$, based on past monthly or seasonal observations of various EOF orders.

Table 7 shows two sets of candidate predictors for the prediction of season EOF 1 (i.e., December-January-February). The first set was obtained by systematically studying the crosscorrelation between 'easonal EOF 1 and all possible lagged valued EOF components and by keeping all terms lying above the 2σ noise level (i >., $\sigma = \sqrt{1/n}$). It is interesting to note that very few short lags appear with the exception of the last monthly (i.e., November) and the last seasonal (i.e., September-October-November) EOFs of order 1. The second set of candidate predictors was assembled simply by listing various components and lags that seemed to make sense from a meteorological point of view. Actually, all previous (nonoverlapping) seasons were incorporated up to a lag of 27 months, which corresponds approximately to the well-known quasi-biannual oscillation phenomenon.

A large number of experiments were then conducted. We describe here results for the forecast of projections onto seasonal EOF 1. Parameters for the various models were estimated without using the last 48 observations (i.e., from 1977 to 1980) so that we could have an independent validation set which was used to evaluate competing forecasts. The types of models considered were as follows:

AVERAGE FIELD FOR MONTH 12

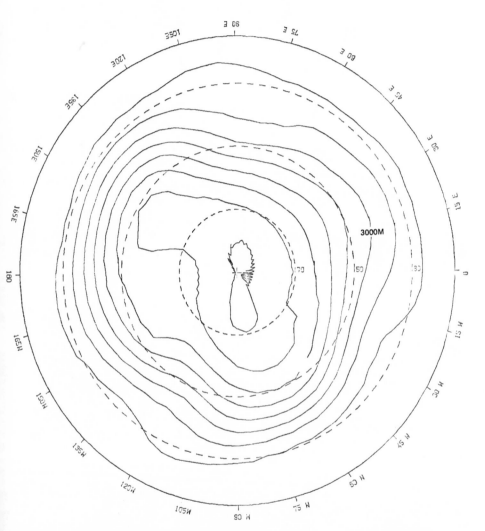

Fig. 7 Mean 700-mbar altitude field for December. Contour lines are 60 m apart; 3000-m contour is labeled.

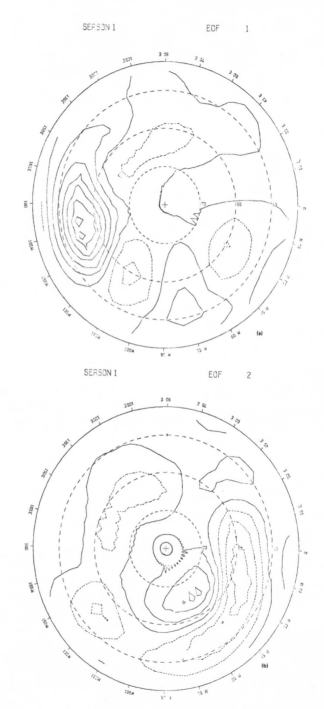

(a)

(b)

SERSON 1   EOF   3

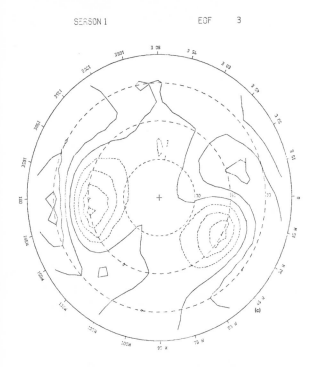

Fig. 8 Typical winter season EOF shapes.

1. GMDH using PRESS as a selection criterion.
2. GMDH using mean-squared error over the checking set as a selection criterion.
3. Linear model: In this model a stepwise linear regression is performed over the set of candidate predictors.
4. Zero model: For this model, our estimate is simply the mean of the series. Since the data were centered during preprocessing, this amounts to predicting 0 for each series element.
5. Persistence model: Our definition of persistence is $\hat{Y}(t) = Y(t - 3)$. A lag of 3 is the shortest lag for which consecutive seasons do not overlap.

In all of the model types listed, lags of less than three seasons were not allowed because overlapping seasons were used. For model types 1 to 3, the data were divided into a training set and a checking set using the DUPLEX algorithm. Actually, model types 1 and 3 made no use of the checking set, but for consistency we wanted all regression models to be evaluated over exactly the same data. The training set contained 80% of the observations (not including those in the validation set) and the checking set 20%. The GMDH models were carried to two levels because higher levels

YEAR 1966   MONTH 12   ACTUAL FIELD

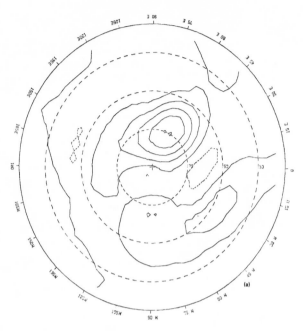

YEAR 1966   MONTH 12   1 EOF

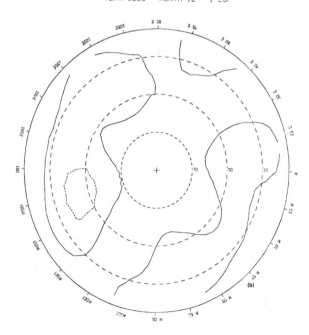

YEAR 1956   MONTH 12   6 EOFS

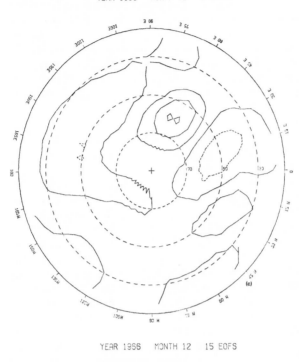

YEAR 1956   MONTH 12   15 EOFS

Fig. 9 Anomaly field and its reconstruction using 1, 6, and 15 EOFs. Contour lines are 60 m apart; dashed lines are negative contours; zero contour is labeled.

Table 7 List of Candidate Predictors for Seasonal EOF 1[a,b]

Set 1			Set 2		
Monthly EOF	1	Lag 2	Seasonal EOF	1	Lag 3
Seasonal EOF	1	Lag 3	Seasonal EOF	1	Lag 6
Seasonal EOF	1	Lag 20	Seasonal EOF	1	Lag 9
Seasonal EOF	1	Lag 21	Seasonal EOF	1	Lag 12
Seasonal EOF	1	Lag 22	Seasonal EOF	1	Lag 15
Seasonal EOF	1	Lag 23	Seasonal EOF	1	Lag 18
Seasonal EOF	3	Lag 23	Seasonal EOF	1	Lag 21
Seasonal EOF	4	Lag 13	Seasonal EOF	1	Lag 24
Seasonal EOF	4	Lag 25	Seasonal EOF	1	Lag 27
Seasonal EOF	4	Lag 26	Seasonal EOF	2	Lag 3
Seasonal EOF	4	Lag 27	Seasonal EOF	2	Lag 6
Seasonal EOF	6	Lag 21	Seasonal EOF	2	Lag 9
Seasonal EOF	10	Lag 26	Seasonal EOF	2	Lag 12
Seasonal EOF	12	Lag 6	Seasonal EOF	2	Lag 15
Seasonal EOF	13	Lag 20	Seasonal EOF	2	Lag 18
Seasonal EOF	13	Lag 21	Seasonal EOF	2	Lag 21
Seasonal EOF	14	Lag 28	Seasonal EOF	2	Lag 24
Seasonal EOF	15	Lag 17	Seasonal EOF	2	Lag 27
Seasonal EOF	15	Lag 26	Seasonal EOF	3	Lag 3
Seasonal EOF	15	Lag 27	Seasonal EOF	3	Lag 6

[a]Lags are given in months.
[b]Entries in set 1 for EOF 1 were suggested by correlation analysis.

did not offer improvements in terms of the selection criteria. Model types
were compared by looking at the means and standard deviations of the forecast errors over the validation set.

Table 8 indicates some results obtained for the forecast of seasonal
EOFs 1, 2, and 3. Figure 10 shows plots of some of the forecasts over the
validation set. The best model for forecasting EOF 1 is the persistence
model, followed by the zero model. GMDH and the linear regression models
are the worst. We found little difference between the two sets of predictors
and little difference between the GMDH nonlinear models and the linear
regression model. The second set of predictors offers a small but statistically insignificant improvement over the set of predictors selected by
correlation analysis. Nonlinear terms in the GMDH models are significant
in that they were accepted by the stepwise regression procedure, but empir-
ically the nonlinear coefficients are small and make little difference in the
models. Although we had hoped that nonlinear models would offer great
improvement, the plots of GMDH models and the linear models look very
much alike. If one model predicts a peak well, the others tend to do the same.

Table 8 Mean and Standard Deviations of Forecast Errors for Seasonal EOF 1 over the Validation Set

| EOF number | GMDH using press | | GMDH using check MSE | | Model | | | | | |
| | | | | | Linear model | | Persistence model | | Zero model | |
	Mean	St. dev.	Mean	St. dev.	Mean	St. dev.	Mean	St. dev.	Mean	St. dev.
EOF 1										
First set of predictors	27.30	30.33	27.94	29.71	26.72	31.08	-1.15	45.62	14.20	31.88
Second set of predictors	23.92	32.56	22.92	28.86	20.53	35.58				
EOF 2	-0.94	13.17	-0.67	13.00	-2.29	13.62	-0.30	17.15	-1.85	12.00
EOF 3	0.59	12.11	0.25	11.61	0.82	10.87	0.31	17.07	-0.94	10.78

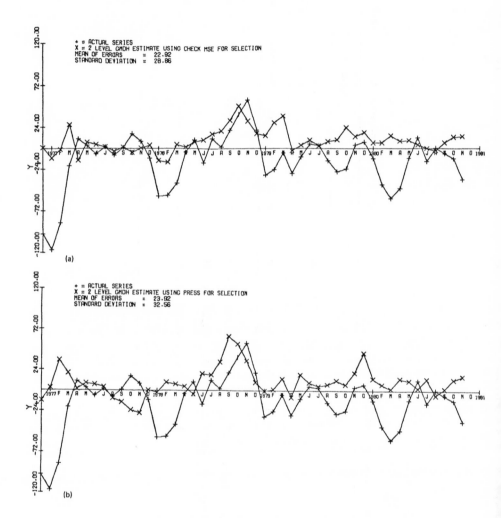

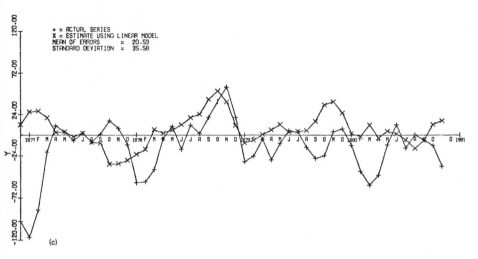

(c)

Fig. 10 Forecasts of validation set for EOF 1 using models identified from predictor set 2.

In all instances, GMDH had the opportunity to select persistence as the preferred models since the persistence variable was included in all the sets of candidate predictors. This was not done, yet the persistence model was better in forecasting over the validation set. Part of the problem is that the validation set is short and may represent an atypical region of data (in fact, the first element of the validation set is January 1976, one of the coldest months in memory along the East coast), but this is also an indication that there is not a strong underlying climatic model sitting out there waiting to be uncovered. Rather, GMDH is finding weak and inconsistent structures which may or may not be artifacts of the data set.

3. Conclusion and Future Area of Research

The main thrust of this preliminary study was to investigate the feasibility of applying GMDH to identify nonlinear dynamical features in the 700-mbar geopotential data. It turns out that these data, at least when compressed by projection over EOFs, do not appear to have sufficient structure to be identified by regression techniques (whether linear or nonlinear like GMDH). We feel that our lack of success is not related to the GMDH algorithm, but rather to the choice of data compression techniques and to the limited size of available data.

Since this study, several papers have appeared in recent literature on map-typing techniques (e.g., Refs. 14 and 15), which tend to indicate that considerable improvement can be observed when compressing meteorological data maps by using some form of "rotated EOFs" derived by factor analysis techniques. Since these techniques are based on correlation fitting ideas, we expect that they might reproduce large-scale, spatially clustered, atmospheric features better than the variance fitting techniques used here which reproduce the total energy of the signal but not its phase. We will be investigating these factor analysis techniques in the near future, as well as the usefulness of augmenting the 700-mbar data set with sea surface temperature (SST) data.

IV. SUMMARY

We have described an implementation of the basic GMDH algorithm which features stepwise regression, the DUPLEX algorithm for data splitting, and the PRESS selection criterion. We have described our experience with GMDH models in areas of crop yield and climate forecasting. We have had encouraging results in models for Canadian wheat yield. We have had moderately good results in the forecasting of localized meteorological time series (precipitation). Our analysis of hemispheric 700-mbar geopotential altitudes has not shown significant improvement, although we remain confident that GMDH will be a viable technique for modeling these data as soon as adequate preprocessing techniques are found to preserve the information content of large-scale meteorological data.

ACKNOWLEDGMENT

Part of this study was supported by Contract NA-81-SAC from the NOAA Climate Analysis Center.

REFERENCES

1. N. Wiener, Nonlinear Problems in Random Theory, Wiley, New York, 1958.
2. J. J. Duffy and M. A. Franklin, A Learning Identification Algorithm and Its Application to an Environmental System, IEEE Trans. Syst. Man Cybern. SMC-5(2):226-240 (1975).
3. S. Ikeda, M. Ochiai, and Y. Sawaragi, Sequential GMDH Algorithm and Its Application to River Flow Prediction, IEEE Trans. Syst. Cybern. SMC-6(7):473-479 (1976).
4. R. Snee, Validation of Regression Models: Methods and Examples, Technometrics 19(4):415-428 (1977).
5. R. W. Kennard and L. A. Stone, Computer Aided Design of Experiments, Technometrics 11:137-148 (1969).
6. D. M. Allen, The Prediction Sum of Squares as a Criterion for Selecting Predictor Variables, University of Kentucky, TR 23, 1971.
7. H. V. Rice, Application of the GMDH Algorithm to Climate/Crop Forecasts, Am. Soc. Agric. Eng., Winter Meet., New Orleans, La., 1979.

8. LACIE, NOAA-CEAA Technical Note 76-2, LACIE-00433, JSC-11658, Columbia, Mo., Oct. 1976.

9. P. Toldalagi, W. Lebow, and R. K. Mehra, Pilot Study of a Nonlinear Statistical System Identification Technique for Model Identification and Prediction of 700mb Data, Final Report Contract NA-81-SAC-00630, Scientific Systems, Inc., Cambridge, Mass., 1981.

10. C. E. Duchon and A. J. Koscielny, ARIMA Models for Forecasting Meteorological Time Series, 7th Conf. Prob. Stat. AMS, Monterey, Calif., 1981.

11. A. Koscielny, Multivariate Autoregressive Modeling of Monthly Averaged 700 MB Height Anomalies, 7th Conf. Prob. Stat. AMS, Monterey, Calif., 1981.

12. E. N. Lorenz, Empirical Orthogonal Functions and Statistical Weather Prediction, M.I.T. Statistical Forecasting Project, Report 1, Contract AF19-604-1566, 1956.

13. D. L. Gilman, Empirical Orthogonal Functions Applied to Thirty-Day Forecasting, M.I.T. Sci. Rep. 1, Contract AF19-604-1283, 1957.

14. M. B. Richman, The Retrieval of Meteorological Patterns via Rotated Principal Components, 7th Conf. Prob. Stat., AMS, Monterey, Calif., 1981.

15. J. E. Walsh and M. B. Richman, Seasonality in the Associations Between Surface Temperatures over the United States and the North Pacific Ocean, Monthly Weather Rev. 109:767-783 (1981).

7

Nonlinear Prediction Models for River Flows and Typhoon Precipitation by Self-Organizing Methods

SABURO IKEDA/Institute of Socio-Economic Planning, University of Tsukuba, Sakura, Ibaraki, Japan

I. INTRODUCTION

This chapter is concerned with the applications of the GMDH to prediction models of river flow and typhoon precipitation. Both applications are associated with the modeling of environmental, ecological, and social systems which have a great deal of uncertainty and unknown interactions between elements and parameters. Most forecasting models, for example, of river flow are based on hydrological and topological characteristics of the river basins such as "areal precipitation," "evapotranspiration," or "underground storage" [1,2]. These models are certainly appropriate for a qualitative analysis of the runoff mechanisms, but their applicability to a forecasting or an operational use may be limited. It is partly because these physical models are possibly formed only with many assumptions and simplifications with respect to the real environmental, ecological, and social processes, and partly because the introduced elements and factors involved in their models are all under uncertainty in terms of their quantitative measurement. The prediction of typhoon precipitation is far more difficult than that of river flow, since it is harder to find such a sufficient physical structure that explains an exact quantitative relationship between the precipitation and meteorological variables in connection with typhoons.

Another approach to quantitative modeling is "time-series analysis" of the input/output data: namely, prediction models are identified from the given data by the use of various linear statistical methods, such as "maximum likelihood," "least squares," and "Kalman filtering" [3,4]. However, those linear time-series analyses have some difficulties not only in treating the nonlinearity of the runoff mechanism, but also in dealing with cases where the sample size is relatively small, a situation often encountered when we are to construct models of environmental, ecological, or social systems for prediction or operational purposes. In this chapter attention will be focused on modeling of the nonlinear phenomena of runoff by rainfall or precipitation when a typhoon is approaching affected river basins but only a small number of data sequences are available for the model building.

II. AN ADAPTIVE PREDICTION MODEL
OF RIVER FLOW

This section gives a brief description of an adaptive prediction scheme by a modification of Ivakhnenko's basic GMDH algorithm. To simplify illustration of our model, let us first describe a river basin, the Karasu River basin, in the upper reaches of the Tone River, located north of Tokyo. Figure 1 shows the Karasu River basin, for which the daily river flow $q(t)$ (m^3/day) was measured at the mouth of the river, and the daily mean areal precipitation $r(t)$ (mm/day) was given as an averaged value of the four observation points denoted by squares.

Instead of introducing defined variables with physical meanings such as "evapotranspiration" or "underground storage," we shall consider a general model as follows:

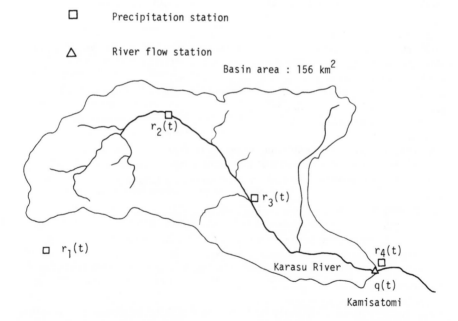

Fig. 1 Karasu River basin in Gunma Prefecture in Japan. (From Ref. 6.)

$$q(t) = F(r(t-h), r(t-2h), \ldots, r(t-mh), q(t-h), q(t-2h), \ldots, q(t-nh)) \qquad (1)$$

The problem is to determine the unknown function F from the available past data of $r(t-mh)$ and $q(t-nh)$, where h is a sampling interval, assumed to be 1 day, and m and n represent the number of past days to be included in the prediction equation (1). The function F is assumed to form a "Ivakhnenko polynomial,"

$$y = a + \sum_i b_i x_i + \sum\sum c_{ij} x_i x_j + \cdots \qquad (2)$$

where

$$y = q(t), \quad x_1 = r(t-h), \quad x_2 = r(t-2h), \quad \ldots, \quad x_{m+1}$$

$$= q(t-h), \quad x_{m+2} = q(t-2h), \quad \ldots$$

Figure 2 displays the tracking behavior of the actual flow $q(t)$ together with the predicted value $q_1(t)$ by a simple GMDH model:

$$q_1(t) = -0.2001 + 0.9846q(t-1) + 0.2245r(t-1) - 0.0180q^2(t-1)$$
$$+ 0.0040r^2(t-1) - 0.0048q(t-1)r(t-1) \qquad (3)$$

This model was derived by utilizing Ivakhnenko's basic GMDH and the data for the period May 1 to August 31, 1964. The heuristics that were used in applying the basic GMDH algorithm to our prediction equation (1) are as follows [5]:

1. Alternate separation of the original data into training and checking sequences
2. Prehistory of 10 days with respect to each input, $r(t)$ and $q(t)$
3. Partial description (intermediate variable) of second-order polynomial
4. Five partial descriptions to be passed into the next selection layer

It is surprising that the simple nonlinear model (3) gives good estimations for the whole period. However, it is also true that it underestimates some peak points from May to June. The root-mean-square error for this interval becomes

$$\sigma_1 = \left\{ \frac{1}{60} \sum_{t=5/1}^{t=6/30} [q(t) - q_1(t)]^2 \right\}^{\frac{1}{2}} = 3.87 \qquad (4)$$

To improve the prediction capacity for the nonlinear character of the runoff process, that is, for the prediction model to follow abrupt increases

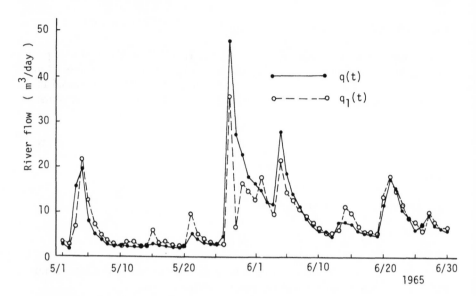

Fig. 2 Tracking behavior of predicted value $q_1(t)$ together with the actual flow $q(t)$.

of river flow, an adaptive identification method is presented as a modified version of the basic GMDH. The major modifications are seen in both the generation rule for intermediate variables and the selection rule for intermediate variables to the next layer. The procedure for this modified GMDH algorithm is summarized as follows:

A. Modified GMDH Algorithm

Step 1: Select input variables. First, determine the interval of pre-history with respect to the candidate variables, and then select N "useful" input variables according to the selection rules, such as correlation criteria between input and output variables, or simply by physical considerations.

Step 2: Divide the original data into training and checking sequences. The separation rule is very heuristic. The role of the two sequences may be changed to avoid a bias due to the difference of two sequences.

Step 3: Suppose that we have the intermediate variables $Z_i^{(k-1)}$ (i = 1, ..., M) in the (k - 1)th layer. We determine the intermediate variables in the kth layer according to the following two-stage procedure:

a. Transform $Z_i^{(k-1)}$ by the second-order polynomial

$$\hat{Z}_i^{(k)} = a_{io} + a_{i1} Z_i^{(k-1)} + a_{i2} (Z_i^{(k-1)})^2, \quad i = 1, 2, \ldots, M \quad (5)$$

where the coefficients a_{ij} ($j = 0$, 1, 2) are determined by using the training data so as to minimize the mean-squared error,

$$\overline{(y - \hat{Z}_i^{(k)})^2} \tag{6}$$

The overbar denotes the sample mean. For $k = 1$ we skip this procedure (5) and set $Z_i^{(1)} = x_i$.

b. Combine the transformed intermediate variables $Z_i^{(k)}$ and $Z_j^{(k)}$ in the following manner:

$$Z_s^{(k)} = b_{s0} + b_{s1}\hat{Z}_i^{(k)} + b_{s2}\hat{Z}_j^{(k)} + b_{s3}(\hat{Z}_i^{(k)})^2$$
$$+ b_{s4}(\hat{Z}_j^{(k)})^2 + b_{s5}(\hat{Z}_i^{(k)})(\hat{Z}_j^{(k)})$$

$$i, j = 1, 2, \ldots, M; \quad i \neq j; \quad s = 1, 2, \ldots, \frac{M(m-1)}{2} \tag{7}$$

where the coefficients b_{sj} ($j = 0$, 1, $\ldots$, 5) are determined by using the training data so as to minimize the mean-squared error,

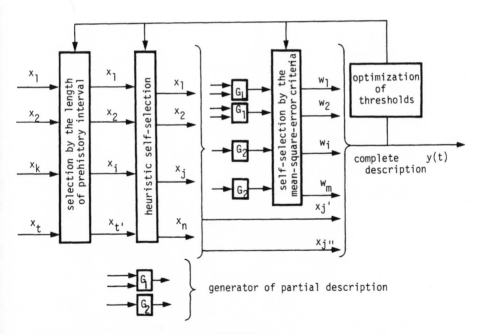

Fig. 3 Structure of the modified GMDH.

$$e_s^2 = \overline{(y - Z_s^{(k)})^2} \tag{8}$$

<u>Step 4</u>: Select intermediate variables among $Z_s^{(k)}$, $s = 1, \ldots, m(M-1)/2$, which give M_1 smallest mean-squared errors e_s^2 for the checking data. Here the same transformation (5) and the combination (7) should be applied to the checking data. In addition, we add M_2 original inputs on this selection stage, if necessary, to increase the variety of the polynomial terms (see Fig. 3).

<u>Step 5</u>: Replace k by k + 1 and go to step 3. Repeat steps 3 and 4 until the mean-squared error e_m^2 of the best predictor Z_m in the present layer does not exceed that of the best one in the preceding layer over the prescribed bound of approximation.

It should be noted that the resultant complete description (final model) by this procedure is more complex than that of the basic GMDH on the same number of selection layers: namely, we can obtain a higher-order polynomial with a small number of layers. Meanwhile, addition of the original input variables into each layer gives more multiple features of the system

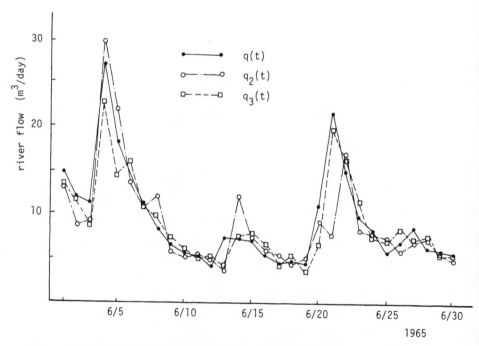

Fig. 4 Tracking behaviors of the predicted values by the adaptive scheme with the modified GMDH $q_3(t)$ and by the basic GMDH $q_2(t)$.

Table 1 Partial Description for $q_3(t)$ on June 5, 1965

Input variables	$x_1 = q(t - 1), \quad x_2 = r(t - 1), \quad x_3 = r(t - 2)$

First layer

$$w_1^{(1)} = -0.711 + 1.181x_1 + 0.015x_2 - 0.022(x_1)^2 + 0.002(x_2)^2 + 0.0101x_1x_2$$

$$w_2^{(1)} = 2.009 + 0.454x_2 + 0.071x_3 - 0.0007(x_2)^2 + 0.0006(x_3)^2 - 0.0103x_2x_3$$

Second layer

$$\hat{w}_1^{(2)} = -0.00008 + 1.0000w_1^{(1)} - 4.0 \times 10^{-7}(w_1^{(1)})^2$$

$$\hat{w}_2^{(2)} = 0.0380 + 0.9922w_2^{(1)} + 0.00017(w_2^{(1)})^2$$

$$w_1^{(2)} = -0.3488 + 0.8171\hat{w}_1^{(2)} + 0.2833\hat{w}_2^{(2)} - 0.0084(\hat{w}_1^{(2)})^2 - 0.0240(\hat{w}_2^{(2)})^2 + 0.0304\hat{w}_1^{(2)}w_2^{(2)}$$

. . . .

Final model $q_3(t) = w_1^{(2)}$

that might once have been discarded at an earlier stage. This whole procedure is illustrated in Fig. 3.

An adaptive scheme of the prediction model is constructed by making
use of the modified algorithm above in order to improve the forecasting
ability of fitting the peak values of river flow with a limited amount of input/
output data. For example, let us suppose that the data from only the preceding 30 days are used each day. The identified structure of the model (1)
is not always the same each day, since the content of the data sequence is
renewed day by day. In other words, the model structure will be changed
adaptively to the variation involved in each data sequence.

Figure 4 shows the tracking behavior of the predicted values $q_2(t)$ and
$q_3(t)$ of the Karasu River flow for the period from June 1 to June 30, 1965,
where $q_2(t)$ is the prediction by the basic GMDH and $q_3(t)$ is that by the
adaptive scheme of the modified GMDH with the data sequence of 30 days.
In this calculation, the number of the "useful" input variables and the number of the intermediate variables are optimally chosen to be 10 and 5,
respectively. As is seen in Fig. 4, the adaptive prediction model, in particular with the modified GMDH, indicates a remarkable improvement of
fitness to the peak values, such as those on June 4 and June 21. In fact, the
root-mean-square errors defined by

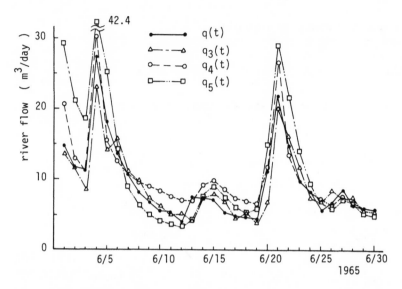

Fig. 5 Tracking behaviors of the predicted values by the modified GMDH $q_3(t)$ and by the hydrological models, $q_4(t)$ and $q_5(t)$.

$$\sigma_i = \left\{ \frac{1}{30} \sum_{t=5/1}^{t=6/30} [q(t) - q_i(t)]^2 \right\}^{\frac{1}{2}} \tag{9}$$

are

$$\sigma_2 = 3.27 \quad \text{and} \quad \sigma_3 = 1.86 \tag{10}$$

Table 1 displays the final model of $q_3(t)$ on June 5, 1965. Shown in Fig. 5 is a comparison between the predicted values by the adaptive scheme $q_3(t)$ and those by elaborate hydrological models such as the "tank model," $q_4(t)$, and the "storage function model," $q_5(t)$, which are taken from Refs. 2 and 6. The root-mean-square errors of q_4 and q_5 for the same period are

$$\sigma_4 = 2.23 \quad \text{and} \quad \sigma_5 = 5.19 \tag{11}$$

III. A SEQUENTIAL PREDICTION MODEL
OF RIVER FLOW

One of the arguments involved in the adaptive prediction scheme is certainly a problem of "stability" in terms of the model structure. Since the identified structure varies day by day, it may cause unnecessary overfitting to such measured data that happen to have a considerable degree of noise or error due to a higher degree of polynomials used in the constructed model. To stabilize the structure of the prediction model, a sequential formula of the modified GMDH algorithm was developed [7]. First, we determine an optimal nonlinear structure of the input/output relation (1). Next, whenever new measurement data are obtained, the coefficients of intermediate variables (5) to (7) are changed sequentially according to the following procedure, with the model structure unchanged.

A. Sequential GMDH Algorithm

Consider the following equation for each intermediate variable:

$$A_p X = Y_p \tag{12}$$

where A_p is a $p \times q$ matrix, X is a $q \times 1$ coefficient vector, and Y_p is a $p \times 1$ output vector. The integer q is 3 for (5) and 6 for (7), respectively. When A_p has a maximum rank, the least-squares solution of (12) is given by

$$X^* = (A_p^t A_p)^{-1} A_p^t Y_p \tag{13}$$

where $(A_p)^{-1}$ and A_p^t denote the inverse and transpose of matrix A_p, respectively.

Now, suppose that we get a new observation. Then we have an additional equation,

$$BX = Z_{p+1} \tag{14}$$

where B is a $q \times 1$ vector matrix provided by a new measurement, and Z_{p+1} is the latest output. Then, combining (12) and (14), we have

$$A_{p+1} X = Y_{p+1} \tag{15}$$

where

$$A_{p+1} \triangleq \begin{bmatrix} A_p \\ B \end{bmatrix}, \quad Y_{p+1} \triangleq \begin{bmatrix} Y_p \\ Z_{p+1} \end{bmatrix}$$

For (15), the least-squares solution becomes

$$X^*_{p+1} = (A_{p+1}^t A_{p+1})^{-1} A_{p+1}^t Y_{p+1} \tag{16}$$

Furthermore, if we define

$$P_p = (A_p^t A_p)^{-1} \tag{17}$$

from (15) we have

$$P_{p+1}^{-1} = A_{p+1}^t A_{p+1} = A_p^t A_p + BB^t = P_p^{-1} + BB^t \tag{18}$$

Now, let us establish a recursive formula of (16) in order to simplify the computation of the least-squares solution X_{p+1}^* with the new measurement data. In accordance with a "matrix inversion lemma" [11], P_{p+1} can be calculated by means of P_p and other terms. Namely, if matrices P_{p+1}, P_p, H_{p+1}, and R_{p+1} satisfy the equation

$$P_{p+1}^{-1} = P_p^{-1} + H_{p+1}^t R_{p+1}^{-1} H_{p+1} \tag{19}$$

then P_{p+1} is given by

$$P_{p+1}^{-1} = P_p - P_p H_{p+1}^t (H_{p+1} P_p H_{p+1}^t + R_{p+1})^{-1} H_{p+1} P_p \tag{20}$$

where P_{p+1}^{-1}, P_p^{-1}, and R_{p+1}^{-1} are nonsingular and H_{p+1} is of maximum rank. From (18) and (19), we set $H_{p+1}^t = B$ and $R_{p+1} = I_q$ (a $q \times q$ identity matrix) in (19); then we can obtain the equation

$$P_{p+1} = P_p - P_p B(B^t P_p B + 1)^{-1} B^t P_p \tag{21}$$

Thus, substitution of (21) for the least-squares solution (16) makes the recursive formula

$$X^* = P_{p+1}(A_p^t Y_p + BZ_{p+1}) = X_p^* + P_p B(B^t P_p B + 1)^{-1}(Z_{p+1} - B^t X_p^*) \tag{22}$$

Since initial values X_p and P_p are given by the predetermined model, we can compute only the inverse of a scalar value, $B^t P_p B + 1$, instead of computing the inverse of a $(p + 1) \times (p + 1)$ matrix in (16). Whenever a new measurement is made, this procedure is repeated to generate each intermediate variable with the predetermined structure on each layer.

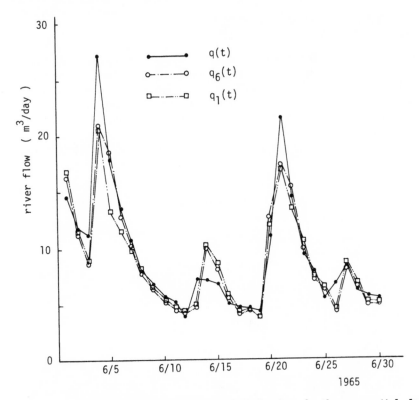

Fig. 6 Tracking behaviors of the predicted values by the sequential algorithm q_6(t) and by the basic GMDH q_1(t).

This sequential GMDH algorithm is effective when the input/output
relation of the system changes gradually according to nonstationing elements
of other factors that have not been taken into consideration in the model.
Only the parameters of the obtained structure are gradually adapted to the
newly observed data. Let us see how this sequential algorithm works in the
case of the Karasu River basin.

First, as a stable structure of the initial prediction model, we employ
(3) derived from either the basic or the modified GMDH algorithm, because
the model (3) is a product at the first layer. Next, the sequential formula
(22) is applied to the 1-day-ahead prediction on each day for the period from
June 1 to June 30, 1965. The tracking behaviors of the actual flow q(t) and
the predicted values q_1(t) and q_6(t) are shown in Fig. 6, where q_1(t) is the
predicted value by the model (3) and q_6(t) is the predicted value by the sequential algorithm (22) with the use of the model (3). The root-mean-square
errors, calculated by (9), are given as follows:

$$\sigma_1 = 2.07 \quad \text{and} \quad \sigma_6 = 1.84 \tag{23}$$

IV. NONLINEAR PREDICTION MODEL
OF TYPHOON PRECIPITATION

For flood control, it is extremely important to estimate the amount of precipitation in due time when typhoons are approaching the basin areas concerned. It is therefore highly desirable to develop prediction models of
typhoon precipitation which give reliable and accurate forecast with respect
to the amount of rainfall on the basis of the meteorological data. We have
developed a nonlinear GMDH model for the prediction of typhoon rainfall
which attempts to provide quantitative information on the anticipated amount

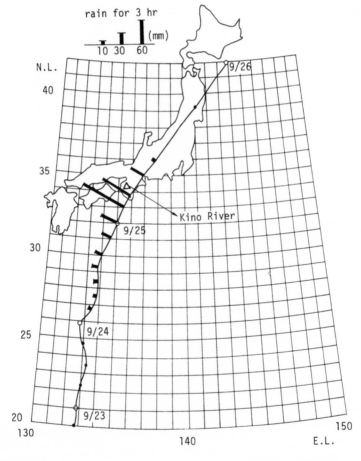

Fig. 7 Course of the No. 1 typhoon and precipitations for 3 hours at Kino
River basin in Japan. (From Ref. 9.)

Table 2   Data List of Typhoon Precipitation at Kino River Basin

Data group number	Number of samples (serial no.)	Name (typhoon no.)	Date
1	16 (1-16)	13	9/24/1953
2	10 (17-26)	15	9/25/1956
3	28 (27-54)	17	8/23/1958
4	13 (55-67)	15	9/24/1959
5	13 (68-80)	24	9/16/1965
6	13 (81-93)	20	9/16/1972
7	13 (94-106)	14	9/17/1963
8	10 (107-116)	22	9/25/1958
9	22 (117-138)	11	8/10/1960
10	22 (139-160)	16	8/28/1960
11	26 (161-176)	18	9/15/1961
12	10 (177-186)	34	10/26/1967

of typhoon rainfall for the operational use of flood control in the Kino River basin area [8]. The Kino River basin is one of the basins that is frequently affected by typhoons in Japan.

Figure 7 shows the typical course of a typhoon (No. 13, 1953) and the observed precipitation in 3 hours at a point in the Ohtaki district of the Kino River basin (denoted by a triangle), when the center of the typhoon was located at each point (denoted by a black bar) along its landward course. The original data consist of 13 groups of data on typhoons that brought a considerable amount of precipitation to this river basin (see Table 2).

A. Model Structure and Input Variables

Based on the following major physical and meteorological factors that cause typhoon precipitation:

1. Relative distance between the river basin and the active area of the typhoon
2. Amount of vapors carried by the typhoon
3. Scale of the typhoon potential

the following 11 elements were taken as the input variables:

x_1: the north latitude of the typhoon center
x_2: the east longitude of the typhoon center
x_3: the atmospheric pressure at the typhoon center
x_4: the moving speed of the typhoon center

x_5: the radius of the typhoon area
x_6: the distance from the typhoon center
x_7: vapor pressure at the Shionomisaki observation point (S)
x_8: the atmospheric pressure on the sea surface at the point S
x_9: the humidity at the point S
x_{10}: the atmospheric temperature at the point S
x_{11}: precipitation for 3 hours in one period before the past

(The point S is located at the nearest seacoast among the meteorological observatories.)

The one-step-ahead prediction model of typhoon precipitation,

$$y = f(x_1, x_2, \ldots, x_{11})$$ (24)

will be identified in the form of the partial polynomials of second order by the GMDH algorithm:

$$y_k = a_0 + a_1 x_i + a_2 x_j + a_3 (x_i)^2 + a_4 (x_j)^2 + a_5 (x_i x_j)$$ (25)

$$i, j = 1, 2, \ldots, 11; \quad k = 1, 2, \ldots$$

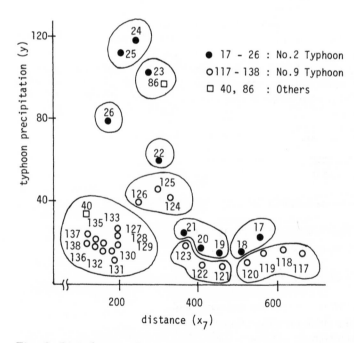

Fig. 8   Distribution of the samples in the two-dimensional space ($y - x_7$).
(From Ref. 10.)

B. A GMDH Algorithm with Clustered Input/Output Data

The sample size, 186, does not seem to be too small compared with that of the input variables, 11, when we consider the application of the GMDH algorithm to this typhoon model. However, if we look at the distribution of the samples in the data space of 11 dimensions in which the prediction (interpolation) is to be conducted, it is easily seen that the typhoon data are sparsely scattered and sometimes concentrated in a few points. Figure 8 illustrates an example of such sample distribution in the two-dimensional space of y (precipitation for 3 hours) and x_7 (distance from the typhoon center). It is therefore important to take the distribution structure of the input/output data into account, because the model identified is strongly biased by being fitted to the particular narrow domains.

In this respect, we introduce a "clustering technique" as the heuristics of the division of data into training and checking sequences. The original data samples are clustered into a given number of clusters by using a measure of similarity or dissimilarity between data samples. Figure 8 shows one example of clustered data by means of a "farthest neighbor method." The largest cluster in the left side of the domain $(y - x_7)$ contains 13 samples, but points 22 and 26 belong to the "isolated cluster." The procedure of data division is as follows:

1. The whole samples M are clustered into N (N < M).
2. One sample from each cluster is included in the training and checking sequences, respectively, if the cluster has more than two samples. The rest of the samples in the cluster are used for the evaluation of the identified model.
3. The "isolated cluster" sample is put into the training sequence.

In our case, 186 samples are clustered into 70 clusters, in the following manner:

1. Clusters of two or three samples: 49
2. Clusters of over three samples: 11
3. Clusters of one sample: 10

Figure 9 shows the tracking behavior of the estimated or predicted values of typhoon precipitation y together with those of a physical model that is taken from Ref. 9. This physical model was constructed by using the location of the typhoon center (x, y):

$$\log \frac{R}{K} = a_1 + a_2 x + a_3 y + a_4 x^2 + a_5 y^2 + a_6 xy \qquad (26)$$

where R is the typhoon precipitation and K represents the supplementary coefficients of rainfall derived from the vapor and atmospheric pressures.

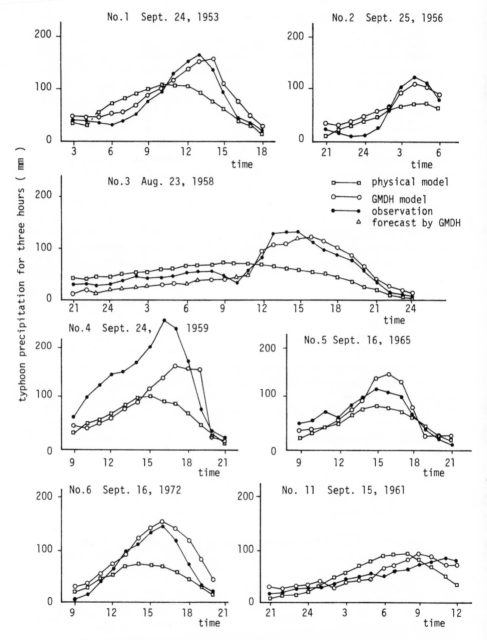

Fig. 9 Tracking behaviors of the predicted values by the GMDH model with data clustering and by the physical model. (From Ref. 9.)

Table 3 Prediction Model for Typhoon Precipitation

Intermediate variables	Input variables		Coefficients of the GMDH model (25)					
			a_0	a_1	a_2	a_3	a_4	a_5
First layer								
z_1	x_1	x_7	2.77	-0.56	-4.63	3.48	-1.98	0.77
z_2	x_1	x_6	-8.95	19.24	1.70	4.03	-12.48	-3.39
z_3	x_4	x_8	-2.44	8.33	8.06	8.34	-0.01	-10.52
z_4	x_6	x_{11}	-0.64	6.00	-4.23	0.04	-3.91	3.00
Second layer								
w_1	z_1	z_3	0.18	-0.11	-1.31	6.29	-1.53	2.35
w_2	z_2	z_4	0.02	0.93	0.11	3.65	-3.13	0.25
. . .	. . .	. . .	. . .	. . .	. . .	. . .	. . .	. . .
Complete description								
y	w_1	w_2	0.02	0.06	0.66	3.86	-0.94	-1.99

Coefficients a_1 to a_6 are to be identified from meteorological data from the typhoon. Table 3 lists the variables and coefficients of the prediction model. The variables selected seem to match meteorological knowledge of the typhoon's precipitation.

V. CONCLUSION

There is no methodology that covers all aspects of prediction models for hydrological and meteorological applications. The GMDH algorithms have, of course, both strengths and weaknesses compared with other prediction models. Our heuristic self-organization method, GMDH, has the specific features of algorithmic simplicity, short computation time, and applicability to a limited amount of data. However, it is also acknowledged that these models may not fully explain the physical characteristics of the processes concerned.

In this chapter, several modified versions of the basic GMDH algorithm, such as the adaptive scheme of the modified GMDH, the sequential GMDH algorithm, and the GMDH algorithm with clustered input/output data, were presented. They reflect methodological characteristics of the GMDH that have great potential for utilizing the knowledge of specialists versed in both the hydrological field and systems science.

REFERENCES

1. P. Eagleson, Dynamic Hydrology, McGraw-Hill, New York, 1970.
2. M. Sugawara, On the Analysis of Runoff Structure About Seven Japanese Rivers, Jpn. J. Geophys. 2(4):1-76 (1961).
3. G. Box and G. Jenkins, Times Series Analysis—Forecasting, Holden-Day, San Francisco, 1970.
4. M. Hino, Runoff Forecasts by Linear Predictive Filter, J. Hydraul. Div. ASCE 96:681-701 (1970).
5. S. Ikeda, S. Fujishige, and Y. Sawaragi, Nonlinear Prediction Model of River Flow by Self-Organization Method, Int. J. Syst. Sci. 7(2):165-176 (1976).
6. Tone-River-Control-Office, Report on the Runoff Analysis of Low Flows in Upper Reaches of the Tone-River, July 1967 (in Japanese).
7. S. Ikeda, M. Ochiai, and Y. Sawaragi, Sequential GMDH Algorithm and Its Application to River Flow Prediction, IEEE Trans. Syst. Man Cybern. SMC-6(7):473-479 (1976).
8. S. Ikeda and H. Ann, Nonlinear Prediction Model for Typhoon Precipitation, J. Meas. Control (Jpn. Soc. Control Eng.) 19(7):733-737 (1980) (in Japanese).
9. Wakayama Construction Bureau, Min. of Construction, Report of Flood Control Management for Kino-River Basin, 1977 (in Japanese).

10. D. Wishart, An Algorithm for Hierarchical Classification, Biometrics
    25:165-170 (1969).
11. A. Sage, Optimum Systems Control, Prentice-Hall, Englewood Cliffs,
    N.J., 1968.

8

Let's Ask GMDH What Effect the Environment Has on Fisheries

HUGH A. BROOKS*/ Virginia Institute of Marine Science, Gloucester, Virginia

THOMAS H. PROBERT† / MITRE Corporation, McLean, Virginia

I. INTRODUCTION

Fishermen have long been aware of the influence of environmental changes on the abundance and distribution of fishes. Short-term environmental fluctuations affect the distribution of fishes. For example, the seasonal change from spring to summer marks the return of many commercially important species from winter offshore areas to estuaries such as the Chesapeake Bay. Early Long Island and Virginia fishermen predicted the spring arrival of shad (Alosa) and weakfish (Cynoscion) with environmental indicators such as the flowering of the shad bush and the dogwood (Austin and Ingham, 1978). Long-term climatic changes are known to cause fish stocks to appear and disappear. In 1942, Ottestad reported a high correlation between 85 years of cod catches in the Vest Fjord, in northern Norway, and widths of pine tree rings, which were used as indicators of fluctuations in the climate of the area. Recent laboratory and field investigations by fisheries scientists have resulted in an increasing awareness of the impact of the environment on the stock size of commercially valuable species. If the environmental factors causing fluctuations in stock size were identified, this knowledge could lead to accurate predictions of where and when to fish to maximize a fisherman's catch.

Models of the influence of the environment on fisheries were based initially on a relationship between one environmental parameter and fish abundance. Pearson's 1948 study related fluctuations in blue crab density to discharge volume in Chesapeake Bay tributaries. Sette (1943) noticed larval mortality in the Atlantic mackerel was affected by unfavorable winds.

Present affiliations:

*Systems Development Corporation, Research Triangle Park, North Carolina
†Institute for Defense Analyses, Alexandria, Virginia

169

Murphy and Shomura (1955) have shown that the degree of aggregation of yellowfin tuna in the zones of convergence and divergence of the Pacific equatorial currents depends on the temperature structure of the water. When the zone of divergence is narrow, the fish are tightly packed, but when cool water is spread over a wide area, there is little aggregation.

Since biological systems are often nonlinear, the influence of environmental fluctuation on fish abundance may be extremely complex. The complexity has forced researchers to use statistical techniques such as multiple regression and multivariate time-series modeling to model the effect of several environmental variables on catch data. To illustrate GMDH as an appropriate tool to analyze these complex interactions, short-term seasonal shark catch data and a 22-year time series of blue crab landings were each related to environmental data by GMDH.

II. PREPROCESSING, PROCESSING, AND POSTPROCESSING

The first step in the analysis was the preprocessing of data. The shark data set was collected by monthly sampling in an eelgrass bed located in the lower Chesapeake Bay during 1979 and 1980 (Brooks and Lascara, 1982). A 200-ft section of 5-in. gill net was set in a shallow northern section of the eelgrass bed, in a southern section of the same eelgrass bed near a 20-ft dredged channel, and outside the eelgrass bed. Each of the three nets was checked for juvenile sandbar sharks, Carcharhinus plumbeus, every 3 hr over a 24-hr period. The environmental data collected during sampling included year, tide stage, water depth, time of day (dawn, day, dusk, night), temperature, salinity, dissolved oxygen, and location of the gill net. The data set contained 238 observations. Since the sandbar shark predominately ate blue crabs, the density of blue crabs collected in 1981 from the same eelgrass bed (Weinstein and Brooks, 1982) was used as a monthly relative index of food supply. The number of adult female sandbar sharks (Colvocoresses, 1981) captured by offshore longline sets was used as a biological descriptor of season for the juvenile sandbar sharks.

Once the shark data set was constructed, relationships between shark catch and environmental variables were analyzed by constructing a simple correlation matrix between all possible combinations of variables (Table 1). This analysis indicated that the environmental variables were not highly correlated with each other and therefore would not create a singular matrix in the GMDH procedure. Log transformation of the catch data greatly increased the correlation values between environmental variables and the shark catch.

Constructing the 22-year time series of blue crab, Callinectes sapidus, data was more involved than building the shark data set. Approximately 30 environmental variables were gathered by Harris and Van Engel (1981) to study the relationship between climatic variation and population size of the Chesapeake Bay blue crab. Some of the variables used were: water temperature, salinity, stream flow, solar radiation, sunspot activity,

Table 1 Correlation Matrix Between All Possible Combinations of Variables in the Shark Data Set[a]

	Sharks	Year	Crabs	Location	Depth	Temp.	Salinity	Dissolved oxygen	Tide stage	Time of day	Parent
Sharks	1.0000	0.0424	0.5636	-0.1278	0.0878	0.5066	0.4735	-0.1800	-0.0237	0.2406	0.3032
Year	0.0424	1.0000	-0.2291	-0.0184	0.1557	-0.0764	0.3281	-0.3284	0.0577	-0.0044	0.1251
Crabs	0.5636	-0.2291	1.0000	-0.0191	0.1096	0.6843	0.4919	-0.2335	0.0131	0.0500	0.2509
Location	-0.1278	-0.0184	-0.0191	1.0000	0.2405	-0.0512	0.0174	0.0676	-0.0489	-0.0113	-0.0197
Depth	0.0878	0.1557	0.1096	0.2405	1.0000	0.2553	0.1949	-0.1874	-0.1173	0.0108	0.1409
Temperature	0.5066	-0.0764	0.6843	-0.0512	0.2553	1.0000	0.3620	-0.3710	-0.0010	0.0702	0.3938
Salinity	0.4735	0.3281	0.4919	0.0174	0.1949	0.3620	1.0000	-0.3033	-0.0349	-0.0227	0.3418
Dissolved oxygen	-0.1800	-0.3284	-0.2355	0.0676	-0.1874	-0.3710	-0.3033	1.0000	0.0002	0.0320	-0.3365
Tide stage	-0.0237	0.0577	0.0131	-0.0489	-0.1173	-0.0010	-0.0349	0.0002	1.0000	-0.0121	-0.0019
Time of day	0.2406	-0.0044	0.0500	-0.0113	0.0108	0.0702	-0.0227	0.0320	-0.0121	1.0000	0.0854
Parent	0.3032	0.1251	0.2509	-0.0197	0.1409	0.3938	0.3418	-0.3365	-0.0019	0.0854	1.0000

[a]Sharks, log of shark catch; Crabs, blue crab abundance in the eelgrass bed; Location, location of gill net; Depth, water depth (when the net was emptied); Parent, number of adult female sharks.

Table 2 Correlation Matrix Between All Possible Combinations of Variables in the Crab Catch Data Set[a]

	Landings	Mayair	Streamfl.	Skycover	Survey	Julyair	Northjan.	Eastjan.	Northsep.	Eastsep.
Landings	1.0000	0.3892	-0.4454	0.2107	0.3816	-0.3273	0.2535	-0.1454	0.4081	0.1430
Mayair	0.3892	1.0000	0.0392	0.2431	0.1543	0.0468	0.1863	-0.2640	0.2247	-0.1310
Streamfl.	-0.4454	0.0392	1.0000	0.1256	-0.2531	0.1791	0.0847	-0.3576	-0.2911	-0.0213
Skycover	0.2107	0.2431	0.1256	1.0000	0.2760	-0.1786	0.3170	0.1582	-0.0522	-0.3928
Survey	0.3816	0.1543	-0.2531	0.2760	1.0000	0.0524	0.3503	0.0168	0.3706	-0.0153
Julyair	-0.3273	0.0468	0.1791	-0.1786	0.0524	1.0000	-0.1320	0.0486	-0.0881	0.1100
Northjan.	0.2535	0.1863	0.0847	0.3170	0.3503	-0.1320	1.0000	0.1380	0.4783	-0.0145
Eastjan.	-0.1454	-0.2640	-0.3576	0.1582	0.0168	0.0486	0.1380	1.0000	-0.0061	-0.2135
Northsep.	0.4081	0.2247	-0.2911	-0.0522	0.3706	-0.0881	0.4783	-0.0061	1.0000	0.1628
Eastsep.	0.1430	-0.1310	-0.0213	-0.3928	-0.0153	0.1100	-0.0145	-0.2135	0.1628	1.0000

[a]Landings, crab catch; Mayair, May air temperature lagged 2 years; Streamfl., October stream flow through mouth of Chesapeake Bay lagged 1 year; Skycover, total July sky cover lagged 2 years; Survey, York River crab catch lagged 2 years; Julyair, July air temperature lagged 2 years; Northjan., January N–S wind vectors; Eastjan., January E–W wind vectors; Northsep., September N–S wind vectors; Eastsep., September E–W wind vectors.

Ekmann transport, sky cover, precipitation, cooling degree-days, and wind velocity. Bivariate correlations were performed between each environmental variable and blue crab landings in Virginia. The variables were also lagged up to 24 months (the typical age of a marketable blue crab) and again compared to the landings data. Correlations were also performed between all environmental variables to avoid possible variable combinations that would cause a singular matrix to arise during the GMDH procedure. The environmental variables that were most highly correlated (Table 2) with blue crab landings were used in the GMDH analysis. The variables selected were: total July sky cover lagged 2 years, October stream flow through mouth of Chesapeake Bay lagged 1 year, September and January wind-velocity vectors, May through August York River crab catch (from VIMS blue crab survey) lagged 2 years, and May and July air temperature lagged 2 years.

After preprocessing, a GMDH program was written in Pascal. Pascal was selected rather than FORTRAN because Pascal required structured programming and would allow variable-size stacks. These stacks could expand and contract with the number of equations saved at each iteration of the procedure. Pascal would therefore allocate only the memory necessary to store the results of each iteration. The basic program involved three stacks. One stack saved equations from the previous iteration whose predicted catch correlated most closely to the actual catch. Another stack saved the "best" equations from the current iteration. The final stack kept a history of the procedure so that the "best" polynomial could be rebuilt and used as a model to investigate the effects of environmental variables on the catch data. The actual procedure closely followed Farlow's (1981) description of the GMDH method.

Initial program runs indicated that double precision was necessary to obtain accurate GMDH polynomials. Since most Pascal compilers do not allow double-precision operations and data types, an external FORTRAN program was called to solve the set of simultaneous equations for each iteration. After the "best" polynomial was selected, a FORTRAN postprocessor procedure searched the GMDH history file to reconstruct the selected polynomial.

This description of GMDH preprocessing, processing, and postprocessing has been included to illustrate the rather large amount of work necessary to use GMDH. The algorithm is relatively uninvolved compared to the large amount of preprocessing of the data as well as programming considerations to use the technique. Time lags and data transformations are still necessary to allow GMDH to find the optimum pattern in the data set.

III. INTERPRETATION OF THE RESULTS OF GMDH

The correlation between the predicted values of the shark model created by the GMDH postprocessor and the actual shark abundance data was 0.892. The GMDH procedure stopped after eight iterations. The environmental

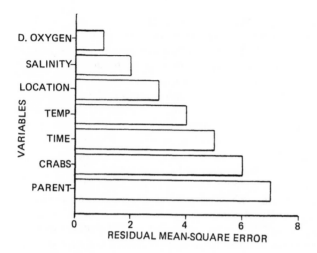

Fig. 1 Residual mean-squared errors for 2.5% standard deviation of variable.

variables used in the model were: number of adult female sharks, blue crab density in the eelgrass bed, time of day, temperature, salinity, dissolved oxygen, and location of the gill net. Year, tide stage, and water depth (when the net was emptied) were not significant variables in predicting shark abundance. Although one now knows which variables may be affecting shark abundance, the next question to be addressed is the relative order of importance

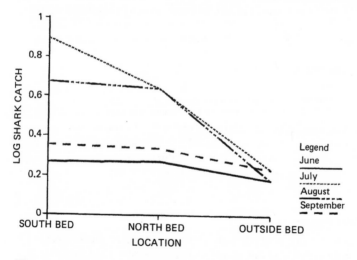

Fig. 2 Location versus log shark catch.

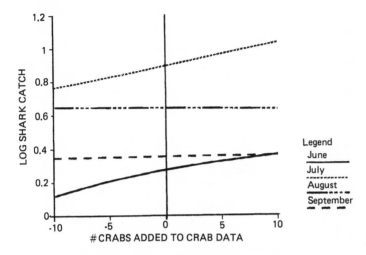

Fig. 3 Crab abundance versus log shark catch.

of the variables. For instance, does a typical fluctuation in temperature have more of an effect on shark catch than a fluctuation in salinity or dissolved oxygen? Because of the complexity of the polynomial, one cannot look at coefficients in the model and compare the relative importance of the variables.

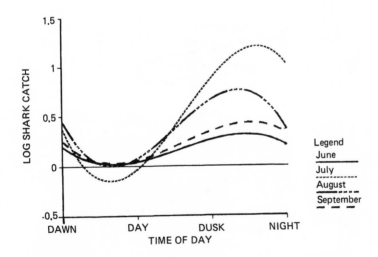

Fig. 4 Time versus log shark catch.

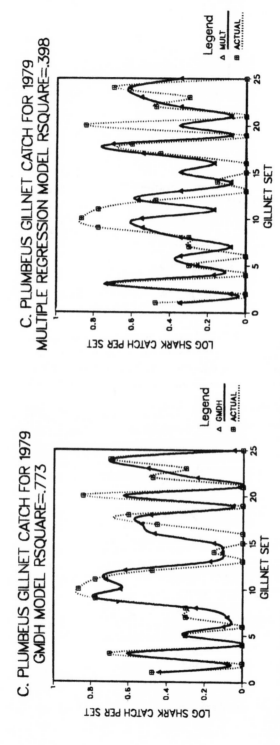

Fig. 5 Comparison of predicted landings and actual recorded landings for a GMDH model and a multiple regression model.

A standard fluctuation was defined as 2.5% of the standard deviation of an environmental variable in the data set. The importance of each variable was measured by calculating the mean-squared error between the predicted values from the modified data set (where the "standard fluctuation" had been added to a variable) and the predicted values of the unmodified data set. The larger the mean-squared error, the more impact a "standard fluctuation" in a variable had on the shark catch. Figure 1 illustrates the residual mean-squared errors and importance ranking of variables based on this technique.

Sensitivity analysis of the model can illustrate the effects of a single variable on the shark catch. The parameter of interest is allowed to vary within a specified range while all other model variables are fixed. The predicted catch can then be plotted versus the values of the fluctuating parameter to illustrate the impact that this parameter has on the catch. Figures 2 to 4 were created by using sensitivity analysis on night observations during June, July, August, and September. These observations were during periods when the sharks were most abundant.

Figure 2 illustrates the effect of varying the location variable in the data set. One can see that for all four months the model predicts that more sharks would be captured in the eelgrass bed than in the outside sandbar area surrounding the eelgrass bed. Figure 3 indicates that food supply may be important to the abundance of sharks in June and July when they first enter the eelgrass bed, but not in August and September. Figure 4 shows that most sharks were caught between twilight and night (around 2000 hr) in the eelgrass bed. Figure 5 illustrates how the GMDH model mimics actual shark catch with greater accuracy than the multiple regression model.

Procedures similar to those used in the shark analysis were used to analyze the blue crab time series. The GMDH model was built using environmental and blue crab landings data from 1956 to 1976. The correlation between actual and predicted landings was 0.999. Sensitivity analysis indicated that the GMDH model was two orders of magnitude more sensitive to river discharge than to other model variables. Environmental data from 1977 were entered into the model to predict crab landings for that year. Despite the high correlation between predicted landings and actual landings for 1956-1976, the prediction for 1977 varied greatly from the actual landings. Since the time series contained just 22 points, GMDH may not have been able to detect the entire underlying pattern in the data. With a longer time series more variation in the independent and dependent variables would have been given to GMDH, and other patterns between the variables may have emerged.

IV. SUMMARY AND CONCLUSIONS

GMDH is a valid tool to analyze the complex interactions between environmental variables and fisheries data. GMDH will select the important environmental variables that may be influencing the catch data of important commercial species. Sensitivity analysis of the resulting GMDH model can illustrate the impact of each variable on the catch data.

As with other multivariate techniques, the GMDH procedure does involve a large amount of data preprocessing to find hidden time lags and to avoid highly correlated environmental data. Data transformations may also enhance the relationships that GMDH may find.

The power of the GMDH procedure may become more apparent in the future when fisheries scientists combine population dynamic models, environmental influences, and fishery economic models using GMDH. Since GMDH does not make any assumptions about the data, the linking of these three approaches to studying fisheries should be relatively uncomplicated, and the potential for accurate predictive models seems very high.

REFERENCES

Austin, H., and Ingham, M. (1978). Use of Environmental Data in Prediction of Marine Fisheries Abundance. Proc. Climate Fisheries Workshop, Alton Jones Campus, University of Rhode Island, Mar. 1978, pp. 5.1-5.6.

Brooks, H. A., and Lascara, J. V. (1982). The Role of the Megapredator Carcharhinus plumbeus in an Eelgrass Bed in the Lower Chesapeake Bay. VIMS, Gloucester Point, Va., in preparation.

Colvocoresses, J. (1981). Virginia Institute of Marine Science Long Line Survey. VIMS, Gloucester Point, Va.

Farlow, S. J. (1981). The GMDH Algorithm of Ivakhnenko. Am. Stat. 35: 210-215.

Harris, R. E., and Van Engel, W. A. (1981). Relationship Between the Chesapeake Bay Blue Crab and Its Climatological Environment. Data Rep. 15, VIMS, Gloucester Point, Va.

Murphy, G. I., and Shomura, R. S. (1955). Longline Fishing for Deep-Swimming Tunas in the Central Pacific, Aug.-Nov. 1952. U.S. Fish Wildlife Serv. Spec. Sci. Rep. 137.

Ottestad, P. (1942). On Periodic Variations in the Yield of the Great Sea Fisheries and the Possibility of Establishing Yield Prognoses. Fisk. Dir. Skr. Ser. Havundersok. 7(5):1-11.

Pearson, J. C. (1948). Fluctuations in the Abundance of the Blue Crab in Chesapeake Bay. U.S. Fish Wildlife Serv. Res. Rep. 14.

Sette, D. E. (1943). Biology of the Atlantic Mackerel (Scomber scomberus) of North America. Fish Bull. 50(38):149-237.

Weinstein, M. P., and Brooks, H. A. (1983). Comparative Ecology of Nekton Residing in a Tidal Creek and Adjacent Seagrass Meadow: Community Composition and Structure. Mar. Prog. Ser. 12:15-17.

9

Predictive GMDH Models of Shrimp Catches: Some Practical Considerations

MICHAEL H. PRAGER and SAUL B. SAILA/Graduate School of Oceanography, University of Rhode Island, Narragansett, Rhode Island

I. INTRODUCTION

Measured by the ex-vessel value of its harvest, the U.S. shrimp fishery is the most valuable of our fisheries, averaging about 25% of all fish and shellfish landed, and valued in 1979 at $471,573,000. Of this value, the Gulf of Mexico contributed about 80%. After adding the economic effects of landing, processing, and reselling the shrimp, the economic impact of the Gulf fishery was estimated at $1.2 billion in 1979. At the same time, the fishery employed about 30,000 persons throughout the Gulf region, and provided recreational and bait fishing for tens of thousands more (GMFMC, 1981; Etzold and Christmas, 1977).

The National Marine Fisheries Service, which regulates the offshore fishery, has divided the fishing grounds into 21 statistical areas (Fig. 1). Of the reported commercial catch from these 21 areas—which totals about 120 million pounds (heads-off) annually—brown shrimp (Penaeus aztecus) forms roughly 55%, white shrimp (Penaeus setiferus) 30%, and pink shrimp (Penaeus duorarum) 11% by weight (GMFMC, 1981). Other species make minor contributions. In this chapter we consider catches of brown and white shrimp in statistical areas 17 and 18.

A modeling method that could forecast an area's annual catch several months in advance would benefit fishermen and fishery managers alike. Both fishing and monitoring personnel could be directed to areas of higher yield. Shrimping trips could be planned more efficiently. Processors could adjust their staffs in anticipation of good and bad years.

The year-to-year variation in shrimp abundance has been examined by several investigators. For example, Garcia and LeReste (1981) found that a region's annual shrimp production often varies in relation to climatological

179

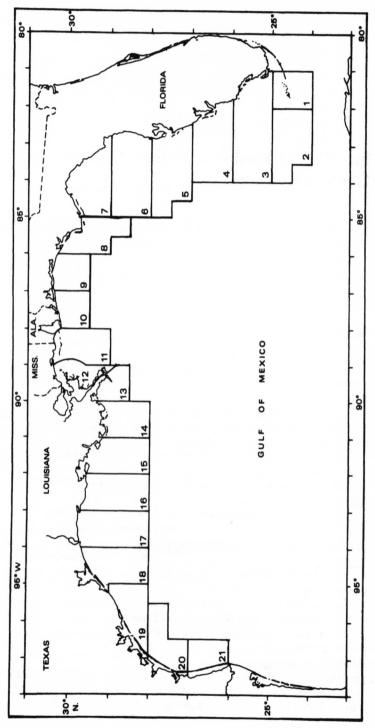

Fig. 1 Gulf of Mexico, illustrating the 21 statistical areas of the fishing grounds.

factors. Saila et al. (1982), using regression on principal components, accounted for most of the year-to-year variability in brown and white shrimp catches in terms of environmental fluctuations. Hunt et al. (1980) developed a successful harvest model for brown shrimp in Pamlico Sound (North Carolina) based on temperature and salinity measurements in late spring. Taking a more general point of view, Stocker and Hilborn (1981) examined seven methods of short-term forecasting in marine fish stocks, and concluded that time-series methods might be useful. Among our reasons for choosing GMDH in modeling the Gulf of Mexico shrimp fishery were that the method's use of interaction terms seemed biologically appealing, and that the method has not had extensive previous application in forecasting, especially in fisheries science.

In undertaking the study, we hoped, first, to develop a useful forecasting methodology for catches of brown and white shrimp. We also wished to gain a clearer perspective on GMDH's usefulness as a fisheries management tool; and finally, we hoped to optimize the technical aspects of our GMDH modeling procedures.

II. METHODS

A. Data

Our data set ran from 1961 to 1979, and included, for each annual catch value, a set of quarterly environmental statistics. As we wanted to develop a model to forecast the year's catch by midyear, we included only the first two quarters' environmental data in each yearly observation. The environmental variables were:

> Variables 1-8: First- and second-quarter means of the north and east components of Ekman transport (wind-driven water circulation) at two locations: 30°N, 92°W, and 30°N, 95°W.
> Variables 9-16: First- and second-quarter totals of discharges of the Mississippi, Sabine, Trinity, and Neches rivers.
> Variables 17-28: First- and second-quarter minima, maxima, and medians of surface-water temperature and salinity at North Jetty (just northeast of Galveston, Texas). Data from North Jetty comprised the longest continuous records of temperature and salinity in the area, but measurements were discontinued after 1974. We therefore estimated values for 1975-1979 by simple multiple regression onto an overlapping data set from Galveston Bay. Mississippi River discharge data were used to improve the fit of the regression equations for estimating North Jetty salinities.

To allow for the influence of the environment on the early developmental stage of the shrimp, we included in each observation the environmental data from the four quarters of the preceding year. (As most shrimp are harvested when they are less than a year old, these data covered more than the entire life history for most individuals.) The number of environmental values per catch record was thus tripled, to 84. To simulate any possible stock-recruitment relationship, we included the preceding year's catch as an additional input variable. (The assumption here was that any year's catch is an indicator of that year's stock size, which in turn is a result of that year's recruitment. Thus the preceding year's catch, as an indicator of spawning stock size, might be correlated to the present year's catch, which resulted from the spawning.) From the total of 85 input variables, we selected for each analysis the 20 variables most highly correlated with catch over time (as measured by the Pearson product-moment correlation coefficient, r).

B. Models

To clarify the properties of our GMDH modeling procedures, and thereby improve our forecasts, we decided to construct a fairly large number of models and compare them in a systematic way. To this end, we used a quasi-factorial experimental plan, varying five factors over two levels each. Thus we ran 2^5 (or 32) experiments (models). Although we do not consider it a true factorial design, as the five factors were not true treatments, we will employ the terms "factor" and "level" for ease of expression. The five factors are summarized in the following table:

The Five Factors and Their Levels

Factor	Level 1	Level 2
Species	Brown	White
Area	Area 17	Area 18
Year predicted	1977	1978
Point division	Random	D(k)
Stepwise subset selection	All points	Checking set

Table 1, part (a), lists all the factor levels of each experiment. The "species" and "area" entries in the table above are self-explanatory. The significance of "year predicted" is that we used data from 1961 to 1976 to predict 1977's catches; similarly, we used data from 1961 to 1977 to predict 1978's catches.

"Point division" refers to the division of the data into a training set and a checking set, which are used (at each iteration of the GMDH process) to estimate parameters and to rank models, respectively. Two divisions were used. In the first, points were randomly assigned to either set. In the second, points were assigned on the value of the modified D(k) statistic,

Table 1 Factor Levels and First-Period Prediction Error for Each Model[a]

GMDH model number	Species		Area		Year		Point division		Stepwise		First-period prediction error for each model (%)
	B	W	17	18	77	78	Random	D(k)	All	Ckg	
7A		A′	A		A		A		A		8.4
7B		B	B		B			B		B	22.0
8A	A				A	A	A		A		-60.2
8B	B				B	B		B		B	-62.2
9A	A		A		A		A		A		-26.8
9B	B		B		B			B		B	-35.6
10A		A	A		A		A		A		10.5
10B		B	B		B			B		B	-52.3
12A	A		A	A			A		A		-2.7
12B	B		B	B				B		B	-34.6
14A		A	A		A		A		A		14.5
14B		B	B		B			B		B	24.4
15A	A		A		A		A		A		-33.4
15B	B		B		B			B		B	-39.3
16A		A	A	A			A		A		21.7
16B		B	B	B				B		B	30.0
17A		A	A			A	A		A		-33.4
17B		B	B			B		B		B	-33.3
18A		A	A		A		A		A		-25.3
18B		B	B		B			B		B	35.1
19A	A				A	A	A		A		-103.0
19B	B				B	B		B		B	-103.0
20A	A				A	A	A		A		149.2
20B	B				B	B		B		B	200.8
21A	A		A		A		A		A		-6.8
21B	B		B		B			B		B	-6.8
22A	A		A		A		A		A		-23.5
22B	B		B		B			B		B	114.8
23A		A	A		A		A		A		-30.2
23B		B	B		B			B		B	-13.8
24A		A	A		A		A	A	A		-61.2
24B		B	B		B			B		B	-58.7

[a] A's and B's are used instead of X's solely to improve legibility.

183

developed by Duffy and Franklin (1973) from the original D(k) of Ivakhnenko (1971). In using D(k), a trend line (a polynomial of degree 4 or less) was established for each variable, and the variable's deviations from the trend line figured. Then D(k) was computed for each data point as

$$D(k) = \sum_{j=1}^{n} \left[\frac{x_j - p(x_j)}{m(x_j)} \right]^2 \tag{1}$$

where n is the number of variables, x_j the jth variable at time k, $p(x_j)$ the value of the trend line of x_j, and $m(x_j)$ the mean of x_j over time. The points with the highest D(k) were put into the training set; the others, into the checking set. The presumption in the use of D(k) is that the values of the variables closer to the trend lines are likely to reflect the basic nature of the system and should be used for ranking the models. Moreover, by putting the points with higher D(k) into the training set, a wide spread of values is assured when estimating parameters. This makes it less likely that future extreme values of variables will drive the model out of range.

"Stepwise selection method" also involves the mechanics of our GMDH procedure. We first ran 16 GMDH analyses (one for each combination of species, area, year, and point division), using the computer program described in Chapter 16. We performed the necessary algebra with the REDUCE-2 computer program (Hearn, 1973), to arrive at a list of potential model terms for each of the 16 models. From each list, we selected a statistically significant subset by stepwise regression (or, in a few cases, all-subsets regression). Performing this subset selection for each of the GMDH analyses in two ways gave us our 32 models. The two ways—(1) on all the data, and (2) on the checking set data only—were the two "levels" of the stepwise factor.

We then calculated model parameters (regression coefficients) by ordinary least squares, and computed a forecast (C*) of the following year's catch (C).

C. Analysis of Results

For each of the 32 models, we calculated the signed error (E1) as a percentage of actual catch by the following formula:

$$E1 = \frac{C^* - C}{C} \times 100\% \tag{2}$$

and additionally defined the unsigned error as the absolute value of E1, that is

$$E2 = |E1| \tag{3}$$

Except where specified otherwise, references to prediction error refer to the unsigned percentage error, E2.

It should be emphasized that our choice of this criterion was more or less arbitrary, and that other measures of forecasting accuracy, such as the magnitude of error ($C* - C$) or squared error ($C* - C)^2$, might have been used. What was not arbitrary was our choice to rank models by prediction error, rather than by their fit to the same data used to develop the model. Our experience with GMDH and other curve-fitting techniques (see Sec. IV and Fig. 2) has convinced us that any level of fit desired can be obtained by modern statistical methods, and furthermore, that this fit is no indication of a workable forecasting method.

We used two-tailed Wilcoxon signed-rank tests (Hollander and Wolfe, 1973) to determine whether either level of the stepwise selection or point division factors had produced models with significantly lower unsigned errors. This nonparametric test makes few distributional assumptions about the data, but does assume independent data values. This assumption may not have been met, as groups of data values had factors (such as year) in common, which may have caused correlations. We felt, nonetheless, that even approximate significance probabilities would be of interest, both to ourselves and to our readers. Hence we have reported P values in several instances. We have not, however, attempted to draw any single line between significant and nonsignificant results.

We also compared the accuracies of our GMDH models to those of predictions made with two simple time-series estimators: the grand mean (of catch) and the 5-year running average (of catch).

III. RESULTS

A. General Results

Figure 2 shows a few of our models, including some of the best- and worst-fitting. It was apparent that good fit to the model-building sequence was not necessarily an indicator of good prediction accuracy: nor was model root-mean-squared error (RMSE) on the checking set significantly correlated with unsigned prediction error ($r = -0.18$, $P < 0.33$). (The GMDH algorithm uses RMSE on the checking set to indicate goodness of fit.) Although the accuracy of prediction varied widely from model to model, several patterns were apparent (and are discussed later).

The short table below gives an overview of the models' first-period prediction errors. In this table and the ones that follow, "S.E." refers to the standard error of the mean; and all errors are given as percentages of actual catch [Eqs. (2) and (3)].

(a)

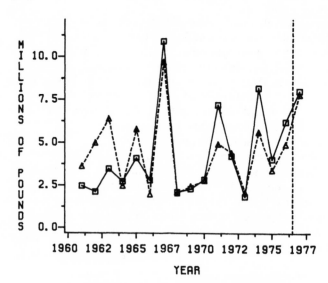

(b)

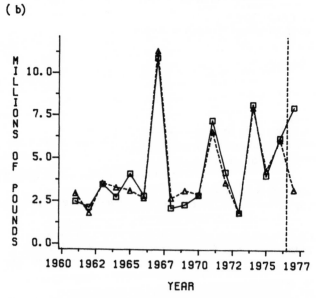

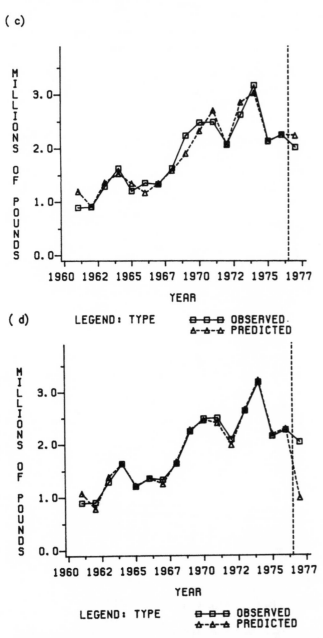

Fig. 2 Four of our GMDH models: (a) experiment 12A; (b) experiment 8A;
(c) experiment 10A; (d) experiment 10B. Data to the left of the dashed line
were used to develop each model.

Overall Statistics of Percentage Error

Error type	n	Minimum error	Maximum error	Mean error	S. E.
Signed (E1)	32	-103.7	200.8	-6.7	11.3
Unsigned (E2)	32	2.7	200.8	46.2	7.8

In addition, Table 1 lists the prediction error of each experiment individually. Histograms showing the distributions of errors make up Fig. 3. In general, prediction error was fairly high, averaging 46%—probably too high for the forecasts to be useful.

B. Effects of the Five Factors

The effects on prediction error of the different values of species, area, year predicted, stepwise method, and point division are described in the following five sections.

1. Species

In 12 (of 16) cases, models predicting white shrimp catch were more accurate than corresponding models predicting catch of brown shrimp.

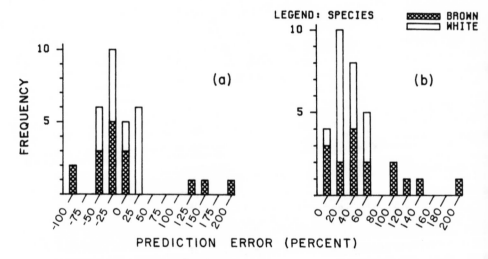

Fig. 3 Distribution of first-period prediction errors of 32 GMDH models: (a) signed error; (b) unsigned error.

Unsigned Error Percentage (E2) Statistics—by Species

Species	n	Minimum	Maximum	Mean	S. E.
Brown	16	2.7	200.8	62.7	14.2
White	16	8.4	61.2	29.7	4.0
Difference	16	-26.6	142.1	33.0	3.0

The "difference" row in this table and the similar tables that follow represents the differences in error between models using the two levels of the factor in question when all other factors are held the same. For example, in this case, since

$$D = E2(\text{brown}) - E2(\text{white}) \tag{4}$$

(where D is this difference variable and E2 the unsigned error [Eq. (3)]), a positive value of D indicates a lower error for the model of white shrimp catch than for the model of brown. As there are, besides species, four factors having two levels each, there are 2^4 (or 16) pairs of experiments differing (within each pair) only by species. Thus the statistics for D are based on a sample size of 16 experiments, as indicated.

2. Area

Catches in area 17 were predicted with less than half the error, on the average, of those in area 18.

Unsigned Error Percentage (E2) Statistics—by Area

Area	n	Minimum	Maximum	Mean	S. E.
17	16	6.8	114.8	29.9	4.4
18	16	2.7	200.8	62.1	13.4
Difference	16	-125.7	30.7	-31.9	11.5

3. Year

The predictions for 1977 were generally much more accurate than those for 1978, even though all predictions were made only one year ahead. The difference in errors can be seen in the following table.

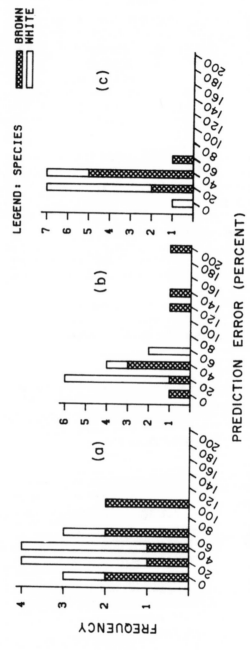

Fig. 4   Error distributions of three modeling techniques: (a) random point division; (b) point division by $D(k)$; (c) average forecast of methods (a) and (b).

Unsigned Error Percentage (E2) Statistics—by Year

Year	n	Minimum	Maximum	Mean	S.E.
1977	16	2.7	62.2	29.9	4.4
1978	16	6.8	200.8	62.4	14.1
Difference	16	-166.2	38.5	-32.5	14.1

4. Stepwise Method

The Wilcoxon test supported ($P < 0.02$) the advantage of using only the checking set for best-subset selection; mean error was less by about 16% of catch. Using the mean forecast of both methods was intermediate in error: superior ($P < 0.02$) to all-points method; inferior ($P < 0.09$) to the checking-set method.

Unsigned Error Percentage (E2) Statistics—by Stepwise Method

Modeling method	n	Minimum	Maximum	Mean	S.E.
All points	16	6.8	200.8	54.2	12.2
Checking set	16	2.7	149.2	38.2	9.8
(Difference)	16	-16.4	91.3	16.0	6.6
Mean forecast	16	4.9	175.0	42.5	10.8

5. Point Division

Although there were large differences between the two methods in individual cases, the advantage of random division was questionable ($P < 0.13$) when all 16 pairs of experiments were considered at one time.

We found that lower prediction error was obtained (for a given species, area, and year) by averaging the predictions of the two models developed with different point divisions ($P < 0.03$). In particular, the mean and standard deviation of the error (E2) were both reduced, so that fewer forecasts had extreme errors, and none had an error higher than 54%. This comparison is illustrated in the following table and in Figs. 4 and 5.

Unsigned Error Percentage (E2) Statistics—by Point Division

Modeling method	n	Minimum	Maximum	Mean	S.E.
Random	16	6.8	103.0	38.0	7.8
D(k)	16	2.7	200.8	54.3	13.6
(Difference)	16	-108.0	57.5	-16.3	10.7
Mean forecast	16	0.9	54.0	29.0	3.9

(a)

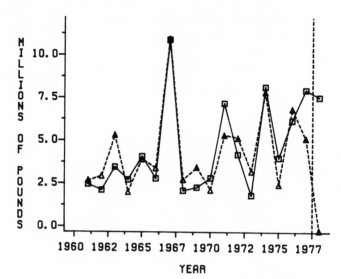

(b)

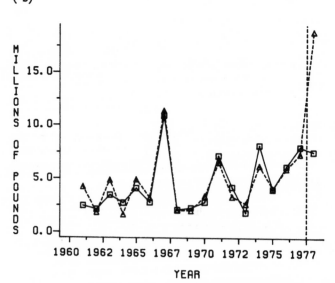

(c)

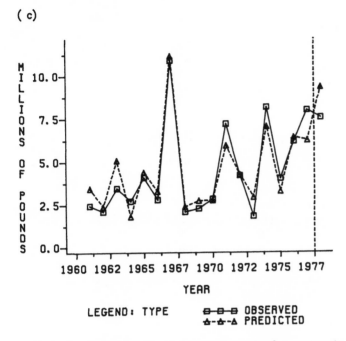

Fig. 5 Example of reduced forecast error when averaging two models.
Models are of 1978 brown shrimp catch in area 18: (a) experiment 19A,
random point division; (b) experiment 20A, D(k) point division; (c) averaged
forecast of models 19A and 20A.

We also obtained this reduced error (by averaging forecasts) when considering only models constructed with stepwise method 2 (use of the checking set), which gave lower mean error to begin with. The mean forecast
error of the eight predictions obtained in this way was the lowest mean error
of any group of models yet considered.

Unsigned Error Percentage (E2) Statistics—by Point Division
(for Models with Stepwise Method 2)

Modeling method	n	Minimum	Maximum	Mean	S.E.
Random division	8	6.8	103.0	34.9	11.5
D(k) point division	8	2.7	149.2	41.4	16.5
Mean forecast	8	11.5	45.7	25.3	4.0

Table 2 Comparison of GMDH to Simple Time-Series Forecasting Methods

Forecasting method[a]	n	Minimum error (%)	Maximum error (%)	Mean error (%)	S. E.[b]
GMDH—subset 3[c]	8	11.5	45.7	25.3	4.0
GMDH—subset 2[d]	16	0.9	54.0	29.0	3.9
Running average	8	13.1	57.2	33.0	5.1
GMDH—subset 1[e]	16	2.7	149.2	38.2	9.8
Mean catch	8	8.3	68.1	40.8	7.9
GMDH—32 models	32	2.7	200.8	46.2	7.8

[a]Methods are ranked by mean error.
[b]S. E.: standard error of the mean.
[c]Stepwise method: checking set only. Average of two point-division forecasts.
[d]Not controlled for stepwise method. Average of two point-division forecasts.
[e]Stepwise method: checking set only. Not controlled for point division.

C. Comparison to Other Forecasting Methods

The results of our comparisons to other forecasting methods are presented in Table 2. GMDH was found to be both the best and the worst forecaster: the best predictions were made by GMDH, using subset selection on the checking set, and averaging the forecasts of the two point-division methods. The worst predictions were attributable to all 32 GMDH models taken as a group; that group had the highest mean error of any group considered. The simple time-series methods' errors were in between.

IV. DISCUSSION

A method of forecasting shrimp catch several months in advance would be a valuable tool in managing the Gulf of Mexico fishery. Our results, which demonstrate forecasting accuracy with roughly 25% error (Table 2), suggest that GMDH is such a method. A pilot program in the fishery could confirm or falsify this belief, as could further experiments along the lines explored here.

The superiority of prediction for white shrimp we attribute to its lower year-to-year variability in catch. The simple time-series methods also predicted catches of white shrimp more accurately than those of brown. We have no explanation for the consistently higher errors of prediction in 1978.

The improved results by averaging predictions of the models from two point divisions emphasize the sensitivity of GMDH models to this factor.

Different point divisions have, in our experience, consistently led to models that fit and predicted differently. This problem has been discussed in the literature (Yurachkovskiy and Groshkov, 1980; Vysotskiy, 1976) but has not yet been resolved. We have found no clearly superior method of point division, but recommend averaging the forecasts of at least two point divisions. We recognize that this can be an expensive and time-consuming practice. In the case of the shrimp fishery, at least, the results appear to be worth the expenditure.

It appears that stepwise selection on the checking set may produce more accurate forecasts by reducing overfitting. We recommend its consideration whenever GMDH modeling is performed with an algorithm (such as ours) which requires taking a subset of the final model. Of course, those using overspecified GMDH models, which have more parameters than data points, will not need this technique.

Our GMDH algorithm ranked models by checking set RMSE (a function of squared errors). In contrast, we adopted prediction error (a percentage) as the criterion of model quality. We suspect that in future work, prediction error of the final model might be reduced by using prediction error itself, instead of RMSE, as the internal GMDH model ranking criterion.

The structures of the model equations may provide some insight into stock-recruitment relationships of white and brown shrimp. Of the 16 models of white shrimp catch, 14 model equations included lagged catch, a pattern that is consistent with some form of stock-recruitment relationship. This pattern, however, can also be explained by the persistence of year classes over annual boundaries (i.e., January 1), a phenomenon expected in this summer-spawning species. For brown shrimp, the picture was quite different. In area 18, lagged catch was virtually uncorrelated to current catch, and was therefore not used as an input variable. In area 17, where lagged catch was used as an input variable, only two of the eight brown shrimp models retained it. We infer from these results that the stock-recruitment relationships of brown shrimp in these two statistical areas may be weaker or nonexistent. We emphasize that our data are from a small geographic area, and that the geographic limits of the true breeding area are not known. A similar study of brown shrimp over a larger area might find different patterns.

In conclusion, we feel that we have demonstrated the practicality of GMDH modeling in the Gulf shrimp fishery; that we have accumulated some evidence of its superiority to at least the simple time-series models considered here; and that we have brought to light some practical considerations which may reduce first-period prediction error considerably.

ACKNOWLEDGMENTS

We would like to thank the staff at the National Marine Fisheries Service Galveston Laboratory, particularly Dr. Charles Caillouet, Dr. Edward

Klima, and Dennis Koi. The National Marine Fisheries Service and the Texas Parks and Wildlife Department supplied the data that made this work possible.

The manuscript was reviewed by Dr. Charles Caillouet, John Hoenig, Roy Johnson, Cynthia Jones, Dr. Edward Klima, and Henry Walker, all of whom suggested improvements.

Much of the analysis was supported by a National Science Foundation graduate fellowship to one of the authors (M. H. P.). In addition, the University of Rhode Island and its Graduate School of Oceanography contributed extensive computer time.

REFERENCES

Duffy, J., and Franklin, M. (1973). A Case Study of Environmental System Modeling with the Group Method of Data Handling. Proc. Joint Automatic Control Conf., Columbus, Ohio, pp. 101-111. Distributed by Nat. Tech. Inf. Serv., U.S. Dept. of Commerce, as Rep. AFOSR-TR-73-2220.

Etzold, D. J., and Christmas, J. Y. (1977). A Comprehensive Summary of the Shrimp Fishery of the Gulf of Mexico United States: A Regional Management Plan. Gulf Coast Res. Lab., Ocean Springs, Miss. (Tech. Rep. 2, Part 2.)

Garcia, S., and LeReste, L. (1981). Life Cycles, Dynamics, Exploitation and Management of Coastal Penaeid Shrimp Stocks. FAO Fisheries Tech. Paper 203. Food and Agriculture Organization of the United Nations, Rome.

GMFMC (1981). Draft Update of Fishery Management Plan for Shrimp: Gulf of Mexico. The Gulf of Mexico Fishery Management Council, Tampa, Fla.

Hearn, A. C. (1973). REDUCE 2 User's Manual, 2nd ed. University of Utah, Salt Lake City.

Hollander, M., and Wolfe, D. A. (1973). Nonparametric Statistical Methods. Wiley, New York.

Hunt, J. H., Carroll, R. J., Chinchilli, V., and Frankenberg, D. (1980). Relationship Between Environmental Factors and Brown Shrimp Production in Pamlico Sound, North Carolina. Spec. Sci. Rep. 33; N.C. Dept. of Natural Resources, Moorehead City.

Ivakhnenko, A. G. (1971). Polynomial Theory of Complex Systems. IEEE Trans. Syst. Man Cybern. SMC-1:364-378.

Saila, S. B., Walker, H. A., Lorda, E., Kelly, J., and Prager, M. H. (1982). Analysis of Data on Shrimping Success, Shrimp Recruitment, and Associated Environmental Variables. Vol. I in W. B. Jackson (Ed.) Shrimp Population Studies: West Hackberry and Big Hill Brine Disposal Sites Off Southwest Louisiana and Upper Texas Coasts, 1980-82. NOAA/NMFS final report to DOE. (University of Rhode Island GSO Tech. Rep. 82-4.) Distributed by Nat. Tech. Inf. Serv., U.S. Dept. of Commerce, Springfield, Va.

Stocker, M., and Hilborn, R. (1981). Short-Term Forecasting in Marine
    Fish Stocks. Can. J. Fish. Aquat. Sci. 38:1247-1254.
Vysotskiy, V. N. (1976). Optimum Partitioning of Experimental Data in
    GMDH Algorithms. Sov. Autom. Control 9(3):62-65.
Yurachkovskiy, Yu. P., and Groshkov, A. N. (1980). Optimal Partition of
    Initial Data into a Training Sequence and a Checking Sequence by Analy-
    sis of the Distribution Function of the Criterion. Sov. Autom. Control
    13(2):1-7.

10

GMDH Forecasting of U.S. Interest Rates

KENICHI OHASHI/The World Bank, Washington, D. C.

I. INTRODUCTION

This study is an attempt to develop GMDH forecasting models for two key U.S. interest rates; the federal funds rate and the yield on the 5-year U.S. Treasury note (see Secs. V.B and V.C for definitions of the two rates). It relies on the results obtained in an earlier project conducted by H. Kanemasu in collaboration with this author and others in which multivariate ARIMA models were developed for several U.S. financial and economic variables (see Kanemasu, 1980). Kanemasu found that several time series with certain lags influenced the monthly changes in the federal funds rate and the 5-year Treasury note yield. The current study tries to extend the earlier results by exploiting the ability of GMDH to capture the nonlinearities in the relationships among the various time series in the model. Although in the models by Kanemasu, the effects of those series on the two interest rates were in some cases through the autoregressive side and in other cases through the moving-average side, this study will only consider the fact that those series affected the interest rates with certain lags. Therefore, the GMDH models developed here are quite different from the original Kanemasu models.

This project was done in two stages. In the first stage, 83 monthly observations up to September 1980 were used to develop models. In this exercise, the last 12 observations were set aside to conduct an ex post forecast performance test. It was found that two models for the federal funds rate did quite well and none of the models for the 5-year Treasury note yield produced satisfactory results. In the second stage of the project, the true forecasting performance of the two federal funds rate models was monitored in the succeeding months.

199

II. DEFINING A MODEL

One of the unusual features of the models studied here is that the model structure changes from month to month to keep up with the changing environment of the financial market. In other words, there is no model identified, in a conventional econometric sense, as the one that explains how the two interest rates are determined and can be projected. New models are developed every month using the most up-to-date information to forecast the interest rates for the following month.

This poses a problem for defining a model meaningfully. When GMDH is applied to the new set of observations in each period, the model builder could easily find himself choosing different ways each time to proceed from one GMDH level to another or different ways to define the fit and check sets, depending on recent experiences with the GMDH approach. This kind of flexibility and the amalgamation of more rigid modeling and personal experiences are perhaps essential in a successful application of GMDH. From the standpoint of evaluating the usefulness of GMDH, however, this creates an insurmountable problem, because it becomes impossible to distinguish how much of the forecast performance is attributable to the model itself and how much to the personal skill of its user.

In this project it was decided from the outset that no element of subjectivity should enter the process of model building. This meant that the methods of dividing observations into a fit and a check set, choosing the equations for higher modeling levels, and determining when to stop needed to be clearly spelled out so that any person would produce exactly the same result by following them. Indeed, only such a set of rules may be called a well-defined model in this study.

III. ECONOMIC BACKGROUND OF THE MODELS

The time series under study here are two of the key interest rates in the U.S. financial system, although they are by no means of exclusive importance. Some elements of the institutional and theoretical backgrounds of these rates are explained in this section. In particular, the federal funds rate will be discussed in greater length, because it is a more specialized rate and also because more of the forecasting effort was concentrated on this rate.

A. The Federal Funds

The federal funds market is one of the pivotal markets in the U.S. financial system. More than 14,000 commercial banks and other participants trade immediately available funds, mostly on an overnight basis. The federal funds rate is the interest rate charged in such an overnight transaction. The original need for the market arose from the reserve requirements

imposed by the Federal Reserve System on various financial institutions.*
Required reserves (i.e., certain percentages of deposit liabilities specified
by Regulation D) must be held in a combination of vault cash and non-interest-
bearing reserve balances at a Federal Reserve Bank. Since reserves do not
earn any interest, banks try to minimize their holding of excess reserves
(i.e., reserves in excess of what is required by the Fed). Although banks
only need to meet the requirements on a weekly average basis,† unexpected
changes in assets or liabilities can easily create some shortfall or excess
every week. This gives rise to a market in which the excess funds are pur-
chased by banks with reserve deficiencies.

However, this function of smoothing out the reserve funds distribution
alone does not justify all the attention that this market receives. The special
importance of the federal funds market is attributable to two other factors.
First, the central bank constantly applies a varying degree of pressure on
this market to implement its monetary policy objectives. Second, many large
banks have come to rely on overnight money as a more permanent source of
their funds, and consequently made the federal funds availability a particularly effective tool of monetary policy.

B. The 5-Year Treasury Note

U.S. Treasury notes are the interest-bearing obligations of the U.S. government. They are issued for initial maturities of 2 years to 30 years. Together with the 3-month to 12-month Treasury bills, which are issued at
discount instead of having semiannual interest payments, the Treasury notes
are used to finance shortfalls in government revenue. Although the yield for
any maturity is important, the 5-year yield is one of the most important,
because the financial market considers it the representative intermediate
rate and the issue in that maturity range is actively traded in the bond market. Unlike very short-term interest rates, which may be influenced heavily
by technical and temporary factors in the financial markets, intermediate-
term to long-term rates are presumably determined by the long-run cost
of credit and the long-term expectations about inflation, and are therefore
less volatile than shorter maturity rates.

C. Theoretical Basis for Selecting Explanatory Variables

Interest rates are the prices paid to obtain liquidity (or money) on various
terms in regard to the length of time for which liquidity is made available

---

*Commercial banks, thrift institutions, U.S. branches and agencies of
foreign banks, and Edge Act Corporations (i.e., international banking corporations established in the United States by a U.S. bank) are all subject
to reserve requirements of Regulation D.
†This was changed to a biweekly system in February 1984.

and the risk involved. They are determined in the general process through which an economy as a whole attains an equilibrium. The mechanism of this process has been a subject of much controversy in economics. Broadly speaking, however, there is little doubt that some variables play essential roles in this general equilibrium system. The level of economic activity, money supply, inflation, and investment demand, in addition to the interest rates, are considered to constitute the basic elements of the system in the classical tradition of economics. Economic activities and investment demand influence the demand for liquidity. Money supply interacts with the demand for liquidity to determine interest rates. Inflation affects the real (or inflation-adjusted) demand for liquidity. For intermediate- and long-term rates, the expected inflation rate influences the nominal interest rate, as the latter is considered to be the total of the real interest rate and the expected rate of inflation. Exchange rates may be important as well, for foreign capital markets are closely linked to the U.S. counterpart through foreign exchange markets.

The time series chosen for this study are closely related to one of these factors, with particular emphasis on their potential ability to forecast the future movements in interest rates. Therefore, certain variables which are theoretically relevant for the determination of interest rates may not appear in the models developed here, because of the lack of predictive power. For example, the long-term interest rate and the overnight rate should influence each other, as investors would always compare the relative attractiveness of different rates. Yet they are not used to forecast each other, because their mutual influences seem to be only contemporaneous.

IV. THE MODEL STRUCTURE

A. Time Series Involved

Before proceeding further, all the series that are used in this study are defined in detail here.

> FFR: monthly change in the average federal funds rate.
> 5YN: monthly change in the 5-year Treasury note (constant maturity) yield.
> M1A: monthly change in the 12-month growth in logarithm of seasonally unadjusted M1A (i.e., the sum of currency in circulation and conventional demand deposits), or
>
> $$[\ln(M1A(T)) - \ln(M1A(T-12))] - [\ln(M1A(T-1)) - \ln(M1A(T-13))]$$
>
> Beginning in January 1981, M1A was adjusted for the nationwide introduction of NOW accounts.*

---

*With the nationwide introduction of NOW (negotiable order of withdrawal)

EX: monthly change in the trade-weighted index of the U.S. dollar exchange rates.

BS: monthly change in the 12-month growth in seasonally unadjusted total business (or manufacturing and trade) sales.

CPI: monthly change in the seasonally unadjusted Consumer Price Index for all urban consumers.

CIL: monthly change in the 12-month growth in logarithm of the amount of commercial and industrial loans outstanding at the weekly reporting (i.e., large) commercial banks. Seasonally unadjusted.

The numbers after these variable codes will indicate the number of months the series are lagged. This study will use 15 different combinations: FFR, FFR1, 5YN, 5YN1, M1A1, M1A3, M1A5, EX1, EX4, EX6, BS5, CPI1, CIL6, CIL7, and CIL8.

B. Selection of the Series

The selection of particular time series with certain lags as explanatory variables in this project relied on the findings of H. Kanemasu (1980). He found that FFR is affected by FFR1, M1A1, EX1, EX4, EX5, CPI1, CIL6, and CIL7; and that 5YN is affected by 5YN1, M1A1, M1A3, M1A5, EX4, EX6, BS5, and CIL8. The current study takes these relationships as given and concentrates on discovering the best nonlinear combinations of these series for forecasting the two interest rates. Thus, for FFR, models involving all the eight series were built as the most general models. Similarly, for 5YN, the relevant eight series were used to build the most general models. In addition, for each dependent variable, models using only a subset of the respective eight series were examined. These subsets correspond to the series used in the best performing models for FFR and 5YN in the Kanemasu study.

---

accounts in January 1981, large sums of money which were formerly in the M1A category shifted into the newly available NOW accounts. This caused a temporary drop and a slower growth in M1A and a compensating rise and a faster growth in "other checkable deposits (OCDs)," which are mostly NOW accounts. To remove the effects of shifting in funds and to maintain a continuity in the basic money supply measure, the so-called "shift-adjusted M1A" was estimated by the Federal Reserve. As there were funds flowing into the OCD category from outside M1A as well as from M1A, a certain portion of growth in OCD was estimated to have come from of M1A, and was added back to the observed M1A to arrive at the shift-adjusted M1A. In this study, a similar procedure was used for every month since January 1981. The proportion of the incremental OCD that was attributed to M1A was 80% in January, 72.5% in February through April, and 50% thereafter.

C. Meaning of the Selected Series

In view of the discussion in Sec. III.C, BS is an indicator of the overall economic activities; CIL is a proxy for the liquidity demand; CPI is a measure of inflation; EX is the indicator of the potential influence of the foreign capital markets; and M1A is money supply. Ideally, the model for 5YN should have a measure of expected inflation, but there is no proven indicator of expected inflation.

D. Choosing the Fit and Check Sets

Given the set of time series to work with, much of the model-building effort was focused on the choice of the fit and check sets. Of the initial data set covering the 85-month period from September 1973 to September 1980, the data from the last 12 months were set aside for an ex post test of forecasting performance of the models. For each of the last 12 months [e.g., month (T + 1)] all the data up to month T were divided into a fit and a check set according to various methods and GMDH applied to produce forecasts for the interest rates in month (T + 1). Since it is impractical and too costly to test every possible way of constructing the two sets, a few approaches were selected that seemed to represent different philosophies about the underlying dynamic system of the interest rate determination.

The first approach was based on the idea that the system was stable over time. This implied that a randomized separation of observations would be perfectly adequate. However, this was not successful, as the system was apparently not very stable over time. The result is quite consistent with the fact that substantial changes have occurred in U.S. financial markets since 1970s. (The gradual deregulation of interest rates, the introduction of NOW accounts and money market mutual funds, and the adoption of the new monetary control procedure by Federal Reserve in 1979 are just a few of these changes.)

A different approach was needed to find models that were capable of explaining the latest behavior of the two rates. Since GMDH selects the models on the basis of their forecast performance over the check set, it was decided to use only recent data points as a check set. Thus the second method was to use the last X number of observations as the check set, while fitting the models on the first (T - X) periods. This was attempted with X = 12 and X = 1, but neither produced reasonable results. Without the data from the latest 12 months in the fit set, the models were apparently blinded from the newest developments. On the other hand, using only one period as a check set seemed to bias the models too much toward explaining that particular period.

Finally, a reasonable balance appeared to be struck when the collection of observations from every other month of the last 12 months was used as the check set and the rest were included in the fit set, in addition to the first (T - 12) observations. This method allows the models to incorporate the changes in the most recent months and adapt to the structural changes in the financial system. At the same time it retains a check set which is sufficiently up to date so that choosing the models on the basis of a good forecasting performance over the set is likely to ensure a good forecast for the next month. The detailed rules that were finally selected for building the models are described in the next section.

E. The Rules for Model Building

The rules that were followed in building GMDH models for FFR and 5YN are explained in detail here. For actual computation, this project used the GMDH program provided by Data Resources, Inc. (DRI), on its software system, EPS.

1. For predicting FFR or 5YN for period (T + 1), the first T observations are used.
2. The first T observations are split into a fit set with (T - 6) observations and a check set with six observations. The first (T - 12) observations are always in the fit set. Of the 12 most recent observations, every other one is included in the fit set and the remainder constitute the check set. There are two ways to divide the last 12 observations.

   Method A: Check set includes T, (T - 2), ..., (T - 10); fit set includes (T - 1), (T - 3), ..., (T - 11). In other words, the most recent observation from period T is in the check set.
   Method B: The reverse of method A; that is, the most recent observation is in the fit set.

3. The maximum GMDH level used is three. (This actually proved not to be restrictive, because none of the modeling process that involved the third GMDH level produced a new model at the fourth level.)
4. From each level of modeling, a certain number of equations with the lowest check mean-squared errors (CMSE) are brought to the next level. Any redundant equations are omitted (i.e., those equations that contain only a constant term and those that are simply linear combinations of the equations that are already included). The number of equations to be carried to the next level depends on the number of independent variables. The following rule was used:

Number of independent variables for level 1	Number of equations for level 2	Number of equations for level 3
5	4	3
6	5	3
8	6	4

5. The equation with the least CMSE among all the equations is chosen as the final model for period (T + 1).
6. The criterion of variable inclusion or exclusion for stepwise regressions at each stage is the t-test; and the tolerance value is 1.5 for inclusion and 1.0 for exclusion.

Given a particular set of observations, an identical GMDH model will be constructed no matter who builds the model, as long as these rules are followed.

F. Specification of the Models

The following table outlines the structure of the three different models for FFR and the same number of models for 5YN that were examined at length in this study. The first two models for each dependent variable correspond to the "best" (in terms of the smallness of variance of residuals over a specified period) model in Kanemasu (1980). The distinction between models A and B derives from the two different methods of defining the fit and check sets. (A and B correspond to methods A and B discussed in Sec. IV.E.) The last model for each interest rate is built from all the variables that were considered relevant. It turned out that the most general models (i.e., FFR-C and 5YN-C) did not seem superior to the more parsimonious counterparts. As a result, only one of the two methods of selecting the fit and check sets was applied to these models.

Model name	Dependent variable	Independent variables	Last observation in:
FFR-A	FFR	FFR1, M1A1, EX1, EX4, BS5	Check set
FFR-B	FFR	FFR1, M1A1, EX1, EX4, BS5	Fit set
FFR-C	FFR	FFR1, M1A1, EX1, EX4, BS5, CPI1, CIL6, CIL7	Check set
5YN-A	5YN	5YN1, M1A1, M1A5, EX4, EX6, BS5	Check set
5YN-B	5YN	5YN1, M1A1, M1A5, EX4, EX6, BS5	Fit set
5YN-C	5YN	5YN1, M1A1, M1A3, M1A5, EX4, EX6, BS5, CIL8	Check set

V. OBSERVATIONS ON THE MODELS AND THEIR FORECAST PERFORMANCE

The structure of the final FFR and 5YN models and their forecasts for the initial test period (i.e., October 1979 through September 1980) are summarized in Tables 1 to 4 and Fig. 1. In addition, for models FFR-A and FFR-B, which were eventually chosen as the best models, similar information is shown for 12 months of 1981 as well.

Table 1 GMDH Model Structure: Federal Funds Rate

Model	Independent variables	Last Observation in:
FFR-A	FFR1, M1A1, EX1, EX4, BS5	Check set
FFR-B	FFR1, M1A1, EX1, EX4, BS5	Fit set
FFR-C	FFR1, M1A1, EX1, EX4, BS5, CPI1, CIL6, CIL7	Check set

		GMDH level of final model[a]			Independent variables[b]							
Year	Month	A	B	C	FFR1	M1A1	EX1	EX4	BS5	CPI1	CIL6	CIL7
1979	October	1	2	1	ABC	B	AB					C
	November	2	1	2	ABC		AB		A	C		
	December	1	2	1	B	AC	BC		B			
1980	January	1	1	2	AC	B	ABC		C	C		
	February	1	2	1	B	AC	BC		B			
	March	2	1	2	AC	B	ABC		AC			
	April	2	2	2	B	ABC	ABC		B			
	May	2	1	2		ABC			AC			
	June	3	2	3	ABC	ABC	AC		AC	C		
	July	2	1	3	BC	ABC		A	A	C	C	
	August	2	2	1	AC	ABC	A	AB	AB			
	September	2	2	2	BC	ABC	AB	A	AB		C	
	October	2	2		A	AB			AB			
	November	2	3		AB	AB	B		AB			
	December	2	3		AB	AB	A		AB			
1981	January	2	2		AB	AB			AB			
	February	2	2		AB	AB			A			
	March	3	2		AB	AB	B		A			
	April	2	2		AB	AB			A			
	May	1	2		AB	A		B				
	June	3	2		AB	AB	B		A			
	July	1	3		AB	AB			B			
	August	3	3		AB	AB			A			
	September	1	3		AB	AB						
	October	1	1		AB	AB						
	November	1	2		AB	AB	B		B			
	December	1	2		AB	B	B					

[a]A, B, and C correspond to models FFR-A, FFR-B, and FFR-C.
[b]For each month, the independent variables used in the respective models
are indicated by A, B, and C.

Table 2 GMDH Forecasts: Federal Funds Rate (%)

Year	Month	Actual figures Monthly average	Actual figures Monthly change	Predicted change FFR-A	Predicted change FFR-B	Predicted change FFR-C
1979	September	11.43	0.47			
	October	13.77	2.34	0.96	0.21	0.32
	November	13.18	-0.59	-0.14	1.50	2.35
	December	13.78	0.60	0.04	-0.31	-0.01
1980	January	13.82	0.04	0.43	0.17	0.50
	February	14.13	0.31	0.21	0.12	0.24
	March	17.19	3.06	0.17	0.16	0.16
	April	17.61	0.42	0.22	-1.91	0.22
	May	10.98	-6.62	-1.02	-0.47	-1.15
	June	9.47	-1.51	-15.11	-7.47	-11.33
	July	9.03	-0.44	0.07	-0.50	0.89
	August	9.61	0.58	0.07	0.06	-0.11
	September	10.87	1.21	0.12	0.41	1.00
	October	12.81	1.94	1.41	2.32	
	November	15.85	3.04	0.29	0.90	
	December	18.90	3.05	0.75	0.78	
1981	January	19.08	0.18	-1.49	0.89	
	February	15.93	-3.15	0.17	-0.01	
	March	14.70	-1.23	0.41	-1.40	
	April	15.72	1.02	-0.05	-0.29	
	May	18.52	2.80	3.17	2.08	
	June	19.10	0.58	1.03	1.11	
	July	19.04	-0.06	-1.78	-1.33	
	August	17.82	-1.22	-0.35	-0.25	
	September	15.87	-1.95	-0.68	-1.90	
	October	15.08	-0.79	-2.16	-1.39	
	November	13.31	-1.77	-0.75	-0.65	
	December	12.37	-0.94	-0.61	-0.80	

Table 3 GMDH Model Structure: 5-Year Treasury Note Yield

Model	Independent variables	Last observation in:
5YN-A	5YN1, M1A1, M1A5, EX4, EX6, BS5	Check set
5YN-B	5YN1, M1A1, M1A5, EX4, EX6, BS5	Fit set
5YN-C	5YN1, M1A1, M1A3, M1A5, EX4, EX6, BS5, CIL8	Check set

		GMDH level in final model[a]			Independent variables[b]							
Year	Month	A	B	C	5YN1	M1A1	M1A3	M1A5	EX4	EX6	BS5	CIL8
1979	October	1	2	1	ABC	B		ABC				
	November	2	1	2	ABC						AC	
	December	2	1	2	ABC	A		A			B	C
1980	January	3	1	3	ABC	ABC			AC		AC	C
	February	1	2	1	B	ABC		AC			B	
	March	2	1	3	AC	ABC					AC	C
	April	1	2	1	ABC	AB					B	
	May	2	1	2		ABC	A				AC	C
	June	1	2	2		ABC	C			B	BC	
	July	2	1	1	AB	ABC	AC					C
	August	1	2	3	AB	ABC	C	C		C	C	C
	September	2	1	1	AB	ABC					A	C

[a] A, B, and C correspond to models 5YN-A, 5YN-B, and 5YN-C.
[b] For each month, the independent variables used in the respective models are indicated by A, B, and C.

Table 4 GMDH Forecasts: 5-Year Treasury Note Yield (%)

		Actual figures		Predicted change		
		Monthly	Monthly			
Year	Month	average	change	5YN-A	5YN-B	5YN-C
1979	September	9.41	0.35			
	October	10.63	1.22	-0.04	-0.25	-0.04
	November	10.93	0.30	0.19	0.41	0.19
	December	10.42	-0.51	0.11	0.12	0.15
1980	January	10.74	0.32	-0.12	-0.10	-0.10
	February	12.60	1.86	0.11	0.18	0.11
	March	13.47	0.87	0.55	0.25	1.43
	April	11.84	-1.63	0.24	-0.01	0.38
	May	9.95	-1.89	0.16	-0.65	-0.05
	June	9.21	-0.74	0.11	0.20	0.25
	July	9.53	0.32	-0.17	-0.22	-0.02
	August	10.84	1.31	0.10	0.10	-0.02
	September	11.63	0.78	0.49	0.73	0.45

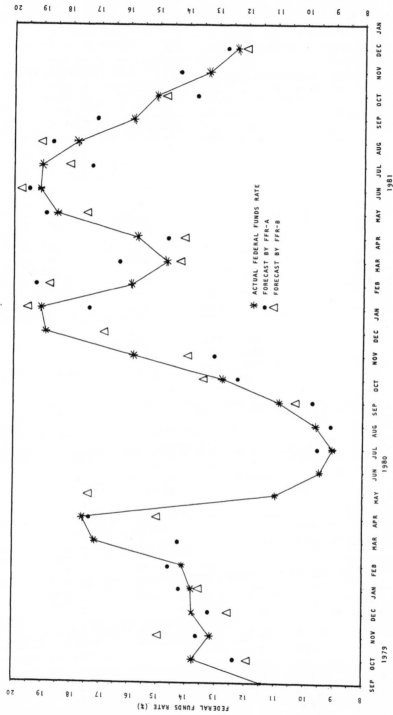

Fig. 1 GMDH forecasts: federal funds rate.

A. The General Model Structure

The time series actually used in the final models changed significantly each month. Indeed, there was not a single variable that was consistently present in at least one of the six models throughout the 12 months of the test period. (Even M1A1 was missing from all six models in one of the months.) The GMDH level of the final models, which corresponds roughly to the degree of nonlinearity in the models, varied from one to three over the test period. In no case, however, would the modeling have reached the fourth level, even without the third-level limit imposed in this study. (Every time the modeling reached the third GMDH level, all the equations from the third level were taken to the next level, but they never produced a nontrivial model.) Most of the models included a relatively limited number of variables. For FFR, the maximum number of variables used was five out of five in FFR-A, four out of five in FFR-B, and five out of eight in FFR-C. For 5YN, the maximum was four out of five in 5YN-A, three out of five in 5YN-B, and six out of eight in 5YN-C.

B. The Federal Funds Rate

The forecasts of FFR models are shown in Table 1. Unfortunately, it is obvious that the models perform rather poorly in predicting the magnitude of the changes (i.e., FFR). However, in the U.S. financial market, where interest rates show a high degree of volatility, one would often benefit greatly even if he or she knew only whether the federal funds rate would rise or fall. In other words, models that project the correct sign of FFR would still be quite useful, especially if they catch the major turning points in the rate. In this respect, these models appear to do much better. In the initial 12-month period of performance testing, FFR changed its sign four times, making it a more difficult period in which to forecast the correct signs of FFR. Of the 12 signs, FFR-A missed only one, FFR-B missed three, and FFR-C missed four.

It was noted in the model selection process that the most general model, FFR-C, did not seem to predict with greater accuracy than did the more parsimonious models, despite the larger amount of information it used. Therefore, it was decided to monitor only the first two models in the succeeding months. It was also noted that there appeared to be some tendency for these models to "overreact" to their forecast errors. This is particularly evident in the forecast for June 1980, which followed an unprecedented drop of 6.62% in one month. All the models seriously underestimated the size of the decline in May, and in a seeming reaction to it, they all anticipated an oversized drop for the following month. (Actually, two of the models forecast the overnight rate to fall below zero.)

C. The 5-Year Treasury Note Yield

The forecast performance of 5YN models are disappointing even by the correct-sign criterion. The best record was 5 misses out of 12 by 5YN-B, and 5YN-A and 5YN-C models registered 7 misses out of 12. Moreover, there was no recognizable pattern in errors (such as overreaction to the previous error) which could make the forecast still useful in some way. As a consequence, these models for 5YN were abandoned.

VI. FURTHER EXPERIENCE WITH THE FEDERAL FUNDS RATE MODELS

Two of the federal funds rate models (i.e., FFR-A and FFR-B) were continually monitored in the months following the initial study. Their projections and the information pertaining to the basic structure of the models are shown in Tables 1 and 2 and Fig. 1.

A. Structural Changes

The structure of the models seemed to change significantly over time. In the 15 months following the initial test period, FFR1 and M1A1 became almost fixed parts of the two models, while the frequency of the use of the EX series seemed to diminish substantially. The following table shows the percentage of times that each of the variables was used in the two models during the first 12 months and the next 15 months (i.e., altogether 24 models in the first period and 30 models in the second period).

	FFR1	M1A1	EX1	EX4	BS5
October 1979 to September 1980	58.3%	70.8%	66.7%	16.7%	50.5%
October 1980 to December 1981	96.7%	93.3%	20.0%	3.3%	50.0%

The increase in the influence of the federal funds rate and the money supply, and the corresponding decline in the influence of the exchange rates, agree with common perceptions in the financial market regarding the relative importance of these variables. It is widely felt that starting in about the spring of 1980, the money supply became the focus of monetary policy and market concern and that U.S. short-term rates began to influence exchange rates rather than the latter affecting the former. Business sales appear to have maintained a similar degree of usefulness as an explanatory series in the two periods, although the mere presence of the series in the model is not necessarily a good measure of the explanatory power.

B. Forecast Characteristics

It has been recognized that whenever FFR changed its sign, there was a tendency for the two models to differ in the sign of their forecasts for that month or the one immediately preceding. In other words, whenever the two models point in opposite directions, there has been a strong tendency for FFR to reverse its sign in that month or in the following month.* FFR-A and FFR-B produced forecasts of opposite signs in 7 of the 27 months, and in every one of those seven cases, FFR changed its sign either in that same month or one month later (it did both in one case). Perhaps the two models as a group expressed their uncertainty about the near future when a turning point is imminent. This characteristic enhances the usefulness of forecasts produced by these two models.

With the interpretation that the contradicting forecasts by the two models signal a turning point, the forecast record of these models has been quite good, as indicated in the following table. Even in the case where both models erroneously predicted a fall in the federal funds rate, the divergent forecasts for the preceding month implied an increase.

Forecasts	Frequency	Correct	Incorrect	Sign changed in: Same month	Next month
Both up	11	11	0		
Both down	9	8	1		
Disagree	7	7[a]	0[a]	3	5

[a]Correct in predicting a turn in the same month or one month later.

VII. CONCLUSIONS

In this study, GMDH models were built for the federal funds rate and the yield on the 5-year Treasury note. Two of the models for the overnight rate proved reasonably successful, especially for predicting the direction of the change. However, none of the models for the 5-year Treasury note yield produced reliable forecasts. The unsuccessful results of the 5YN models may be related to the difficulty of quantifying long-term expectations of the participants in the financial markets, which are crucial determinants of long-term rates.

In the course of this project, it became clear that the systems under examination could not be assumed stable over time. The two most successful models for the federal funds rate (or procedures for building them)

---

*This observation was initially made by William Dellal, who at the time was affiliated with the World Bank.

attempted to cope with this situation by carefully constructing the fit and check sets. Observations from every other month of the latest 12 months were used as the check set, while the remainder of the observations from those 12 months, in addition to all observations from the earlier periods, were used as the fit set. This method seemed to achieve a balance between the need for a fit set that incorporates the latest structural changes in the models and the need for a check set that distinguishes the most effective models for forecasting the next period.

The importance of accumulating experiences with the models was recognized in interpreting the forecasts. In the case of the two federal funds rate models, their tendency to project changes in opposite directions shortly before a turning point in the overnight rate significantly enhanced the information value of the forecasts. The gradual shift in the time series included in the final models also seems to identify some additional variables that may prove effective in the models. Extension of the models along this line, however, is left for the next stage of the project.

ACKNOWLEDGMENTS

The author is greatly indebted to Dr. Jacques Rolfo and Dr. Hiromitsu Kanemasu of the World Bank, Dr. William Dellal of Morgan Guaranty Trust Co., and Dr. Baldwin Hui of Data Resources, Inc., for their invaluable comments and suggestions. This project was funded by the Investment Department of the World Bank.

REFERENCE

Kanemasu, H. (1980). Multivariate Stochastic Modelling of U.S. Monetary and Economic Variables (Preliminary Report), July, the World Bank.

11

A Method for Predicting Sales Amount by the Use of IWSM and GMDH

JUNJI NOMURA/Matsushita Electric Works, Ltd., Osaka, Japan

I. INTRODUCTION

In inventory management for a production firm, one of the most important but difficult problems is how to determine production quantities and ordering points for the next period [1, 2]. To optimize these problems, it is necessary to improve the prediction accuracy of sales quantities, because a prediction error will cause a shortage of goods or excessive stock. It is also a difficult problem because decisions should include consideration of the uncertainties of the future trend of the market, sales effort, quantities in stock, and so on. Prediction of future demand on the basis of economic indicators, identification of an equation for prediction by using multiregression analysis, and other factors have been investigated. But those approaches are not as suitable for sales prediction, because an actual sales amount is not related directly to economic indicators, and sales trends change gradually with the passage of time.

Thus in this chapter we propose the combined use of interpretive weighted structural modeling (IWSM) [3] and the GMDH algorithm. Through this process, important prediction factors are selected from among factors examined by persons in charge of sales prediction, and a future sales amount is predicted on the basis of these factors. A real-world problem of sales prediction of architectural goods has been dealt with successfully by the present procedure.

II. DIFFERENCE BETWEEN DEMAND PREDICTION AND SALES PREDICTION

The subject of demand prediction relates to the demand for goods in a specific social area in which all firms are considered. Sales prediction,

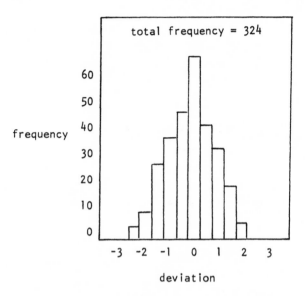

Fig. 1 Distribution of errors in sales predictions.

on the other hand, is concerned with the demand for goods in a market
occupied by a specific firm. The predicted sales amount S is given by

$$S = r \times D \tag{1}$$

where r is the occupation ratio of the firm and D is the total predicted demand amount for goods. If occupation ratio r is constant, the value of S is
obtained from Eq. (1) by predicting demand amount D. But, in practice, r
often undergoes a change because of uncertainties of sales, particularly
with regard to such human factors as sales policy and sales effort. So the
approach for sales prediction has to be discussed differently from that of
demand prediction.

There are numerous approaches to the prediction of the future sales
amount: for example, multiple regression, ARMA, ARIMA, and others.
In our firm, the Winter exponential smoother has been used. Figure 1
shows the distributions of errors in sales prediction observed in past data.
The results of predictions for each item using these methods are not bad,
but for principal items or for a group of items, the accuracy is not as
satisfactory. As a consequence, we held some meetings to discuss procedures and methods for prediction, and decided to adopt a combined IWSM
and GMDH method.

III. SALES PREDICTION USING IWSM AND GMDH

The sales prediction procedure using IWSM and GMDH is shown in Fig. 2. Sales of principal items or a group of items is not influenced directly by macroeconomic indicators, but primarily by factors peculiar to goods. So the most important step is to select prediction factors from among factors examined by persons in charge of sales prediction every month, as shown at step 2 in Fig. 2. More detailed steps of this procedure are as follows.

Step 2.1: To make clear contextual relations among factors listed in step 1, we question each person in charge regarding three types of relations: relative importance, influencing relation, and quantifiability among factors. For example, we ask each person: "Is factor a_i more important than factor a_j?" The person is requested to reply in one of the following four degrees: "much more important," "more important," "equally important," and "less important." After repeating such pair-wise questionnaires among factors, by using IWSM we construct a hierarchical digraph representing relative importance among factors. Step 2.2: From the digraph of relative importance, we list higher-ranked factors. Step 2.3: From the digraph of quantifiability, we pick out lower-ranked factors among factors listed in step 2.2.

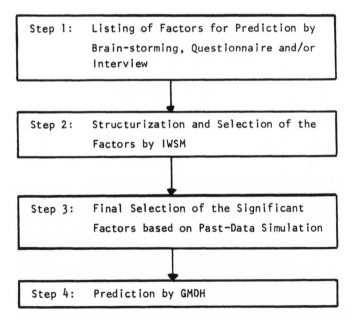

Fig. 2 Procedure of predicting sales amount by IWSM and GMDH.

Step 2.4: From the digraph of the influencing relation, we find appropriate substitutes for factors picked out in step 2.3.
Step 2.5: Following the steps outlined above, we select a set of factors
for sales prediction.

IV. APPLICATION TO A REAL PROBLEM

The real-world problem considered here is concerned with monthly sales
prediction for architectural goods.

A. Analysis and Selection of Prediction Factors

At the outset of this procedure, 44 factors for prediction are developed by
brainstorming among seven persons in charge of sales prediction. Eliminating redundant factors, 22 factors were selected, as shown in Table 1.
Next, we analyzed, by use of IWSM, the relative importance, the influencing

Table 1 Prediction Factors

Number	Factor
1	Sales quantity in the past
2	Policy of management
3	Annual sales plan of the operational division
4	Sales plan in the next period
5	Achievement ratio of the operational division's sales plan
6	Achievement ratio of the total sales plan
7	Achievement ratio of the group sales plan
8	Group sales policy
9	Sales quantity for the operational division
10	Assumed demand
11	Price of merchandise
12	Price policy
13	Price of other firm's merchandise
14	Price of the material
15	Price of the raw material
16	Plan for new merchandise
17	Sales plan for new merchandise
18	Sales effort on new merchandise
19	Sales effort on principal merchandise
20	Sales campaign
21	Shortage of goods
22	Ratio of returned goods

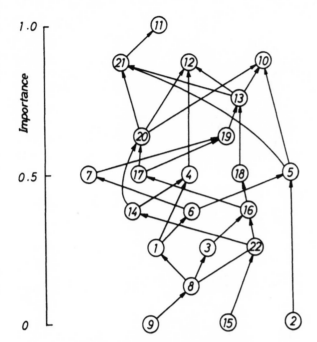

Fig. 3 Relative importance among factors.

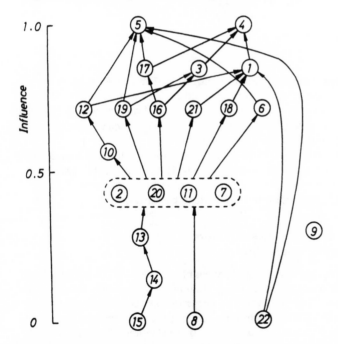

Fig. 4 Influencing relations among factors.

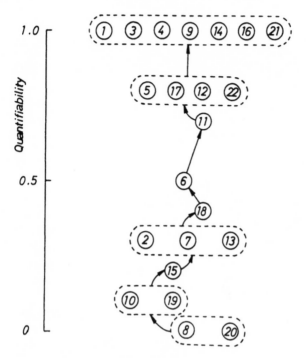

Fig. 5 Quantifiability of factors.

relation, and the quantifiability among factors as shown in Figs. 3 to 5, respectively. In Fig. 3 the upper configurated factors are more important than the lower ones. In Fig. 4 the factor at the end of an arrow is influenced by the factor at the beginning of the arrow. In Fig. 5 the upper configurated factors are more quantifiable than the lower ones.

Applying the procedure outlined in steps 2.2 to 2.5, using Figs. 3 to 5, six factors were selected for sales prediction: 4, 5, 11, 12, 14, and 21. Of these, factors 11 and 14 were selected as substitutes for factors 10 and 13, respectively—important but difficult to be measured or quantified.

B. Final Selection of the Significant Factors
    Based on Past-Data Simulation

In order to select the appropriate input variables to the GMDH method from the factors, we have practiced past-data simulation of 2-month-future sales prediction using the GMDH method. The program we use was developed on the basis of the revised GMDH algorithm using AIC as a criterion for model selection [4]. The past sales data used for simulation are shown in Fig. 6. The procedure for simulation is briefly as follows.

First, after identifying the model based on factor data for 18 months by use of GMDH, sales for the next month were predicted. Next, regarding these predicted sales as the actual ones, sales 2 months in advance were predicted in the same way. As a criterion for addition or elimination of a factor, the following prediction error was defined:

$$E_t = \frac{|P_t - A_t|}{A_t} \qquad (2)$$

where P_t is the prediction amount at time t and A_t is the actual amount at t. To minimize the mean prediction error for the preceding year, a set of factors were selected for sales prediction: 3, 4, 5, 11, and 21.

For reference, we compared the result of this procedure with that of another procedure, as follows:
Factors were added one after another as input variables to the GMDH method according to the correlation coefficients between factors and the actual sales amount.

Figure 7 shows a comparison of two prediction procedures on the basis of a mean prediction error for one year. Based on the results shown, it is

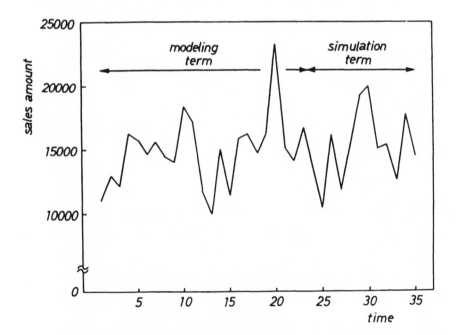

Fig. 6 Actual sales figures used for simulation.

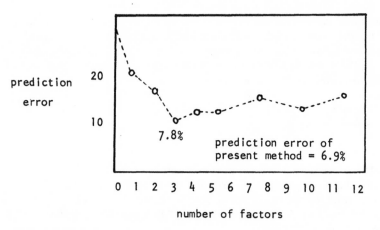

Fig. 7 Transition of prediction error by number of factors.

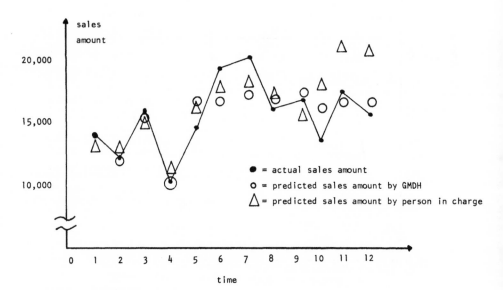

Fig. 8 Comparison of prediction by person and by the GMDH algorithm.

clear that it is reasonable to select prediction factors using this combined procedure.

C. Result of Simulation

By way of example, Fig. 8 shows a comparison of prediction by the GMDH algorithm with that by a person who had been predicting sales for about 10 years. This is a typical example which illustrates the effectiveness of the present procedure.

V. CONCLUDING REMARKS

We have quantified sales predictions for inventory management using IWSM and GMDH. As a result of real-world applications, we have found that the procedure proposed works well. The principal findings are as follows:

1. The IWSM procedure is very useful for structuring and selecting the prediction factors.
2. The GMDH method makes the operation of a sales plan systematic and efficient.

The procedure has been used in a real sales prediction problem with decidedly satisfactory results.

ACKNOWLEDGMENTS

The authors are indebted to Dr. Y. Nishikawa, Professor, Kyoto University; Dr. H. Tamura, Assistant Professor, Osaka University; and S. Sakai, Director of Systems Laboratory, Matsushita Electric Works, Ltd., for their excellent advice and cooperation during this work.

REFERENCES

1. Y. Nishikawa and J. Nomura, Multiobjective Optimization of Inventory Management by Use of Utility Theory, <u>Proc. First Int. Symp. Policy Analysis Inf. Syst.</u>, 1978, pp. 460–465.
2. Y. Nishikawa, J. Nomura, K. Sawada, and H. Takada, Multiobjective Optimization of Ordering Point Decision in Inventory Management by Use of Utility Theory, <u>Proc. 8th IFAC Symp.</u>, Sess. 50.3, Kyoto, Japan, 1981.
3. Y. Nishikawa and J. Nomura, Interpretive Weighted Structural Modeling (IWSM) and Its Applications to a Scheduling and a Budget Allocation, TIMS 24th Int. Meet., 1979.

4. T. Kondo and H. Tamura, Revised GMDH Algorithm of Self-Selecting
    Optimal Intermediate Polynomials Using AIC, <u>Trans. Soc. Instrum.
    Control Eng.</u> 15: 466–471 (1979).

12

On Revised Algorithms of GMDH With Applications

HIROYUKI TAMURA and TADASHI KONDO*/Osaka University,
Osaka, Japan

I. INTRODUCTION

GMDH (group method of data handling), originated by Ivakhnenko (1968), is
a useful data analysis technique for the identification of nonlinear complex
systems, especially when few data are available. The primary advantage
of GMDH is that it self-selects the structure (degree of nonlinearity) of the
model without using a priori information on the relationship among input/
output variables. In the original GMDH (Ivakhnenko, 1968, 1971; Ivakhnenko
et al., 1969), the concept of regularization is used to avoid overfitting the
model to past data. In this concept, available input/output data are divided
into two sets: training data for estimating unknown parameters in the partial
polynomials, and checking data for selecting the intermediate variables in
each layer. Much research was done within Ivakhnenko's group (Ivakhnenko
et al., 1979) on the best method of dividing the data into two data sets. In
the original GMDH we need the following heuristics:

H1: Predetermination of the structure of the partial polynomials
H2: Division of the original data into two sets; training data and
    checking data
H3: Predetermination of the number of intermediate variables selected
    in each layer

The first half and second half of this chapter follow closely the development
in Tamura and Kondo (1980) and Kondo and Tamura (1979), respectively.
* Present affiliation: Toshiba Corporation, Tokyo, Japan.

These heuristics are to be changed so as to find their optimal combination. Therefore, the computational procedure of the original GMDH must be repeated many times, but the final model obtained is rarely for the optimal combination of the heuristics. Furthermore, the results identified depend heavily on the heuristics adopted.

In this chapter we present two revised GMDH algorithms:

RG1: Heuristics-free GMDH which generates optimal partial polynomials.
RG2: Heuristics-free GMDH which generates optimal intermediate polynomials.

The intermediate polynomials in RG2 represent the direct relationship among the original input/output variables, while the partial polynomials in RG1 represent the relationships among the intermediate variables and output variable, just as in the original GMDH of Ivakhnenko. In the original GMDH the polynomials of fixed form are used as the partial polynomials in all the layers, whereas in RG1 the optimal partial polynomials are generated automatically for each pair of variables in each layer, and the polynomials as such are used to construct a final model in the multilayered structure. Therefore, the results do not depend on the heuristics of determining the structure of the partial polynomials, and much better flexibility in constructing a final model can be obtained compared with the original GMDH.

In both RG1 and RG2 the artificial differentiation between training data and checking data is eliminated. Instead of dividing the available data into two sets, all the data are used as the training data and at the same time as the checking data, where instead of using the mean-squared error for the checking data, the prediction sum of squares (PSS) of Allen (1974) or Akaike's information criterion (AIC) (Akaike, 1972-1974) calculated from all the data is used as a criterion for generating optimal partial polynomials or intermediate polynomials, for selecting intermediate variables, and for stopping the multilayered calculation. Therefore, the final model identified fits all the data uniformly well and the results do not depend on the heuristics of dividing the data into two sets.

In both RG1 and RG2 the number of intermediate variables is preferred to be as large as possible to minimize PSS or AIC. That is, the number of intermediate variables is determined not by the heuristics but by the upper limit of the memory capacity of the computer. As described above, since no heuristics of H1, H2 and H3 are needed in RG1 and RG2, we do not need to repeat the computational procedure for finding a better combination of heuristics.

It is difficult to find a physically meaningful model using the original GMDH or RG1 in this chapter. However, by using RG2 a physically meaningful model could be identified if the characteristics of the system are well embedded in the measured data, and these data are noise-free.

II. PARTIAL POLYNOMIALS USED IN PREVIOUS GMDH ALGORITHMS

Among many kinds of mathematical models, such as polynomials, trigono-
metrical functions, and Bayes' formulas, the Kolmogorov-Gabor polynomial

$$\phi = a_0 + \sum_i a_i x_i + \sum_i \sum_j a_{ij} x_i x_j + \sum_i \sum_j \sum_k a_{ijk} x_i x_j x_k + \cdots \qquad (1)$$

is widely used as a complete description of the system model. In the original GMDH, equation (1) is constructed by combining the following second-
order partial polynomials of two variables in multilayers:

$$y_k = b_0 + b_1 x_i + b_2 x_j + b_3 x_i x_j + b_4 x_i^2 + b_5 x_j^2 \qquad (2)$$

In the block diagram of the original GMDH shown in Fig. 1, the multilayered
structure of GMDH is illustrated. The detail of this original GMDH is described in Chap. 1.

In what follows we show some polynomials that have been used before
as the partial polynomials.

A. First-Order Polynomial

$$y_k = b_0 + b_1 x_i + b_2 x_j \qquad (3)$$

By using this partial polynomial, we can only construct a first-order complete polynomial.

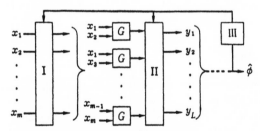

```
I  : Division of the original data
II : Self-selection of the intermediate variables
III: Optimization of the threshold
G  : Generator of a partial polynomial
```

Fig. 1 Block diagram of the original GMDH.

B. Second-Order Polynomial

$$y_k = b_0 + b_1 x_i + b_2 x_j + b_3 x_i x_j \tag{4}$$

$$y_k = b_0 + b_1 x_i + b_2 x_j + b_3 x_i x_j + b_4 x_i^2 + b_5 x_j^2 \tag{2}$$

By using these second-order partial polynomials, we construct a 2^Tth-order polynomial after passing the Tth selection layer. As a second-order polynomial, which contains a smaller number of unknown parameters than Eqs. (4) and (2), the partial polynomial (Ihara, 1975)

$$X_k = w x_i + (1 - w) x_j, \quad w = \text{weight} \tag{5a}$$

$$y_k = b_0 + b_1 X_k + b_2 X_k^2 \tag{5b}$$

has been proposed.

C. High-Order Polynomial (Ikeda et al., 1976)

$$X_i = b_0 + b_1 x_i + b_2 x_i^2 \tag{6a}$$

$$X_j = b_0' + b_1' x_j + b_2' x_j^2 \tag{6b}$$

$$y_k = c_0 + c_1 X_i + c_2 X_j + c_3 X_i X_j + c_4 X_i^2 + c_5 X_j^2 \tag{6c}$$

By using this partial polynomial, we construct a 4^Tth-order polynomial after passing the Tth selection layer.

D. Optimal Partial Polynomials

1. Optimal partial polynomials in which parametrically unstable terms contained in Eq. (2) are eliminated (Endo, 1978).
2. Optimal partial polynomials in which unnecessary terms contained in Eq. (2) are eliminated by applying a stepwise regression method (Draper and Smith, 1966) using a significant test for the training data (Duffy and Franklin, 1975).

By using optimal partial polynomials as such, we can construct a final model of various order between the first and 2^Tth order after passing the Tth selection layer.

The relationship between the number of selection layer and the order of the final model obtained is shown in Fig. 2. The order of the final model

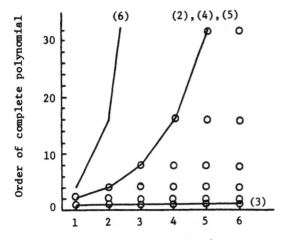

Fig. 2 Order of complete polynomials obtained for different partial polynomials. (From Tamura and Kondo, 1980.)

constructed by using optimal partial polynomials in (1) and (2) is shown by
a circle. By using partial polynomials of Eqs. (2) to (6), the system, which
is described by a polynomial having many input variables with low order,
cannot be identified, because the order of the polynomials is doubled in
each selection layer. By using optimal partial polynomials in (1) and (2), a
system as such could be identified. That is, much broader kinds of systems
can be identified by using optimal partial polynomials compared with using
predetermined polynomials for all possible combinations of two variables.
Furthermore, the number of terms contained in a final model can be decreased by using optimal partial polynomials. But in (1), a lot of computation
time is needed to construct an optimal partial polynomial, and in (2), it is
difficult to find the optimal standard value of variable selection, and furthermore the valuable information contained in the checking data cannot be used
to construct the partial polynomials.

In order to cope with the disadvantages contained in the previous partial
polynomials as described above, we present in the following section a revised GMDH which generates in each selection layer an optimal partial polynomial which minimizes the prediction error evaluated from all the data.

III. HEURISTICS-FREE GMDH FOR GENERATING
OPTIMAL PARTIAL POLYNOMIALS (RG1)

In a GMDH algorithm, PSS or AIC calculated from all the data can be used
as a criterion for generating optimal partial polynomials in each selection
layer, for selecting intermediate variables, and for stopping the multilayered

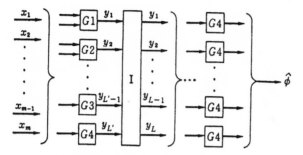

I : Self-selection of intermediate
    variables

(a) Block diagram

$$\left\{\begin{array}{l} y_k = b_0 + b_1 x_i + b_2 x_j + b_3 x_i x_j + b_4 x_i^2 + b_5 x_j^2 \\[2mm] y_k = b_0 + b_1 x_i + b_2 x_j + b_3 x_i x_j \\[2mm] \vdots \end{array}\right.$$

$$y_k = b_0 + b_1 x_i + b_2 x_j$$

$$y_k = b_0 + b_1 x_i + b_2 x_i^2$$

$$y_k = b_0 + b_1 x_i$$

(b) Generators of the optimal partial polynomials

Fig. 3 Block diagram of the heuristics-free GMDH for generating optimal partial polynomials. (From Tamura and Kondo, 1980.)

iterative computation. The significant advantage of using PSS or AIC for model selection is that it is not necessary to divide the available data into training data and checking data. All the data can be used for constructing the model and at the same time for evaluating the prediction error, since PSS and AIC have the ability to evaluate the prediction error incurred by the model. Therefore, the results do not depend on the heuristics for dividing the data into training data and checking data, as is true in the original GMDH.

The block diagram of the heuristics-free GMDH algorithm which generates optimal partial polynomials is shown in Fig. 3, where it is assumed that the complete description of the system model (a final model) can be written as the Kolmogorov-Gabor polynomial, Eq. (1). The algorithm is constructed by the following four procedures.

A. Generating Optimal Partial Polynomials in Each Selection Layer

The optimal partial polynomials are generated through the polynomial generator G1, G2, G3, or G4 as shown in Fig. 3, applying a stepwise regression procedure to the second-order polynomial of Eq. (2). In this stepwise regression procedure PSS or AIC is used as a criterion for selecting dominant variables in Eq. (2).

PSS is defined as (Allen, 1974)

$$PSS = \sum_{\alpha=1}^{n} (\phi_\alpha - \hat{y}_{k\alpha}^*)^2 \tag{7}$$

where

$$\hat{y}_{k\alpha}^* = b_{0\alpha} + b_{1\alpha}x_{i\alpha} + b_{2\alpha}x_{j\alpha} + b_{3\alpha}x_{i\alpha}x_{j\alpha} + b_{4\alpha}x_{i\alpha}^2 + b_{5\alpha}x_{j\alpha}^2$$

$$\alpha = 1, 2, \ldots, n$$

Here n denotes the data length, ϕ_α is the αth observed value for the output variable, $x_{i\alpha}$ is the αth observed value for the input variable x_i, and $\hat{y}_{k\alpha}^*$ is the αth estimated value obtained by a multiple regression analysis of all the data except the αth datum. To compute PSS of Eq. (7), the multiple regression analysis must be repeated n times, and the amount of computation increases exponentially with the number of data. For this reason, when there are many data, it is not practical to compute PSS in the form of Eq. (7). PSS of Eq. (7) can be reduced to (Okuno et al., 1976)

$$PSS = \sum_{\alpha=1}^{n} \left[\frac{\phi_\alpha - \hat{y}_{k\alpha}}{1 - \underline{x}_\alpha^T (X^T X)^{-1} \underline{x}_\alpha} \right]^2 \tag{8}$$

where

$$\hat{y}_{k\alpha} = b_0 + b_1 x_{i\alpha} + b_2 x_{j\alpha} + b_3 x_{i\alpha}x_{j\alpha} + b_4 x_{i\alpha}^2 + b_5 x_{j\alpha}^2$$

$$\underline{x}_\alpha^T = [1, x_{i\alpha}, x_{j\alpha}, x_{i\alpha}x_{j\alpha}, x_{i\alpha}^2, x_{j\alpha}^2]$$

$$\alpha = 1, 2, \ldots, n$$

$$X^T = [\underline{x}_1, \underline{x}_2, \ldots, \underline{x}_n]$$

$\hat{y}_{k\alpha}$ is the αth estimated value obtained by a regression analysis of all the data. With this procedure we need not repeat the regression analysis n times.
AIC is defined as (Akaike, 1974)

$$\text{AIC} = -2 \ln (\text{maximum likelihood}) + 2k$$

where k denotes the number of parameters in the model to be adjusted to attain the maximum likelihood. The information-theoretic justification of the use of AIC in stepwise regression analysis can be found in Akaike (1972, 1973). For linear regression analysis, AIC is reduced to (Akaike, 1972)

$$\text{AIC} = n \ln S_k^2 + 2k + C \tag{9a}$$

$$S_k^2 = \frac{1}{n} \sum_{\alpha=1}^{n} (\phi_\alpha - \hat{y}_{k\alpha})^2 \tag{9b}$$

where C is a constant. Here it is assumed that the noises contained in the model are mutually independent and normally distributed.

The normal equation for the partial polynomial of Eq. (2) can be written as

$$X^T \underline{X} \underline{B} = X^T \underline{Y} \tag{10}$$

where $\underline{B} = [b_0, b_1, \ldots, b_5]^T$ and $\underline{Y} = [\phi_1, \phi_2, \ldots, \phi_n]^T$. For solving normal equation (10), the 7×13 matrix

$$
\begin{bmatrix}
X^T X & \vdots & X^T \underline{Y} & \vdots & I \\
\cdots & & \cdots & & \cdots \\
\underline{Y}^T X & \vdots & \underline{Y}^T \underline{Y} & \vdots & \underline{0}^T
\end{bmatrix}
=
\begin{bmatrix}
\Sigma 1 & \Sigma x_{i\alpha} & \cdots & \Sigma x_{j\alpha}^2 & \vdots & \Sigma \phi_\alpha & \vdots & \\
\Sigma x_{i\alpha} & \Sigma x_{i\alpha}^2 & \cdots & \Sigma x_{i\alpha} x_{j\alpha}^2 & \vdots & \Sigma \phi_\alpha x_{i\alpha} & \vdots & I \\
\vdots & \vdots & & \vdots & & \vdots & & \\
\Sigma x_{j\alpha}^2 & \Sigma x_{j\alpha}^2 x_{i\alpha} & \cdots & \Sigma x_{j\alpha}^4 & \vdots & \Sigma \phi_\alpha x_{j\alpha}^2 & \vdots & \\
\cdots & & & & & & & \\
\Sigma \phi_\alpha & \Sigma \phi_\alpha x_{i\alpha} & \cdots & \Sigma \phi_\alpha x_{j\alpha}^2 & \vdots & \Sigma \phi_\alpha^2 & \vdots & \underline{0}^T
\end{bmatrix}
$$

is constructed, where I denotes an identity matrix, $\underline{0}^T$ denotes a zero vector, and the seventh row is supplemented for computing the RSS (residual sum of squares), which expresses the accuracy of fitting to all the data. Using this matrix we can select the dominant input variables contained in Eq. (2) as follows. When the mth variable in Eq. (2) is to be entered in the partial polynomial, the (m + 1)th column is reduced to the unit vector of the (m + 8)th column using pivoting operations. On the other hand, when the mth variable in Eq. (2) is to be deleted from the partial polynomial, the (m + 8)th column is reduced to the unit vector of the (m + 1)th column using pivoting operations. These variable selection procedure are repeated alternately based on the PSS of Eq. (8) or the AIC of Eq. (9), where the dominant input variables are selected so as to minimize PSS or AIC. Optimal partial polynomials can be constructed as either G1, G2, G3, or G4. The generator G4 generates the same intermediate variable as that generated in the preceding layer.

B. Selecting the Intermediate Variables

The L intermediate variables, which gives L smallest PSS or AIC, are selected from all the intermediate variables. The number L is preferred to be as large as possible within the limit of the memory capacity of the computer. That is, L is not determined in a heuristic manner. When we proceed to the second layer, the intermediate variables y_1, y_2, $\cdots$, y_L are regarded as the input variables of the second layer.

C. Stopping the Multilayered Iterative Computation

When all the generators of the partial polynomials in the selection layer become G4, the iterative computation of the revised GMDH is terminated, because PSS or AIC cannot be further improved.

D. Computation of the Predicted Values

The final model describing a relationship among input/output variables is obtained as a weighted average of the intermediate variables remaining in the final layer. Since we can compare the predicted values obtained from several intermediate variables in the final layer, it is possible to exclude any abnormal one(s) before we obtain the final model as a weighted average. Therefore, a stable prediction model can be obtained.

Since the heuristics-free GMDH algorithm described in this section does not need heuristics, we do not need to repeat the computational procedure for different heuristics.

IV. HEURISTICS-FREE GMDH FOR GENERATING OPTIMAL INTERMEDIATE POLYNOMIALS (RG2)

In this section a revised GMDH is described in which the optimal intermediate polynomials are generated in each selection layer instead of partial polynomials. The intermediate polynomials represent the direct relationship

First layer   Second layer

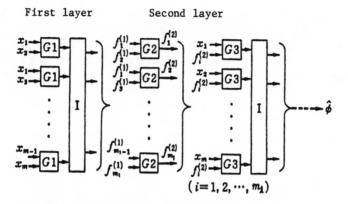

$(i=1, 2, \cdots, m_1)$

I : Self-selection of the optimal intermediate polynomials

G1, G2, G3 : Generators of optimal intermediate polynomials

Fig. 4 Block diagram of the heuristics-free GMDH for generating optimal intermediate polynomials. (From Kondo and Tamura, 1979.)

among the original input/output variables, and the polynomials as such are generated so as to minimize AIC evaluated by using all the data.

The block diagram of the heuristics-free GMDH algorithm which generates optimal intermediate polynomials is shown in Fig. 4, where m_1 denotes the number of intermediate polynomials selected in each selection layer. The algorithm is constructed by the following procedures.

A. Procedure in the First Layer

In the first layer the optimal intermediate polynomials are generated applying a stepwise regression procedure to the following pth-order polynomial:

$$\phi = a_0 + \sum_{i=1}^{2} a_i x_i + \sum_{i_1=1}^{2} \sum_{i_2=1}^{2} a_{i_1 i_2} x_{i_1} x_{i_2} + \cdots$$

$$+ \sum_{i_1=1}^{2} \sum_{i_2=1}^{2} \cdots \sum_{i_p=1}^{2} a_{i_1 i_2 \cdots i_p} x_{i_1} x_{i_2} \cdots x_{i_p} \tag{11}$$

Here Eq. (11) is constructed with L terms. In this stepwise regression procedure, AIC is used as a criterion for selecting dominant variables in Eq. (11). The normal equation for this polynomial, Eq. (11), can be written as

$$(X^T X)\underline{A} = X^T \underline{\Phi} \tag{12}$$

where $\underline{A} = [a_0, a_1, \ldots, a_{L-1}]^T$ and $\underline{\Phi} = [\phi_1, \phi_2, \ldots, \phi_n]^T$. For the normal Eq. (12), the $(L+1) \times (2L+1)$ matrix

$$\begin{bmatrix} X^T X & X^T \underline{\Phi} & I \\ \underline{\Phi}^T X & \underline{\Phi}^T \underline{\Phi} & \underline{0}^T \end{bmatrix}$$

is constructed, where the $(L+1)$th row is supplemented for computing RSS. By using this matrix we can select the combination of the dominant input variables that minimize AIC, and we can construct optimal intermediate polynomials from this combination.

B. Procedure in the Second Layer

In the second layer two kinds of combinations are considered.

1. Combination of Two Intermediate Polynomials
   Selected in the First Layer

Let the ith intermediate polynomial selected in the first layer be

$$\hat{\phi} = f_i^{(1)}(\underline{x}), \quad i = 1, 2, \ldots, m_1 \tag{13}$$

where $\underline{x}$ denotes the input variables, and it is assumed that Eq. (13) contains $K_i^{(1)} (\le L_1 - 1)$ variables. We combine two intermediate polynomials, $f_i^{(1)}$ and $f_j^{(1)}$. Let the equation constructed by all the variables contained in $f_i^{(1)}$ and $f_j^{(1)}$ be

$$\hat{\phi} = f_i^{(1)}(\underline{x}) + f_j^{(1)}(\underline{x}) \tag{14}$$

The normal equation for Eq. (14) can be written as

$$[X_i^{(1)} \mid X_j^{(1)}]^T [X_i^{(1)} \mid X_j^{(1)}]\underline{A} = [X_i^{(1)} \quad X_j^{(1)}]^T \underline{\Phi} \tag{15}$$

where $\underline{A} = [a_0, a_1, \ldots, a_{K_i^{(1)} + K_j^{(1)} + 1}]^T$. For the normal equation (15), the

$(K_i^{(1)} + K_j^{(1)} + 3) \times (2K_i^{(1)} + K_j^{(1)} + 5)$ matrix

$$
\begin{bmatrix}
X_i^{(1)T}X_i^{(1)} & X_i^{(1)T}X_j^{(1)} & \vdots & X_i^{(1)T}\underline{\Phi} & \vdots & I & 0 \\
X_j^{(1)T}X_i^{(1)} & X_j^{(1)T}X_j^{(1)} & \vdots & X_j^{(1)T}\underline{\Phi} & \vdots & 0 & I \\
\hline
\underline{\Phi}^T X_i^{(1)} & \underline{\Phi}^T X_j^{(1)} & \vdots & \underline{\Phi}^T\underline{\Phi} & \vdots & \underline{0}^T & \underline{0}^T
\end{bmatrix}
\tag{16}
$$

is constructed. When $K_i^{(1)}$ variables contained in $f_i^{(1)}(\underline{x})$ are entered into the intermediate polynomial in the second layer, matrix (16) is reduced to

$$
\begin{bmatrix}
I & M_{12} & \vdots & M_{13} & \vdots & M_{14} & 0 \\
0 & M_{22} & \vdots & M_{23} & \vdots & M_{24} & I \\
\hline
\underline{0}^T & M_{32} & \vdots & M_{33} & \vdots & M_{34} & \underline{0}^T
\end{bmatrix}
\tag{17}
$$

where M_{13}, M_{14}, M_{33}, and M_{34} have been already obtained in the first layer as

$$
M_{13} = (X_i^{(1)T}X_i^{(1)})^{-1}X_i^{(1)T}\underline{\Phi}
$$

$$
M_{14} = (X_i^{(1)T}X_i^{(1)})^{-1}
$$

$$
M_{33} = \underline{\Phi}^T\underline{\Phi} - \underline{\Phi}^T X_i^{(1)}(X_i^{(1)T}X_i^{(1)})^{-1}X_i^{(1)T}\underline{\Phi}
$$

$$
M_{34} = -\underline{\Phi}^T X_i^{(1)}(X_i^{(1)T}X_i^{(1)})^{-1}
$$

The remaining parts of the matrix (17) can be obtained as

$$
M_{12} = M_{14}(X_i^{(1)T}X_j^{(1)})
$$

$$
M_{22} = X_j^{(1)T}X_j^{(1)} - (X_j^{(1)T}X_i^{(1)})M_{12}
$$

$$
M_{32} = \underline{\Phi}^T X_j^{(1)} - (\underline{\Phi}^T X_i^{(1)})M_{12}
$$

$$
M_{23} = X_j^{(1)T}\underline{\Phi} - (X_j^{(1)T}X_i^{(1)})M_{13}
$$

$$
M_{24} = -M_{12}^T
$$

Then, by applying a stepwise regression procedure to matrix (17), we can select a combination of the dominant input variables which minimize AIC, and we can construct an optimal intermediate polynomial from this combination.

In this procedure, when the number of selected variables exceeds $(L_1 - 1)$, we try to decrease AIC under the following procedure. First, from $(L_1 - 1)$ variables which have already been contained in the intermediate polynomial, we find the variable that gives the smallest increase in AIC, and delete it from the intermediate polynomial. Then, from the variables that have not been entered in the intermediate polynomial, we find the variable that gives the largest decrease in AIC, and enter it into the intermediate polynomial. We repeat this procedure alternately so as to minimize AIC. When the variable, which is deleted from the intermediate polynomial, is entered into the intermediate polynomial immediately, we terminate the iterative procedure, and construct the optimal intermediate polynomial using the selected variables. The procedure in this part is called generator G2 of optimal intermediate polynomials. From $_{m_1}C_2$ intermediate polynomials generated in this procedure, the m_1 intermediate polynomials, which give the m_1 smallest AIC, are selected.

2. Combination of the Intermediate Polynomial and the Input Variables

Let the ith intermediate polynomial selected in the procedure above be

$$\hat{\phi} = f_i^{(2)}(\underline{x}), \quad i = 1, 2, \ldots, m \tag{18}$$

where $\underline{x}$ denote the input variables and it is assumed that Eq. (18) contains $K_i^{(2)}$ ($\leq L_1 - 1$) variables. We combine the intermediate polynomials $f_i^{(2)}$ ($i = 1, 2, \ldots, m_1$) with the input variables x_j ($j = 1, 2, \ldots, m$). Let the equation constructed by all the variables contained in $f_i^{(2)}$ and $x_j f_i^{(2)}$ be

$$\hat{\phi} = f_i^{(2)}(\underline{x}) + x_j f_i^{(2)}(\underline{x}), \quad j = 1, 2, \ldots, m \tag{19}$$

where $x_j f_i^{(2)}$ contains $(K_i^{(2)} + 1)$ variables. The normal equation for Eq. (19) can be written as

$$[X_i^{(2)} \mid X_{ji}^{(2)}]^T [X_i^{(2)} \mid X_{ji}^{(2)}]\underline{A} = [X_i^{(2)} \mid X_{ji}^{(2)}]^T \underline{\Phi} \tag{20}$$

where $\underline{A} = [a_0, a_1, \ldots, a_{2K_i^{(2)}+1}]^T$. For the normal Eq. (20), the $(2K_i^{(2)} + 3) \times (4K_i^{(2)} + 5)$ matrix

$$
\begin{bmatrix}
X_i^{(2)T}X_i^{(2)} & X_i^{(2)T}X_{ji}^{(2)} & X_i^{(2)T}\underline{\Phi} & I & 0 \\
X_{ji}^{(2)T}X_i^{(2)} & X_{ji}^{(2)T}X_{ji}^{(2)} & X_{ji}^{(2)T}\underline{\Phi} & 0 & I \\
\underline{\Phi}^{T}X_i^{(2)} & \underline{\Phi}^{T}X_{ji}^{(2)} & \underline{\Phi}^{T}\underline{\Phi} & \underline{0}^{T} & \underline{0}^{T}
\end{bmatrix}
\tag{21}
$$

is constructed. Then, by applying the stepwise regression procedure to matrix (21), we select a combination of the dominant input variables which minimizes AIC, and we construct an optimal intermediate polynomial from this combination. The procedure in this part is called generator G3 of optimal intermediate polynomials. From $m_1 \times m$ intermediate polynomials generated from this procedure, the m_1 intermediate polynomials, which give the m_1 smallest AIC, are selected.

The procedure in the second layer is summarized as follows: First, we select the dominant variables from all the variables contained in $f_i^{(1)}$ and $f_j^{(1)}$, and construct the intermediate polynomial $f_i^{(2)}$. Then, we combine the input variables x_j ($j = 1, 2, \ldots, m$) with $f_i^{(2)}$ ($i = 1, 2, \ldots, m_1$), and construct the optimal intermediate polynomial in the second layer.

Instead of using the linear combinations above, we could use a nonlinear combination of $f_i^{(1)}$ and $f_j^{(1)}$ directly, such as a second-order polynomial of two variables, but the number of variables which must be considered in selecting dominant variables becomes very large, and it is not desirable in the practical situation.

C. Procedure in the Third, Fourth, ... Layers

In the third, fourth, ... layers, the procedure used in the second layer is repeated. The multilayered iterative computation will be terminated if one of the following cases is obtained.

1. The AIC is reduced to a very small value and the value of the AIC cannot be decreased further in the next layer.
2. The structures of m_1 intermediate polynomials are the same forms as those of m_1 intermediate polynomials in the preceding layer.

Case (1) indicates that without being disturbed by large noises, the nonlinear relationship among the input and output variables are obtained accurately: that is, it seems to be most probable that a physically meaningful relationship among the input and output variables is obtained. On the other hand, in case (2), the relationship obtained among the input and output variables is not a physically meaningful one. When the multilayered iterative computation is terminated, the intermediate polynomial remained in the final layer is adopted as a complete polynomial of the system.

By using these three procedures as described in Secs. IV.B.1-IV.B.3, we can construct a revised GMDH algorithm which generates optimal intermediate polynomials. The parameters used in this algorithm are as follows:

p: maximum order of the intermediate polynomial in the first layer
L_1: maximum number of the terms in the intermediate polynomials
m_1: number of the intermediate polynomials selected in each layer

These parameter values are preferred to be as large as possible, and are determined not by the heuristics but by the upper limit of the memory capacity of computer.

When we apply the revised GMDH algorithm to the real system, the structure and the parameters of the identified model may be considerably different from those of the real system, because, in general, the measured data of the input and output variables contain noticeable measurement errors. Therefore, we must check the structure of the model based on physical knowledge, and we must check the estimated parameters based on statistical knowledge.

V. APPLICATIONS

The two revised GMDH algorithms

RG1: Heuristics-free GMDH which generates optimal partial polynomials
RG2: Heuristics-free GMDH which generates optimal intermediate polynomials

described above are applied to some real problems.

RG1 is applied to

1. Short-term prediction of air pollution (Tamura and Kondo, 1980)
2. Identification of a dynamic lake model for Lake Ontario (Tamura and Halfon, 1980)

In (1), nonlinear models for predicting SO_2 concentration a few hours in advance are obtained. Inputs are the time-lagged SO_2 concentration, the east-west component, and the south-north component of the wind velocity. The predicted results are compared with the results obtained by a linear regression model, a linear autoregressive model, and an original GMDH model. As a result, a better performance is obtained by the RG1 model than by the linear models and the original GMDH model. In (2), nonlinear dynamic models for water quality dynamics in Lake Ontario are obtained. The state variables in the model are corrected chlorophyll a, volume-averaged over the top 20 m, and soluble reactive phosphorus in three layers, 0 to 20 m, 20 to 40 m, and 40 m to bottom. Inputs are the average quarterly temperature for each layer and the total phosphorus loadings. Two kinds of model are developed: recursive and iterative; they are used to simulate long-term predictions for three different hypothetical rates of phosphorus loading. Results show that RG1 can usefully identify lake models.

RG2 is applied to discovering a physical law (Kondo and Tamura, 1979). The algorithm is applied to noise-free synthetic data obtained in a simple kinetic system where force is applied to an object placed on a perfectly smooth surface, and the displacement and the velocity are observed. Inputs are the mass, force, and displacement and their reciprocals at time t, and the outputs are the displacement and the velocity at time (t + 1). Newton's second law of motion is discovered by RG2, while by using RG1 a physically meaningful model could not be obtained.

VI. CONCLUDING REMARKS

In this chapter two heuristics-free GMDH algorithms, RG1 and RG2, are described. Neither algorithm needs heuristics for dividing the available data into two sets: training data and checking data, for predetermining the structure of the partial polynomials, or for predetermining the number of intermediate variables. As a result, much better flexibility and stability for constructing a final model is obtained, and a much simpler final model can be constructed. Since we do not need to repeat the computational procedure for finding a better combination of heuristics, considerable computational time can be saved. Although it is difficult (probably impossible) to discover a physical law by RG1, RG2 can find a physical law if the characteristics of the system are well embedded in the input/output data and if these data are noise-free.

How should we use RG1 and RG2 properly? If we came across a problem with many variables and complex structure, RG1 would be a more appropriate modeling technique than RG2. This is because it is difficult to find a physically meaningful relationship among the variables when the system contains many variables with a complex relationship. Furthermore, when the observed data contain noise, it is also difficult to find a physically meaningful model by RG2, and therefore RG1 is more appropriate than RG2. By using RG1 we could obtain nonphysical input/output models. Although we need much more computational effort for RG2 than for RG1, we should try RG2 to find a physically meaningful model structure if the characteristics of the system might be well embedded in the observed data.

REFERENCES

Akaike, H. (1972). Automatic Data Structure Search by the Maximum Likelihood. Comput. Biomed. Suppl. Proc. 5th Hawaii Int. Conf. Syst.
Sci., pp. 99-101.

Akaike, H. (1973). Information Theory and an Extension of the Maximum
Likelihood Principle. In Proceedings of the Second Int. Symp. Information Theory. Akadémiai Kiado, Budapest, pp. 267-281.

Akaike, H. (1974). A New Look at the Statistical Model Identification. IEEE
Trans. Autom. Control AC-19:716-723.

Allen, D. M. (1974). The Relationship Between Variable Selection and Data
Augmentation and a Method of Prediction. Technometrics 16:125-127.

Draper, N. R., and Smith, H. (1966). Applied Regression Analysis. Wiley,
New York, pp. 163-216.

Duffy, J. J., and Franklin, M. A. (1975). A Learning Identification Algorithm and Its Application to an Environmental System. IEEE Trans.
Syst. Man Cybern. SMC-5:226-240.

Endo, A. (1978). Identification of a Nonlinear System with the Modified
GMDH (in Japanese). Trans. Soc. Instrum. Control Eng. 14:130-135.

Ihara, J. (1975). Improved GMDH—A Case of Dynamical World Population
Models (in Japanese). Syst. Control 19:201-210.

Ikeda, S., Ochiai, M., and Sawaragi, Y. (1976). Sequential GMDH Algorithm and Its Application to River Flow Prediction. IEEE Trans. Syst.
Man Cybern. SMC-6:473-479.

Ivakhnenko, A. G. (1968). The Group Method of Data Handling, a Rival of
the Method of Stochastic Approximation. Sov. Autom. Control 1:43-55.

Ivakhnenko, A. G. (1971). Polynomial Theory of Complex Systems. IEEE
Trans. Syst. Man Cybern. SMC-1:364-378.

Ivakhnenko, A. G., Koppa, Y. V., Tymchenko, I. K., and Ivakhnenko,
N. O. (1969). Group Handling of Data in Identification of the Static
Characteristic of a Multi-extremal Plant. Sov. Autom. Control 2:30-37.

Ivakhnenko, A. G., Krotov, G. I., and Visotsky, V. M. (1979). Identifi-
cation of the Mathematical Model of a Complex System by the Self-
Organization Method. In Theoretical Systems Ecology, E. Halfon (Ed.).
Academic Press, New York, pp. 325-352.

Kondo, T., and Tamura, H. (1979). Revised GMDH Algorithm of Self-
Selecting Optimal Intermediate Polynomials Using AIC (in Japanese).
Trans. Soc. Instrum. Control Eng. 15:466-471.

Okuno, T., et al. (1976). Multivariate Analysis, Continued. Nikka-Giren,
Tokyo, pp. 66-75 (in Japanese).

Tamura, H., and Halfon, E. (1980). Identification of a Dynamic Lake Model
by the Group Method of Data Handling: An Application to Lake Ontario.
Ecol. Model. 11:81-100.

Tamura, H., and Kondo, T. (1980). Heuristics Free Group Method of Data
Handling Algorithm of Generating Optimal Partial Polynomials with
Application to Air Pollution Prediction. Int. J. Syst. Sci. 11:1095-1111.

13

An Application of the GMDH Algorithm to Economic Modeling

DONALD E. SCOTT and CHARLES E. HUTCHINSON/University of
Massachusetts, Amherst, Massachusetts

I. INTRODUCTION

The concept of applying quantitative model building to economic systems is not new. Several such models has been postulated and used for prediction purposes. The models used to date, however, have been highly heuristic or drastically simplified from a mathematical perspective. The following development presents a procedure for constructing complex nonlinear models using the GMDH technique in a simple and systematic way. A specific preliminary model constructed in this manner is then compared to three other existing models: Pindyck [1], Fed. MIT [2], and St. Louis [3].

The GMDH technique affords the economist-control engineer an increased insight into the inherent structure of complex economic systems. The algorithm "optimally" selects (1) what variables appear in which equations, (2) the degree of lag(s), (3) the optimum degree of nonlinearity of the resulting model equations, and (4) the structure and degree of interaction among variables. The algorithm would appear to be particularly useful in objectively selecting explanatory variables and identifying nonlinear functional relations among economic variables. Consequently, it may be most applicable to those areas of economics where different "schools" of thought have developed concerning the importance of a particular variable or the appropriateness of particular functional relationships.

The authors do <u>not</u> believe that the algorithm will be a panacea for solving all statistical problems within economics, nor that it will reduce the need for economic theory in determining underlying economic relationships. However, the GMDH algorithm may permit further progress toward the development of more generalized models in the economic profession.

One of several interesting properties of the GMDH algorithm is its apparent ability to determine the inherent structure of extremely complex and possibly highly nonlinear systems [4]. In addition, it appears to avoid the potential pitfalls of the regression techniques. It therefore offers a significant advance in the field of mathematical modeling.

Most mathematical models of economic systems are derived using time-series data bases and linear multivariate regression or one of its variations [3,5]. Attempts are sometimes made to include nonlinear relationships in such models. This is usually accomplished by first postulating the form of the nonlinearity and then augmenting the data base with the time series generated by this nonlinearity before proceeding with the regression analysis [2]. There are at least two major and related sources of potential error in such methods: (1) any nonlinear forms are presumed a priori by the investigator rather than being naturally and automatically derived from the data during the modeling process, and (2) although inclusion of postulated nonlinear terms may raise the final correlation between the model's output time series and the actual data, the probability of curve fitting the noise in the data increases toward unity. The first of these errors builds in the investigator's nonlinearity and the second confirms its presence by assigning it a weight.

Another potential error of regression analyses is elimination of terms that do not raise the correlation coefficient between the estimate and the data [6]. If two time series (random variables) have a correlation coefficient close to zero, this does not mean that they are not closely, or even exactly, related in some nonlinear way. The correlation coefficient only measures the amount of clustering that the joint probability density function exhibits about any straight line. For example, two random variables, x and y, related by the deterministic expression

$$y = ax^2 \tag{1}$$

for $a \neq 0$ will have a joint probability density function whose correlation coefficient ρ_{xy} is

$$|\rho_{xy}| < 1 \tag{2}$$

even though the estimator

$$\hat{y} = bx^2 + cx + d \tag{3}$$

will result in identically zero mean-squared error (with $b = a$ and $c = d = 0$). Clearly, both the linear and nonlinear regression techniques described above often represent mutually unattractive alternatives. Choosing the linear regression technique ignored inherent nonlinear and symbiotic relationships among variables [e.g., as in (1)]. Employment of a high-order nonlinear estimator reduces mean-squared estimation error toward zero but tends to curve fit the noise as well as the data.

II. PRELIMINARY GMDH MODEL

A set of economic state variables [1] describing a model of the national economy were tabulated for 26 fiscal quarters (1968 I through 1974 II). These variables were:

C, Personal consumption (billions of constant dollars)
INR, Nonresidential investment (billions of constant dollars)
IIN, Change in industrial inventories (billions of constant dollars)
R, Interest rate on short-term government bills (percent)
RL, Long-term interest rate (percent)
P, Implicit price deflator on GNP (1958 level = 100)
UR, Unemployment rate (percent)
W, Hourly wage rate (constant dollars)
YD, Disposable personal income (billions of constant dollars)
G, Government spending (billions of constant dollars)
DM, Change in the money supply (billions of constant dollars)

With the exception of YD, disposable personal income, and UR, the unemployment rate, all the data above are now available in the Survey of Current Business published by the U.S. Department of Commerce. Each datum in the YD time series is obtained as follows. A generalized measure of taxation t is determined by

$$t = \frac{FR - FTR}{GNP_u} \tag{4}$$

where FR   = federal government receipts, total (billions of current dollars)
       FTR = federal government transfer payments (billions of current dollars)
       GNP_u = gross national product, total (billions of current dollars)

Disposable personal income is then found via

$$YD = (1 - t)GNP_a \tag{5}$$

where GNP_a is gross national product (billions of constant dollars).
The unemployment rate is obtained as

$$UR = \frac{U}{LF} \tag{6}$$

where LF is the civilian labor force (seasonally adjusted millions) and U is the civilian labor force—unemployed (seasonally adjusted, millions).

The variables G and DM are generally thought of as being system input variables, DM being under the control of the Federal Reserve Board and G dependent on congressional action. However, no such distinction between these and the other state variables need be made for the proper functioning of the algorithm. Using the argument k to denote the present quarter, k - 1 the preceding quarter, and so on, the following 20 variables were used as the 20 rows of the data matrix:

$$C(k-1) \quad INR(k-1) \quad IIN(k-1) \quad R(k-1) \quad R(k-2)$$
$$RL(k-1) \quad RL(k-2) \quad RL(k-3) \quad RL(k-4) \quad RL(k-5)$$
$$P(k-1) \quad P(k-2) \quad UR(k-1) \quad UR(k-2) \quad UR(k-3)$$
$$YD(k-1) \quad YD(k-2) \quad YD(k-3) \quad G(k) \quad DM(k)$$

If there are n input variables available, the first layer of the algorithm will compute the six coefficients for each of the $\binom{n}{2}$ possible quadratic Ivakhnenko estimators. The user must then choose how many of these estimators will be saved (whose outputs will be available as inputs to the next layer). Since, eventually, a single estimate vector, $\hat{y}$, is to be generated by the algorithm, it follows that the last layer will contain a single estimator. The numerical choice of how many estimators to save in each intermediate layer is controlled by the user.

At this time no general rule for optimal selection of the number of estimators in each layer is known. The arbitrarily chosen procedure followed to generate the economic model described above was as follows. To estimate the 10 variables, C, INR, ..., YD, a separate GMDH network was developed for each. The computer algorithm was run twice in each case. During the first run the threshold was set in each later (except the last) such that the best 20 (equal to the number of input variables) estimators were allowed to supply input to the next layer. This represents a survival rate for each layer of 20 $\div \binom{20}{2}$, which is approximately 10.5%. Only one estimator was allowed in the last layer. Enough layers were calculated so that the minimum mean-squared error generated by the checking set was observed.

On the second run the algorithm was instructed to generate only the number of layers shown in the first run to result in minimum mean-squared checking set error. Examination of the computer output from the first run shows for each layer which estimators' outputs from the next preceding layer are used. To save time on the second computer run, thresholds were raised selectively in each layer such that only the estimators whose outputs were used by the next layer were recomputed.

Table 1 Optimal Number of Layers

Y	Number of layers	Number of estimators needed for each layer
INR	4	12, 6, 2, 1
IR	3	14, 6, 1
IIN	4	11, 6, 3, 1
R	4	15, 4, 2, 1
RL	4	15, 5, 3, 1
P	1	1
UR	3	19, 5, 1
W	4	8, 4, 3, 1
YD	1	1
C	1	1

It is possible selectively to raise intermediate-layer thresholds above these "natural selection" levels, thus depriving of data later layers in the network. It is not yet clear in general what effects such internal data deprivation has on (1) the mean-squared error (MSE) of the final estimate, (2) the number of layers then needed to reach the minimum (optimum) mean-squared checking set error, or (3) the relative magnitude of this checking set MSE. In the development of the small economic macromodel, no such internal data deprivation was forced to occur.

Table 1 shows how many layers were optimal for each variable. The third column lists the number of estimators required in each layer. For example, in the model for INR, the fourth-layer single estimator used as input data the outputs from third-layer estimators 1 and 2. Those in turn used the outputs, respectively of estimators 1, 6, and 4 of the second layer. This a posteriori method of starting at the output and working back through the developed preliminary network discloses (1) the minimum number of estimators that need to be saved in each layer, and (2) which, if any, original input variables are not used.

In the cases of three variables, P, YD, and C, the algorithm indicated that single-layer networks (single estimator with only two inputs) were optimum. In such cases standard linear regression models using three or more inputs often yield lower mean-squared estimate errors than does the single-layer GMDH estimator. Thus the GMDH method does not in general replace, but rather augments, more standard regression techniques.

Several runs were made, each using a different y variable [C(k), INR(k), ..., YD(k)] as the y vector. The final resulting estimates, $\hat{y}$, compared well with actual data. Figures 1 to 3 show typical results.

The GMDH algorithm automatically selects those combinations of state variables and input variables that yield the most accurate estimate of each

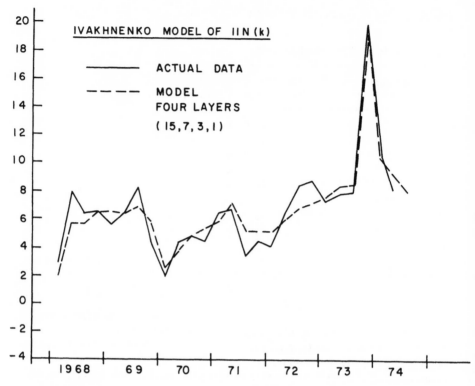

Fig. 1 Change in industrial inventories, IIN(k), in billions of constant (1958) dollars.

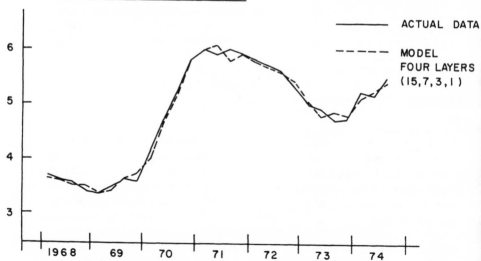

Fig. 2 Unemployment rate, UR(k), in percent.

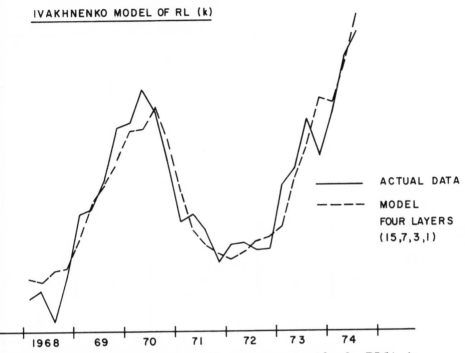

Fig. 3 Long-term interest rate, 10-year government bonds, RL(k), in percent, at annual rate.

output variable one sample time in the future. This property of the algorithm was used to construct a directed graph showing these interdependencies among the model variables.

The matrix shown in Table 2 indicates which variables were selected by the algorithm for inclusion in the prediction equations for each of the variables. Each column contains one of the state variables to be predicted; each row represents one of the known state or input variables. For example, to estimate the unemployment rate, UR, in quarter k, the algorithm selected five input variables: long-term interest rate RL (delayed two and four quarters), unemployment rate in each of the two preceding quarters, and G, government spending in the kth quarter.

This matrix implies a directed graph wherein each node is one of the model variables. A major portion of this graph is shown in Fig. 4. (The degrees of the various lags involved are omitted.) The graph presents the interrelationships among variables as derived by the GMDH algorithm. Only those variables that are affected by, and that affect, other variables are included.

Table 2 General Network Structure

	C	INR	IR	IIN	R	RL	P	UR	W	YD
C(k-1)			X	X						
INR(k-1)	X	X			X	X				X
IIN(k-1)										
R(k-1)			X	X						
R(k-2)		X				X			X	
RL(k-1)				X		X				
RL(k-2)		X						X	X	
RL(k-3)									X	
RL(k-4)			X					X		
RL(k-5)									X	
P(k-1)					X	X				
P(k-2)		X				X	X		X	
UR(k-1)		X	X		X			X	X	
UR(k-2)			X	X	X			X		
UR(k-3)	X		X	X	X	X	X			X
YD(k-1)		X			X					
YD(k-2)				X	X					
YD(k-3)				X						
G(k)		X		X				X		
DM(k)						X				

Variables used in model equations

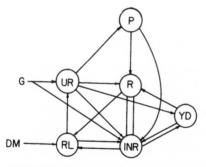

Fig. 4 Statistical effect of input variables government spending, G, and change in money supply, DM, on several of the state variables as derived by the GMDH algorithm.

The statistical effect of monetary action on the unemployment rate is presented as being an indirect one (via a more primary effect on long-term interest rates). This extremely peripheral effect of DM on UR has been referred to by other investigators [6].

Another implication of the directed graph of Fig. 4 is a relatively indirect relationship between DM and R, the short-term interest rate. Although most economic models assume a direct cause-effect relationship between DM and R, the Ivakhnenko model indicates that, at least statistically, this effect is overshadowed by inflation, nonresidential investment demand, and the level of overall disposable income. Monetary policy is, however, observed to affect long-term rates directly and nonlinearly.

To investigate overall model stability and long-range prediction accuracy, the ex post simulation for the GMDH model was allowed to run through the 1974 I period. Generally, the predictions appear to be smoothed, and in some cases, slightly delayed versions of the actual trajectories. No inherent model instabilities were discovered in the model outputs over the entire 25 quarters of this ex post simulation.

The model has, in one sense, "seen the correct answers" in any ex post simulation. A more demanding test is the ex ante simulation, wherein the model predicts values for y over a range of time other than the one on which it has been trained. Any increases over the observed ex post error measurements recorded during such a simulation are typically regarded as being due to changing economic structure.

In the case of R, the yield of U.S. government 3-month bonds, the actual rate dropped from a value of 8.27% in the 74II quarter to an average of 6.72% over the following year. The GMDH model predicted a drop to 7.32%.

In the case of RL, the yield on U.S. Treasury bonds, the actual rate dropped slightly from 7.05% in 74II to an average of 6.97% over the following year. The GMDH model predicted 6.47%.

III. COMPARISONS

For purposes of comparison the values of IIN predicted by Pindyck's model [1] and the Fed.-MIT model [2], are plotted in Figs. 5 and 6. Direct comparison between the GMDH model and the St. Louis model is difficult because of differences in the definitions of the predicted variables, and because of apparent slight discrepancies in the data bases that were used. However, Carlson [2] reports results obtained from an ex post simulation using the St. Louis model. This simulation used model coefficients derived from data through the second quarter 1971, and observed values of the actual money supply and federal expenditures over this period. Initial values of the remaining state variables for 1969 IV were inserted and the resulting predictions for the quarters 1970 I through 1971 II were recorded. The resulting average squared error for three of the resulting St. Louis model

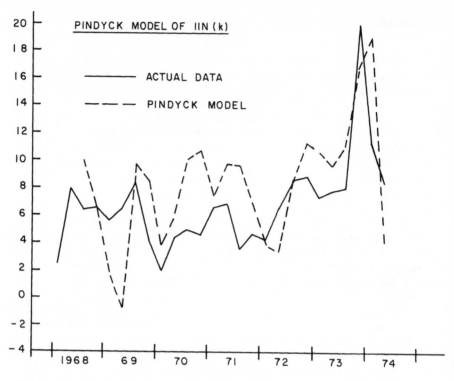

Fig. 5 Projection for IIN(k) derived by Pindyck based on earlier series data.

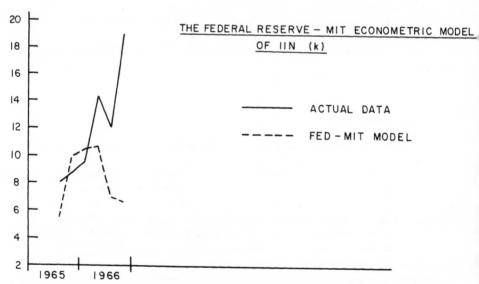

Fig. 6 Projection for IIN(k) as given in the Federal Reserve-MIT econometric model.

variables and for the closest corresponding variables from the same type
of simulation using the GMDH model are shown below.

St. Louis	$\bar{e}^2$	GMDH	$\bar{e}^2$
Unemployment rate	0.1	Unemployment rate	0.109
Corporate A_{aa} rate	0.14	Yield on government bonds	0.0986
Commercial paper rate	1.093	Government 3-month bills	0.385

IV. CONCLUSIONS

The authors believe that the GMDH algorithm may be particularly useful in
the field of economics. Economists who have been concerned with measuring
empirical relationships among economic variables have tended to rely almost exclusively on least-squares regression as the statistical technique
for testing for relationships. To be sure, nonlinear relationships have been
introduced into the classical least-squares model through the nonlinear
transformation of variables, but in such instances the nonlinear form of the
equation being estimated has been determined by the a priori assumption of
the researcher. In recent years, economists have tended to develop more
generalized statistical models which allow the data to play a larger role in
determining the particular functional form that describes the relationship
between the independent and dependent variables: specifically, the development of the constant elasticity of substitution (CES) production function
developed by Arrow et al. [7] and the development by Tsao [8] of a more
generalized form of the Box and Cox transformation [9]. The GMDH algorithm may permit even further progress toward the development of more
generalized models in the economics profession.

   1. In those areas of economics where there is considerable controversy
concerning the proper functional form of basic economic relationships, the
GMDH algorithm may provide an unbiased formulation of the underlying
functional form of the specified economic relationship. Specifically of
interest is the disagreement within the profession as to the proper functional form of both the consumption and production functions. See the
articles by Branson and Klevorick [10], Tsao [8], and Zarembka [11] for a
survey of this literature.

   2. Since the GMDH algorithm tests for nonlinear relationships, it may
provide economists with a means of objectively settling some of the contro-
versies concerning the importance or nonimportance of certain variables.
While this particular problem area is theoretically separable from the
problem of functional form described above, as a practical matter the two
are often interrelated. For example, the argument that an individual's total
real consumption demand is homogeneous of degree zero in prices, money
income, and money wealth leads to the conclusion that aggregate real consumption (for the economy as a whole) should be a function of aggregate

real income and aggregate real wealth, but not the price level. This particular formulation of the consumption function clearly depends on the assumption that consumers are not subject to money illusion. Branson and Klevorick [10] have argued that consumers are in fact subject to money illusion in the short run and hence that prices should be included as a separate independent variable in the aggregate consumption function.

3. In those areas of economics where there is no theoretical basis for specifying the particular functional form of a relationship, the GMDH algorithm may provide an objective basis for determining the most appropriate functional form. For example, economists have frequently transformed their variables into logs or some other nonlinear form because the logs or other nonlinear models provide a better fit of the data. In these cases, there is no theoretical justification for the nonlinear transformation and the use of the GMDH algorithm would appear to be more effective in this instance since it optimally selects the particular nonlinearity that will provide the best fit of the data.

4. At least in certain limited areas within economics, economists have been faced with the problem of virtually an unlimited number of independent variables which may affect an important dependent variable and yet economic theory does not allow one to reduce the number of independent variables to a manageable level. Under these circumstances, factor analysis has been used in lieu of multiple regression analysis. Since the GMDH algorithm not only tests for nonlinear relationships but also selects variables according to their ability to explain the dependent variable, it would appear that the GMDH approach may be more effective than factor analysis in dealing with statistical problems of this nature. Specifically of interest is the attempt by Adelman and Morris to measure the effect of various economic and social variables on the process of economic development [12].

Of the four possible areas of application described above, areas (1) and (2) would be the most fruitful for exploration at this time, primarily because they represent areas of existing controversy within economics where substantial results have already been obtained. Therefore, application of the GMDH algorithm to these particular problem areas would allow comparison with results that have already been obtained by the use of other techniques.

REFERENCES

1. R. S. Pindyck, Optimal Planning for Economic Stabilization, North-Holland, Amsterdam, 1973.
2. F. deLeeuw and E. Gramlich, The Federal Reserve-MIT Econometric Model, Fed. Reserve Bull., Jan. 1968, pp. 11–40.
3. K. M. Carlson, Projecting with the St. Louis Model: A Progress Report, Review—Federal Reserve Bank of St. Louis, Feb. 1972, pp. 20–27.

4. D. E. Scott and C. E. Hutchinson, The GMDH Algorithm—A Technique for Economic Modeling, Proc. 7th Pittsburgh Conf. Modeling Simul., University of Pittsburgh, 1976, p. 729.

5. L. C. Anderson and K. M. Carlson, A Monetarist Model for Economic Stabilization, Review—Federal Reserve Bank of St. Louis, Apr. 1970, pp. 7-25.

6. D. R. Francis, Inflation-Recession—What's a Policymaker to Do? Review—Federal Reserve Bank of St. Louis, Nov. 1974, pp. 3-7.

7. K. J. Arrow, H. B. Chenery, B. S. Minhas, and R. M. Solow, Capital-Labor Substitution and Economic Efficiency, Rev. Econ. Stat. 43: 225-250 (1961).

8. C. S. Tsao, The Linearity Property in the Consumption Function: Estimation, Tests and Some Related Results, Rev. Econ. Stat. 57: 214-220 (May 1975).

9. G. E. P. Box and D. R. Cox, An Analysis of Transformations, J. R. Stat. Soc., Ser. B 26:211-243 (1964).

10. W. Branson and A. Klevorick, Money Illusion and the Aggregate Consumption Function, Am. Econ. Rev. 59:832-849 (Dec. 1969).

11. P. Zarembka, On the Empirical Relevance of the CES Production Function, Rev. Econ. Stat. 52:47-53 (Feb. 1970).

12. I. Adelman and C. Morris, A Quantitative Study of Social and Political Determinants of Fertility, Econ. Dev. Cultural Change 14:129-157 (Jan. 1966).

14

Application of GMDH to Environmental System Modeling and Management

SHIN-ICHI FUJITA / Environmental Pollution Control Center of Osaka Prefectural Government, Osaka, Japan

HIROSHI KOI / Life Environment Department of Osaka Prefectural Government, Osaka, Japan

I. INTRODUCTION

Osaka Prefecture is located in the central portion of Japan, covering most of the Osaka Plain. It faces Osaka Bay on the west and is surrounded by the mountain chains of Rokko-Hokusetsu, Ikoma-Kongo, and Izumi on the north, east, and west. A total of 145 rivers empty into the bay.

The total areas of Osaka Prefecture, extending about 80 km in latitude and about 30 km in longitude, is about 1849 km^2, making it the smallest in Japan. More than 8 million people inhabit this small area, resulting in one of the highest population densities in the country.

Since olden times, Osaka has been the commercial and industrial center of western Japan, and recently a great industrial complex, the Sakai–Senboku industrial area, was constructed with heavy chemical industries and power generating plants along the marine area.

For a decade after 1955, the scale of industrial activities expanded and the number of motor vehicles in Japan increased, thereby accelerating the increase in the concentration of pollutants in urban areas. These conditions, together with an increase in the population density, aggravated the environmental pollutions situation. To cope with this situation, the Air Pollution Control Law and the Noise Regulation Law were enacted by the national government in 1968.

Due to these administrative controls, the concentration of typical pollutants, such as sulfur dioxide and suspended particulates, has been steadily reduced since 1970. However, in addition to controlling these typical pollutants, the public goal is to achieve greater protection of the natural environment and its related amenity value in the 1980s.

To meet these public demands it is becoming more and more necessary to be able to predict the future states of the local environments and to assess the environmental effects of new development projects. Three mathematical models using GMDH were developed for the environmental management of Osaka Prefecture, but these models would also be applicable to other areas.

GMDH, developed by A. G. Ivakhnenko (1968), had some weaknesses in the algorithm which were later improved by various investigators (Duffy and Franklin, 1975; Ivakhnenko et al., 1979). The revised algorithm developed by Tamura and Kondo (1978a,b) using AIC (Akaike's information criterion) or PSS (prediction sum of squares) almost overcomes the pitfalls. So the mathematical models described here use Tamura's revised GMDH (henceforth simply "revised GMDH").

II. SHORT-TERM PREDICTION OF PHOTOCHEMICAL SMOG AND EMISSION SOURCE CONTROL SYSTEM

A. Outline of the Photochemical Smog Control System in Osaka

High concentrations of oxidants (Ox) have been observed on sunny summer days in Osaka. This is called photochemical smog and has caused serious effects, mainly on health. To control this smog, the photochemical smog control system was established by the Environmental Pollution Control Center of the Osaka Prefectural Government.

The main function of the system is to monitor air pollution and meteorological conditions at about 40 monitoring stations. These data are telemetered to the central data processing room in the environmental pollution control center every 10 minutes. Whenever a high Ox concentration is predicted to exceed the emergency level, judged on the weather conditions, a danger warning to reduce the amount of emission gas is issued by the center to about 200 factories, which account for about 80% of the total fuel consumption in Osaka Prefecture (see Table 1).

In such a smog emergency situation the prediction of the Ox concentration level should be made several hours in advance, to provide sufficient lead time for the 200 factories to change their fuel inputs. The procedures for predicting and issuing orders have been systematically developed through skillful experience and carefully gathered information.

The model described here was developed to predict the Ox concentration levels rapidly and accurately so that emergency actions might be carried out easily and smoothly.

B. Short-Term Prediction Model for Ox Concentration

The short-term prediction models of pollutants are classified into two categories. One is a physicochemical model based on the reaction and diffusion of pollutants, and the other is a statistical model based on the time-series analysis of observed data.

Table 1 Emergency Levels and Control of Emission Source on Photochemical Smog Emergency

Emergency levels	Criterion	Control of emission sources
Forecast	Ox > 0.08 ppm	Waiting for prediction
Prediction	Ox > 0.12 ppm	20% reduction of total emission gas
Warning	Ox > 0.24 ppm	20% reduction and waiting for emergency alarm
Emergency alarm	Ox > 0.40 ppm	40% reduction of total emission gas

The former has the advantage of being able to predict the effects of emission source control on the concentration of environmental pollutants, using the contribution rates of individual emission sources; however, the calculation procedure takes too much time to make its use practical for emergency cases.

Statistical models have been studied by many investigators, and discriminant analysis was developed by Lin (1982).

But photochemical smog is formed through complex photochemical reactions which depend on meteorological conditions and are difficult to be predicted by linear statistical models. As a result, a short-term prediction model was developed using revised GMDH which is a nonlinear model, coupled with quantification of qualitative meteorological data (quantification GMDH model).

C. Quantification GMDH Algorithm and Short-Term
    Prediction of Ox Concentration

Since prediction of Ox concentrations should be made more than 3 hours in advance of emergency actions, only 1-hour average values of the on-line monitoring data shown in Table 2 are used for the model.

The applicability of the model for emergency cases was demonstrated in 1979, when from August 11 to 15 typical photochemical smogs were formed in the central part of Osaka Prefecture. During this period actual data from appropriate monitoring stations in the affected area were used in the model (Fujita et al., 1982).

A flow diagram of quantification GMDH calculations is shown in Fig. 1. In the first step, qualitative wind data were quantified followed by the prediction of Ox concentration 3 hours hence using the revised GMDH.

Table 2 On-Line Monitoring Data

Number	Item name	Number	Item name
1	Sulfur dioxide	8	Wind speed
2	Particulates	9	Wind direction
3	Nitrogen oxide	10	Humidity
4	Nitrogen dioxide	11	Temperature
5	Carbon oxide	12	Solar radiation
6	Total hydrocarbon	13	Precipitation
7	Oxidant	14	Traffic volume

1. Quantification of Wind Speed and Direction

The quantification of qualitative data theory developed by Hayashi (1952) was introduced to quantify the wind data as follows:

$$y_i = \sum_{j=1}^{r} \sum_{k=1}^{p_j} X_{jk} \delta_i(jk) + e_i, \quad i = 1, 2, 3, \ldots, n \tag{1}$$

where

$$\delta_i(jk) = \begin{cases} 1 & \text{(ith sample corresponding to kth category on jth item)} \\ 0 & \text{(ith sample not corresponding to kth category on item j)} \end{cases}$$

y_i = objective variable (Ox concentration 3 hours hence)

r = number of items

p_j = number of categories belonging to jth item

X_{jk} = coefficient of kth category belonging to jth item

e_i = error

n = number of samples

As shown in Table 3, the observed values of wind speed and wind direction at the vertical wind monitoring station located in the Sakai-Senboku area, at the monitoring stations in the seaside area of southern Osaka, and at the monitoring stations near Kyoto were chosen as items. The Ox concentration 3 hours hence was taken as the objective variable, and X_{jk} was

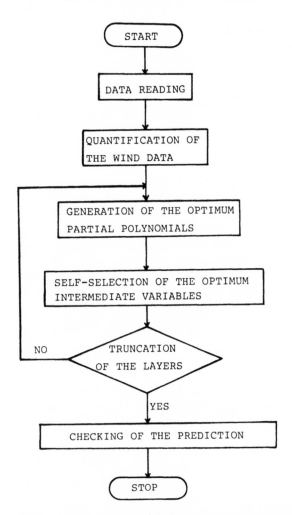

Fig. 1 Flow of quantification GMDH. Wind data were quantified in the first step, and then the Ox concentration 3 hours hence was predicted in the second step using the revised GMDH.

determined so as to minimize the sum of squared differences between the calculated and observed values of the objective variables.

2. Results of Prediction

The Ox concentration 3 hours hence was predicted using revised GMDH (AIC). Input variables are shown in Table 4, and the output variable is the

Table 3 Items and Categories for Quantification of Wind Data

Item	W.D. of seaside station	W.V. of seaside station (m/sec)	W.D. of inland station	W.V. of inland station (m/sec)	W.D. at 65 m height	W.V. at 65 m height (m/sec)	W.D. at 160 m height	W.V. at 160 m height (m/sec)
Category	N	<0.4	N	<0.4	N	<0.4	N	<0.4
	W	0.5-1.4	W	0.5-1.4	W	0.5-1.4	W	0.5-1.4
	S	1.5-2.4	S	1.5-2.4	S	1.5-2.4	S	1.5-2.4
	E	2.5-3.4	E	2.5-3.4	E	2.5-3.4	E	2.5-3.4
		3.5<		3.5<		3.5<		3.5<

Ox concentration 3 hours hence. The results of calculations are shown in
Fig. 2, and the final output equation was the weighted sum of four equations
(see the appendix). The correlation coefficient between the predicted and
observed variables of the Ox concentration was 0.92, and good agreements
were obtained at high concentrations. The error rate for the high-concen-
tration region defined by the equation

$$R = \frac{\Sigma \mid \phi_i - \bar{\phi}_i \mid}{\Sigma \mid \phi_i \mid} \qquad \text{for the regions} > 0.08 \text{ ppm} \qquad (2)$$

where ϕ_i is the observed value of the output variable and $\bar{\phi}_i$ is the calculated
value of the output variable, was 11.7%. These results verify the applicability of the model.

Table 4 Input Variables for Prediction of Ox Concentration
3 Hours Hence

No.	Item name	Time delay	Unit	Sign
1	Oxidant	No delay	ppb	Ox (t)
2	Temperature	2 hr	10°C	T (t-2)
3	Humidity	No delay	%	H (t)
4	Solar radiation	No delay	0.01 μJ/m^2	SR (t)
5	Quantified data	1 hr	ppb	y (t-1)

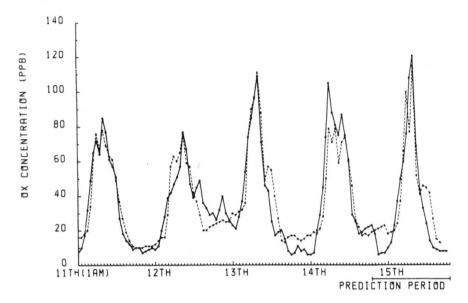

Fig. 2 Result of prediction of Ox concentration. Solid lines indicate the observed Ox concentration, and dotted lines indicate the predicted value of Ox concentration. The equation for prediction was made from 1 A.M. on the 11th to 7 P.M. on the 14th, and predictions were checked from 8 P.M. on the 14th to 7 P.M. on the 15th.

III. ANALYSIS AND PREDICTION MODEL FOR DISTRIBUTION OF COD CONCENTRATION IN OSAKA BAY

Water temperature, pH, transparency, and concentration of COD (chemical oxygen demand), filtered COD, organic and inorganic phosphorus, and nitrogen compounds are monitored regularly at 12 fixed stations in Osaka Bay every month. Among them, COD is regarded as an index of water pollution, and the environmental standards related to COD in Osaka Bay are shown in Fig. 3. Many attempts were made to develop a mathematical model to predict COD levels in Osaka Bay for the control of water pollution in compliance with the standards.

The usual method of calculating the behavior of COD in a bay is the nonreaction-diffusion model based on the diffusion equation of COD concentration (Kaneko et al., 1974). However, this model has the following defects: (1) the slope of the regression line of observed values and calculated values is rather high (usually 3 to 5), (2) the calculated values of COD depend to a large extent on the boundary conditions at the open boundary, and (3) the observed values in summer differ considerably from those in winter; however, the calculated values during both seasons are almost the same.

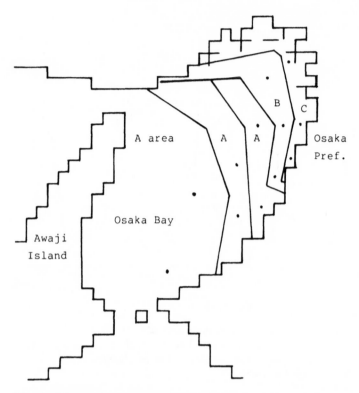

Fig. 3 Monitoring points and environmental standards in Osaka Bay. The
dots indicate the monitoring points in Osaka Bay. The environmental standard
in the A area is 2 ppm, the standard in the B area is 3 ppm, and the standard
in the C area is 8 ppm.

These defects are caused by the changes in COD concentrations due to
biochemical reactions and by the resolution of sedimental deposits. To
overcome these defects, some reaction-diffusion models were proposed
(Tabata, 1977; Dresnak and Dobbins, 1968). However, detailed data such
as diffusion coefficients, rates of biochemical reactions, and resolution of
sedimental deposits in the oceanological conditions were not prepared, so
as to obtain the exact COD level by the reaction-diffusion model. A new
hybrid model was developed, combining a nonreaction-diffusion model and
an eutrophical analysis model using revised GMDH for practical use.

A. Nonreaction-Diffusion Simulation of COD Concentration

The simulation sequence for COD was as follows. The simulation of tidal
flow in Osaka Bay was conducted in the first step, and then the nonreaction-
diffusion of COD was calculated using the results of the first step.

The results showed that the calculated values of COD agreed well with the observed values of the filtered COD. Finally, the relationship between the COD concentration and the filtered one was obtained by using revised GMDH.

1. Simulation of Tidal Flow

The two-dimensional Navier-Stokes equation with latitudinal and longitudinal direction was used for the simulation of tidal flow. The sea depth was divided into two layers where the waters were interchangeable. This is called the two-level model (Rand Corporation, 1973). The equation was approximated by the finite-difference method using a grid size of 4 km^2 and a time interval of 90 sec. A satisfactory numerical solution was obtained by the finite differential equation using the ADI (alternative direction implicit interactive) method (Fujita, 1984).

2. Diffusion Simulation of COD Concentration

The following diffusion equations using the results of the tidal flow simulation and the 3-year average value of COD loads (1976-1978) into Osaka Bay were used (369,441 kg/day) for the diffusion simulation of the COD concentration.

Upper layer:

$$\frac{\partial s}{\partial t} = -u \frac{\partial s}{\partial x} - v \frac{\partial s}{\partial y} - K_i \frac{s - s'}{h} + \frac{1}{h} \frac{\partial}{\partial x}\left(K \frac{\partial s}{\partial x}\right) + \frac{1}{h} \frac{\partial}{\partial y}\left(K_h \frac{\partial s}{\partial y}\right) + s_a + ws_w \tag{3}$$

Lower layer:

$$\frac{\partial s'}{\partial t} = -u \frac{\partial s'}{\partial x} - v \frac{\partial s'}{\partial y} + K_i \frac{s - s'}{h'} + \frac{1}{h'} \frac{\partial}{\partial x}\left(K_h \frac{\partial s'}{\partial x}\right) + \frac{1}{h} \frac{\partial}{\partial y}\left(K_h \frac{\partial s'}{\partial y}\right) - ws_w \tag{4}$$

Boundary conditions:

$$\frac{\partial s}{\partial n} = \frac{\partial s'}{\partial n} = 0 \qquad \text{at the rigid boundaries}$$

$$\frac{\partial s}{\partial n} = a \cdot s \ \text{ and } \ \frac{\partial s'}{\partial n} = a \cdot s' \quad \text{at the open boundaries} \tag{5}$$

where

a   = rate of dilution

u, v = velocity of tidal flow in the latitudinal and longitudinal directions (results of the tidal flow simulation)

s_a = COD load (ppm/hr)

K_i = exchange rate of the density between the lower and upper layers

K_h = diffusion coefficient in the horizontal direction

w   = flow in the vertical direction (when $w \geq 0$, $s_w = s$, and $w > 0$,
         $s_w = s'$)

h, h' = thickness of the upper and lower layers, respectively

These equations were solved numerically by the finite-difference approximation (implicit) method. The diffusion coefficients (K_i and K_h) were determined so as to maximize the correlation coefficient of the calculated COD values with the observed values of filtered COD (i.e., the COD values of seawater from which suspended materials had been removed). The correlation coefficient was 0.92, and the slope of the regression line was 1.17, and rather satisfactory (where $K_h = 2.5 \times 10^6$ m^2/sec, $K_i = 0$, and a = 0.99). The relationship between the calculated values of COD and the observed value of filtered COD is shown in Fig. 4. The values of filtered COD should be converted to the COD values for which environmental standards have been established.

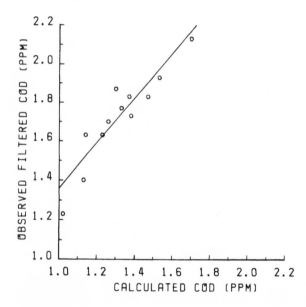

Fig. 4 Results of the nonreaction–diffusion simulation of COD concentration. The relationship between the calculated value of COD and the observed value of filtered COD concentration [the average value for three years (1976–1978)]. The regression equation is COD$_{ob}$ = 0.26 + 1.17 COD$_{cal}$.

3. Eutrophication Analysis Using Revised GMDH

The conversion of the filtered COD values to standard oxidation COD values by the complex eutrophication model is necessary because the COD values vary depending on the biochemical reaction and the resolution and precipitation of substances from and to the sedimental deposits.

So the revised GMDH (PSS) was used for the conversion from filtered COD to standard COD concentration. GMDH has often been used to predict water quality (Duffy and Franklin, 1975; Tamura and Halfon, 1980), but the main use of the model here is for conversion rather than ecological purposes. The conversion by the equation was made for those items that are easy to predict, as shown in Table 5, so that the future prediction for COD values may be carried out easily.

The representative point in C areas in Osaka Bay was selected for calculation by the revised GMDH model. Three years of observed data (1976-1978) were used in the model, and the calculations were checked using the observed data of 1979.

The results of calculations are shown in Table 5, and the final output equation is the weighted sum of four equations (see the Appendix), but in this case the equation expressed approximately one equation, as follows:

$$COD_{cal} = 0.500\ U1 + 0.500\ U2$$

where

$$U1 = 1.74 + 0.137\ Z3\ Z5$$
$$U2 = 1.99 + 0.115\ Z1^2$$

where

$$Z1 = 1.10 + 0.190\ Y3\ Y4$$
$$Z3 = 0.76 + 0.218\ Y4\ Y5$$
$$Z5 = 1.51 + 0.156\ Y2\ Y3$$

where

$$Y2 = 2.20 + 0.029\ X1\ X5 + 0.055\ X5^2$$
$$Y3 = 4.83 - 0.881\ X2 + 0.142\ X4^2$$
$$Y4 = 1.44 + 0.017\ X1\ X3$$
$$Y5 = 5.93 - 0.039\ X2\ X5$$

where Y, Z, and U are the intermediate variables at the first, second, and

Table 5 Input and Output Variables, and Calculated Values of COD Concentration of a Monitoring Point in C Area

Year	Month	Water temp. (°C) X1	Trans- parency (m) X2	DO conc. (ppm) X3	Salt conc. (%) X4	Filtered COD conc. (ppm) X5	COD conc. (ppm)	Calculated COD conc. (ppm)
1976	4	14.9	2.0	9.3	29.8	3.0	3.8	4.0
	5	16.6	1.7	5.8	23.1	3.3	3.7	4.0
	6	21.3	2.1	9.1	24.0	2.2	3.0	4.2
	7	24.3	2.9	7.0	24.1	3.4	4.7	4.2
	8	26.6	1.7	4.8	21.8	2.9	4.1	4.4
	9	23.2	1.3	4.7	28.7	1.6	3.1	3.6
	10	22.2	2.5	4.5	25.6	1.3	3.1	3.0
	11	18.1	2.5	5.9	29.0	2.0	2.5	3.1
	12	13.7	2.8	7.9	25.9	3.3	3.3	3.4
1977	1	7.7	2.6	9.3	29.0	0.8	2.3	2.6
	2	6.9	2.3	10.3	27.5	1.5	3.3	2.8
	3	7.4	1.9	9.4	25.3	3.7	3.8	3.6
	4	11.3	2.0	8.4	25.4	2.2	3.0	3.3
	5	17.6	2.0	8.2	26.4	2.4	2.7	3.8
	6	19.5	2.4	6.5	28.3	1.0	3.3	3.0
	7	21.6	2.5	5.1	24.9	1.3	4.9	3.0
	8	26.5	1.9	6.3	27.3	1.8	4.8	3.9
	9	26.4	2.2	8.1	27.2	2.8	4.4	4.8
	10	23.1	3.8	4.9	26.3	2.2	2.2	2.7
	11	21.1	2.5	5.1	26.5	0.8	1.8	2.9
	12	15.2	2.5	6.5	28.4	1.2	2.9	2.9
1978	1	10.0	3.0	8.6	28.3	1.1	3.1	2.6
	2	8.7	1.9	8.6	22.8	1.8	3.7	3.1
	3	7.5	1.5	9.8	27.8	1.6	2.9	3.1
	4	12.5	1.7	10.0	25.9	2.2	4.7	3.7
	5	17.7	1.1	11.0	27.8	2.5	5.4	5.2
	6	21.4	1.9	8.6	23.0	2.8	5.5	4.7
	7	26.6	2.3	4.8	22.2	2.5	3.7	3.7
	8	28.7	1.1	8.2	28.7	3.8	8.1	8.0
	9	26.7	2.5	8.6	31.3	2.0	4.1	3.9
	10	23.2	2.5	5.4	30.9	1.6	2.9	3.1
	11	19.1	3.0	5.1	30.5	1.7	3.1	2.7
	12	14.2	3.0	7.1	29.1	2.1	2.3	2.8
1979	1	12.5	2.5	7.4	29.3	1.8	2.9	2.9
	2	9.3	2.4	9.4	28.9	2.2	2.8	3.0
	3	9.5	2.5	9.4	30.1	2.1	2.7	2.9
	4	12.9	2.5	9.1	29.2	2.1	4.3	3.1
	5	16.0	2.1	10.0	25.8	2.9	4.9	4.2
	6	19.7	1.6	10.0	22.9	3.0	5.9	5.3
	7	22.8	3.0	5.0	22.2	2.0	3.2	3.0
	8	28.2	1.0	10.0	25.9	3.3	8.4	8.4
	9	24.6	2.8	6.4	30.7	2.1	4.3	3.2
	10	22.2	2.3	5.9	30.1	2.6	3.6	3.6
	11	17.9	2.5	6.7	27.0	2.0	3.0	3.2
	12	14.3	4.0	7.5	28.1	1.5	3.1	2.4
1980	1	7.3	2.0	9.4	26.2	2.5	3.4	3.1
	2	6.6	2.0	11.0	23.8	3.0	4.6	3.3
	3	8.8	2.3	8.7	27.3	2.2	2.9	3.0

third layers, respectively, and the input variables X are as shown in Table 5. For the monitoring station of C area, the error rate was 12.5%.

Phosphorus and nitrogen were not used in the equation because they were determined only twice a year. But if these data were available for use in the model, the result could be significantly improved (Fujita, 1984).

IV. ANALYSIS OF LANDSAT MSS DATA USING REVISED GMDH PREDICTION MODEL OF ENVIRONMENTAL NOISE LEVEL

A. Prediction Model for Environmental Noise Level in Osaka

Environmental noise is caused by various noise emission sources such as motor vehicles and factories, and the noise level is decreased by various obstacles and barriers such as buildings and construct.

Therefore, the prediction of the noise level in the propagation process should be conducted by using the information on regional characteristics, such as traffic flow rate and sizes of buildings.

An environmental noise level prediction model for Osaka Prefecture was developed as follows:

1. The traffic volume in the trunk line was reported by the Construction Ministry and by the Osaka Prefectural Government, and the noise emission level by factories published by the Osaka Prefectural Government is used for the calculations, and these data are averaged over a uniform 500 × 500 square meter mesh, which is called the standard mesh.
2. The model is based on the EPA model (EPA, 1974), assuming that emission source power is uniformly distributed within a mesh. The noise level is assumed to decrease according to the equation $\exp(-mr)/r$, where r is the distance from the source to the receptor and m is the decreasing coefficient depending on the area of buildings in the mesh.
3. It is possible to couple this model with the urban dynamic simulation model for future prediction of environmental noise levels.
4. LANDSAT MSS data are used to obtain the decreasing coefficient.

B. Estimation of the Decreasing Coefficient in Urban Areas by LANDSAT MSS Data

In occupied areas the density of buildings per mesh were estimated by LANDSAT MSS data, using the revised GMDH (AIC) to obtain the decreasing coefficient of noise level due to buildings.

LANDSAT MSS records four spectral bands, and the resolving power is about 80 m on the ground, as shown in Table 6.

Table 6 Intensity of Bands of LANDSAT MSS Data and Observed and Calculated Building Density of Occupied Areas to One Mesh

Mesh	Band 4 (0.5- 0.6μm)	Band 5 (0.6- 0.7μm)	Band 6 (0.7- 0.8μm)	Band 7 (0.8- 1.1μm)	Observed building density (%)	Calculated building density (%)	Error
1	22.9	26.1	27.1	18.2	64.1	72.0	-7.9
2	23.6	26.5	28.3	19.8	64.1	65.6	-1.5
3	23.7	27.3	30.7	22.4	56.5	57.8	-1.4
4	24.1	27.0	30.0	21.1	60.9	61.7	-0.8
5	25.1	28.3	30.8	21.8	68.7	63.0	5.6
6	24.6	27.9	28.9	20.0	65.3	69.0	-3.7
7	25.5	28.9	31.0	22.3	67.6	62.6	5.0
8	25.5	28.7	30.1	20.7	71.6	68.7	2.9
9	24.4	28.4	32.0	23.7	75.8	56.3	19.5
10	23.8	27.7	39.3	33.4	46.3	32.6	13.7
11	23.4	26.7	36.6	30.6	43.9	35.5	8.3
12	22.4	25.0	36.7	31.0	31.6	31.2	0.4
13	25.2	30.0	41.0	34.2	34.3	35.6	-1.4
14	22.4	24.6	41.0	36.6	25.6	22.6	3.0
15	25.0	29.1	40.6	33.9	36.2	34.6	1.5
16	26.4	30.9	38.8	30.9	44.4	43.6	0.8
17	23.2	25.7	33.4	26.6	62.5	41.4	21.1
18	25.6	30.1	39.0	31.4	51.1	41.0	10.1
19	25.7	29.7	36.5	28.3	52.4	47.0	5.5
20	25.6	29.1	37.5	30.0	17.9	41.8	-24.0
21	25.0	28.8	37.0	29.9	35.7	41.4	-5.6
22	25.7	29.8	39.2	31.6	12.7	40.2	-27.4
23	24.3	27.7	37.3	30.8	29.0	37.4	-8.3
24	23.5	26.7	37.7	31.3	24.3	34.2	-9.9
25	23.3	26.1	38.9	33.1	40.0	30.0	10.0
26	23.3	27.0	39.1	33.4	24.1	31.4	-7.3
27	22.4	25.3	37.4	31.6	39.3	30.8	8.4
28	20.3	20.8	37.8	34.9	4.8	17.6	-12.8
29	15.7	14.8	32.9	31.1	0.9	9.9	-9.0
30	12.3	9.9	33.3	33.0	0.0	-1.5	1.5
31	11.9	9.4	30.6	29.3	0.0	-0.4	0.4
32	12.3	9.9	33.4	33.5	0.0	-1.7	1.7
33	11.8	9.4	29.7	28.8	0.0	0.2	-0.2
34	17.9	18.4	47.3	47.2	1.3	4.4	-3.1
35	24.2	28.3	42.3	37.7	2.6	27.6	-25.0
36	25.5	29.0	37.8	31.1	40.1	39.3	0.8
37	23.9	26.4	38.5	32.8	39.8	31.2	8.6
38	18.2	18.2	46.6	46.0	18.2	4.8	13.4
39	18.8	18.7	44.9	43.3	13.4	7.1	6.3
40	19.8	20.8	42.3	39.6	18.1	13.1	5.0
41	17.7	17.7	47.1	47.7	10.8	3.2	7.7
42	18.0	20.3	43.1	40.1	0.0	11.9	-11.9

Source: Koi, 1982.

The building density of occupied areas to one mesh was determined from the aerial photograph of 42 meshes, and the relation between the rate and LANDSAT MSS data was determined using the rate as output variables. There were a total of eight input variables for intensity bands of LANDSAT MSS data and their corresponding inverses. Among the several equations, the following was recognized as being physically significant:

$$S = -20.72 + 64.4 \frac{\text{band 5}}{\text{band 7}} \quad (\%) \tag{7}$$

The equation indicates that density of occupied area per mesh can be estimated by the ratio of band 5 to band 7 (bi-band rate). With this equation a correlation coefficient for the calculated and observed values of 0.90 and a mean-squared error of 10.38 were obtained.

The rate for 63 meshes in which environmental noise levels were observed were calculated by Eq. (7), and the rates were substituted in the equation derived by Takagi (1979), $m = 145 - 1.33(100 - S)/100$ (where S is the rate of buildings per mesh) to obtain the value of the decreasing coefficient. The environmental noise levels (L50) were derived by the following equations:

$$L_a(i,j) = 10 \log J_a(i,j)$$

$$J_a(i,j) = P(i,j)\left[E(mR_0) - E(mR_t) + \frac{50}{a}\exp\left(-m\sqrt{\frac{a}{100}}R_0\right)\right]$$

$$+ \sum_k \sum_1 \frac{P(k,1)}{2\pi d^2}\exp(-md)$$

where

$J_a(i,j)$ = sound energy of a (i,j) mesh

$P(i,j)$ = sound power of a (i,j) mesh

$E(mR)$ = integral exponential function

R_0 = effective radius

$R_t = 500/\sqrt{\pi}$

d = distance between (i,j) mesh and $(k,1)$ mesh

$a = 50$

Figure 5 shows the differences between the calculated and observed values of the environmental noise levels (Koi et al., 1984).

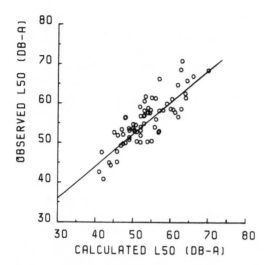

Fig. 5 Result of the prediction of environmental noise level. The regression equation of the calculated L50 value and the observed L50 value was $L50_{ob} = 12.4 + 0.81\ L50_{cal}$, and the correlation coefficient was 0.84, where the number of samples was 63.

V. CONCLUSION

The applicability of GMDH to environmental modeling was demonstrated in three cases.

In case 1, the applicability of GMDH was demonstrated in an emergency situation where a prediction of the Ox concentration was required.

In case 2, GMDH was applied to a reaction–diffusion analysis of COD where there was insufficient information given on the biochemical reaction rates and the resolution of sedimental deposits. However, if there were sufficient a reaction–diffusion model may be more effective than this applied model.

In case 3, the applicability of GMDH for the analysis of LANDSAT MSS data was demonstrated. The results here showed that the efficiency of GMDH was almost the same as that of the linear multiregression analysis (i.e., stepwise regression) when the input data were chosen appropriately.

The results of three cases showed that applicability of GMDH to environmental modeling are fairly good if care is taken in its practical application.

ACKNOWLEDGMENT

The authors wish to thank Professor Tamura and Mr. Kumazawa for their helpful advice.

APPENDIX

In this work, the number of intermediate variables selected was fixed equal to the number of input variables, so there were almost the same number of output equations as the number of input variables. Generally, the final output equation was chosen to be the weighted sum of all the output equations (Tamura and Kondo, 1978a,b).

In this case, Ox concentration 3 hours hence was expressed by the following equation:

$$Ox(t + 3) = 0.25\,\phi1 + 0.25\,\phi2 + 0.25\,\phi3 + 0.25\,\phi4 \quad \text{(ppb)} \qquad \text{(A-1)}$$

where

$$\phi1 = 4.6 - 2.284\,U3 + 3.089\,U5 + 0.027\,U3^2 - 0.025\,U5^2$$

$$\phi2 = -0.4 + 0.545\,U2 + 0.466\,U4$$

$$\phi3 = -0.4 + 0.622\,U2 + 0.389\,U3$$

$$\phi4 = -0.3 + 0.535\,U1 + 0.474\,U4$$

$$\text{(A-2)}$$

where

$$U1 = 3.3 + 0.750\,Z2 + 0.009\,Z2\,Z3 - 0.006\,Z2^2$$

$$U2 = 4.0 - 0.823\,Z1 + 1.60\,Z2 + 0.028\,Z1\,Z2 - 0.025\,Z2^2$$

$$U3 = 9.6 + 0.339\,Z3 + 0.008\,Z3\,Z5 \qquad \text{(A-3)}$$

$$U4 = 3.9 + 0.759\,Z3 + 0.48\,Z1\,Z3 - 0.018\,Z1^2 - 0.027\,Z3^2$$

$$U5 = 3.5 + 0.706\,Z4 - 0.002\,Z4\,Z5 + 0.006\,Z5^2$$

where

$$Z1 = 13.7 + 0.013\,Y1\,Y3$$

$$Z2 = 0.2 + 0.579\,Y1 + 0.284\,Y4 + 0.003\,Y1\,Y4$$

$$Z3 = 0.4 + 0.995\,Y3 + 0.058\,Y2\,Y3 - 0.024\,Y2^2 - 0.033\,Y3^2 \qquad \text{(A-4)}$$

$$Z4 = 1.4 + 0.834\,Y3 + 0.058\,Y2\,Y3 - 0.024\,Y2^2 - 0.033\,Y3^2$$

$$Z5 = Y1$$

where

$Y1 = 137.4 - 3.519\ H(t) - 3.301\ SR(t-1) + 0.77\ H(t)\ SR(t-1) + H(t)^2$

$Y2 = 18.9 + 6.129\ SR(t) - 0.013\ T(t-2)\ SR(t) - 0.018\ SR(t)^2$

$Y3 = 31.9 + 2.326\ SR(t) - 2.523\ y(t-1) - 0.070\ SR(t)\ y(t-1) - 0.025\ SR(t)^2$ $\qquad$ (A-5)

$Y4 = 472.3 + 4.324\ Ox(t) - 2.853\ T(t-2)\ 0.008\ Ox(t)\ T(t-2) - 0.008\ Ox(t)^2$
$\qquad + 0.004\ T(t-2)^2$

where Y, Z, and U were intermediate variables at first, second, and third layers, respectively, and input variables were shown in Table 4.

REFERENCES

Dresnack, R., and Dobbins, W. E. (1968). Numerical Analysis of BOD and DO Profiles. Proc. ASCE SA-5.

Duffy, J. J., and Franklin, M. A. (1975). A Learning Identification Algorithm and Its Application to an Environmental System. IEEE Trans. Syst. Man Cybern. SMC-5:226-236.

Fujita, S. (1984). The Calculation of Pollution of Osaka Bay by a Combined Model of Finite Difference Method and Revised GMDH. (In preparation.)

Fujita, S., Hayama, Y., and Masaki, Y. (1982). The Short Term Prediction of Air Pollutant—Mainly Ox Concentration. Environ. Conserv. Eng. 11: 183-187 (in Japanese).

Hayashi, C. (1952). On the Prediction of Phenomena from Quantitative Data and Quantifications of Qualitative Data from the Mathematical Point of View. Ann. Inst. Stat. Math. 3:69-92.

Ivakhnenko, A. G. (1958). The Group Method of Data Handling, a Revival of the Method of Stochastic Approximation. Sov. Autom. Control 1: 43-55.

Ivakhnenko, A. G., Grotov, G. I., and Visotsky, V. N. (1979). Identification of the Mathematical Model of a Complex System by the Self-Organization Method. In Theoretical Systems Ecology, E. Halfon (Ed.). Academic Press, New York, pp. 325-352.

Kaneko, Y., Horie, T., and Murakami, K. (1974). Tidal Flow and Diffusion of Pollution on Osaka Bay—Studies by Numerical Analysis. Report of the Port and Harbour Research Institute 13:138-149 (in Japanese).

Koi, H. (1982). Application of Remote Sensing Technique to an Environmental System. Environ. Conserv. Eng. 11:75-78 (in Japanese).

Koi, H., Aoi, I., Fujita, S., and Oinishi, T. (1984). The Environmental Noise Level Prediction System in Osaka. (In preparation.)

Lin, G. Y. (1982). Oxidant Prediction by Discriminant Analysis in the South Coastal Air Basic of California. Atmos. Environ. 16:135-134.

Rand Corporation (1973). A Three Dimensional Model for Estuary and Coastal Seas, Vol. 1: Principle of Computation. Rep. R-14170WRR.

Tabata, H. (1977). Simulation of Water Pollution for Enriched Seas. Environ. Creation 7(7).

Takagi, K. (1979). Consideration of the Issues of Estimation of Environmental Noise Level. Environ. Conserv. Eng. 8:1023-1030 (in Japanese).

Tamura, H., and Halfon, E. (1980). Identification of a Dynamic Model by the Group Method of Data Handling: An Application to Lake Ontario. Ecol. Model. 11:81-100.

Tamura, H., and Kondo, T. (1978a). Revised GMDH Algorithm Using Prediction Sum of Squares (PSS) as a Criterion for Model Selection. Trans. Soc. Instrum. Control. Eng. 14:519-524.

Tamura, H., and Kondo, T. (1978b). Nonlinear Modeling for Short-Term Prediction of Air Pollution Concentration by a Revised GMDH. Proc. Int. Conf. Cybern. Soc., IEEE Syst. Man Cybern. Soc., Tokyo and Kyoto, pp. 596-601.

U.S. Environmental Protection Agency (1971). Transportation Noise and Noise from Equipment Powered by Internal Combustion Engines. Natl. Tech. Inf. Serv. Rep. 300.13.

15

A FORTRAN Program for the GMDH Algorithm

STANLEY J. FARLOW / The University of Maine, Orono, Maine

I. GENERAL DESCRIPTION OF THE PROGRAMS

This chapter describes the computer programs GMDH and COMP, which perform most of the computations for the GMDH algorithm. In addition to these two subroutines, a main program is listed which reads the necessary input and calls on these two subroutines. The only restriction is that subroutine GMDH is called before subroutine COMP inasmuch as COMP uses the output of GMDH. The main program then prints out the relevant computed values.

A. Subroutine GMDH(N, M, NT, PI, X, Y, ITREE, TREE, ITER, NITER)

Subroutine GMDH constructs the Ivakhnenko polynomial from the input data arrays X, Y and the numbers N, M, NT, and PI. The parameters X, Y, N, M, and NT have been defined in Chap. 1, and PI is a new parameter which stands for the fractional increase in the number of new variables saved at each level (this is a switch from the procedure Ivakhnenko uses of saving those variables with values $r_j < R$). In other words, if the original number of dependent variables were M (PI must be $0 \leq PI \leq 1$), then after the first level the number of variables would be $M' = M + PI*M$ (and $M'' = M' + PI*M'$ after the next level, etc.).

The output of GMDH is the Ivakhnenko polynomial stored in the arrays TREE and ITREE. ITER is the number of steps it took for GMDH to reach an optimal polynomial (the RMIN curve reached a minimum). For example, if the tree was the one in Fig. 1, then the Ivakhnenko polynomial would be of degree 4 and given by

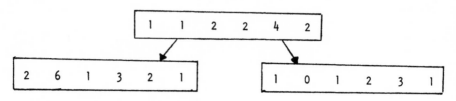

Fig. 1 Simple tree of polynomials.

$$y = 1 + u + 2v + 2u^2 + 4v^2 + 2uv$$

where

$$u = 2 + 6x_1 + x_2 + 3x_1^2 + 2x_2^2 + x_1x_2$$
$$v = 1 + x_4 + 2x_3^2 + 3x_4^2 + x_3x_4$$

Which of the four variables x_1, x_2, x_3, and x_4 of the m variables x_1, x_2, ..., x_m (m might be larger than 4) should be chosen in the polynomial is something the programmer must determine. The array ITREE will show how to construct the tree, while TREE contains the actual coefficients of the polynomial.

B. Subroutine COMP(ZZ, Y, ITREE, TREE, ITER, ITR)

Subroutine COMP uses the tree of polynomials that are stored in ITREE and TREE and evaluates the Ivakhnenko polynomial at the point ZZ (ZZ is an m vector). The output from COMP is the number Y, which is the computed value of the Ivakhnenko polynomial. The array ITR is a work array that need not concern the user.

    A listing of the main program and subroutine GMDH and COMP, together with more detailed descriptions of the subroutine parameters, can be found in the comment statements of the program listings.

II. SOME SAMPLE RUNS

The following data are taken from Draper and Smith's 1966 text, Applied Regression Analysis, page 218. (Draper and Smith took it from a paper by J. T. Wakely.)

x_1 = radiation in relative gram calories/min during the preceding half-day of sunlight (coded by dividing by 100)

x_2 = average soil moisture tension (coded by dividing by 100)

x_3 = air temperature in degrees Fahrenheit (coded by dividing by 10)

x_4 = milligrams of vitamin B_2 per gram of turnip greens

For this problem the variables read by the main program were

M   = number of variables = 3

N   = number of observations = 27

NT = number of observations in the training set = 14

PI = fractional increase in the number of variables saved = 0.0

Y   = the N dependent observations

X   = the N × M array of independent observations

```
THIS DATA IS TAKEN FROM PAGE 218 OF THE FIRST EDITION OF
DRAPER AND SMITH'S BOOK, APPLIED REGRESSION ANALYSIS
(JOHN WILEY AND SONS).

X(1) = RADIATION IN RELATIVE GRAM CALORIES/MIN
X(2) = AVERAGE SOIL TEMPERATURE
X(3) = AIR TEMPERATURE
Y = MILLIGRAMS OF VITAMIN B2.
```

```
---------------------------------------------------------------
NUMBER OF INDEP VAR =       3
NUMBER OF OBS =    27
NUMBER OF VAR IN TRAINING SET =      14
FRACTIONAL INCREASE IN VARIABLES =      0.0

---------------------------------------------------------------
PRINT INPUT DATA
```

OBS	Y	(X(I), I=1,M)		
1	110.40	1.76	0.07	7.80
2	102.80	1.55	0.07	8.90
3	101.00	2.73	0.07	8.90
4	108.40	2.73	0.07	7.20
5	100.70	2.56	0.07	8.40
6	100.30	2.80	0.07	8.70
7	93.70	1.84	0.07	8.70
8	96.60	1.98	0.02	7.60
9	96.20	0.80	0.02	6.70
10	99.00	0.80	0.02	6.20
11	88.40	1.05	0.02	7.00
12	75.30	1.80	0.02	7.30
13	77.10	2.30	0.02	8.20
14	65.70	1.91	0.47	8.30
15	56.80	1.91	0.47	8.20
16	62.10	1.91	0.47	6.90
17	53.20	2.13	0.47	7.60
18	59.40	2.13	0.47	6.90
19	58.70	1.51	0.47	7.50
20	58.00	2.05	0.47	7.60
21	102.80	2.80	0.07	7.40
22	98.90	2.16	0.07	8.80
23	99.40	0.59	0.02	6.50
24	92.00	1.80	0.02	6.50
25	82.40	1.77	0.02	7.60
26	74.00	2.03	0.47	7.60
27	61.00	0.76	0.47	7.40

Fig. 2 Input for the problem. (From Draper and Smith.)

```
LEVEL NUMBER =    1
NO. VARIABLES SAVED =    3
RMIN VALUE (SUMMED OVER CHECKING SET) =  0.52513E-02
MULTIPLE CORRELATION (SUMMED OVER TRAINING SET) =   0.82

GMDH CONVERGED AFTER    1 GENERATION(S)
MULTIPLE CORRELATION (SUMMED OVER TRAINING SET) =   0.82
```

CASE NO.	OBSERVED VALUE	ESTIMATE	ERROR	PERCENT ERROR
1	0.11039999E 03	0.10569897E 03	0.47010193E 01	0.42581682E 01
2	0.10279999E 03	0.99375320E 02	0.34246674E 01	0.33313866E 01
3	0.10101000E 03	0.99375320E 02	0.16246796E 01	0.16085930E 01
4	0.10833999E 03	0.11028885E 03	-0.18888550E 01	-0.17424860E 01
5	0.10070000E 03	0.10191423E 03	-0.12142334E 01	-0.12057924E 01
6	0.10029999E 03	0.10032378E 03	-0.23788452E-01	-0.23717303E-01
7	0.93699997E 02	0.10032378E 03	-0.66237795E 01	-0.70691347E 01
8	0.96599991E 02	0.84281219E 02	0.12318771E 02	0.12752350E 02
9	0.96199997E 02	0.93116394E 02	0.30836029E 01	0.32054062E 01
10	0.99000000E 02	0.98807587E 02	0.19241333E 00	0.19435686E 00
11	0.88399994E 02	0.89970047E 02	-0.15700531E 01	-0.17760782E 01
12	0.75299988E 02	0.87024994E 02	-0.11725006E 02	-0.15571057E 02
13	0.77099991E 02	0.79397507E 02	-0.22975159E 01	-0.29799166E 01
14	0.65699997E 02	0.65700195E 02	-0.19836426E-03	-0.30192430E-03
15	0.56799988E 02	0.65189972E 02	-0.83899841E 01	-0.14771099E 02
16	0.62099991E 02	0.60592392E 02	0.15075989E 01	0.24276953E 01
17	0.53199997E 02	0.62598358E 02	-0.93983612E 01	-0.17666077E 02
18	0.59399994E 02	0.60592392E 02	-0.11923981E 01	-0.20074043E 01
19	0.58699997E 02	0.62244720E 02	-0.35447235E 01	-0.60387077E 01
20	0.58000000E 02	0.62598358E 02	-0.45983582E 01	-0.79282007E 01
21	0.10279999E 03	0.10866942E 03	-0.58694305E 01	-0.57095613E 01
22	0.98899994E 02	0.99839348E 02	-0.93835449E 00	-0.94879121E 00
23	0.99399994E 02	0.95325790E 02	0.40742035E 01	0.40987940E 01
24	0.92000000E 02	0.95325790E 02	-0.33257904E 01	-0.36149874E 01
25	0.82399994E 02	0.84281219E 02	-0.18812256E 01	-0.22830400E 01
26	0.74000000E 02	0.62598358E 02	0.11401642E 02	0.15407622E 02
27	0.61000000E 02	0.61913406E 02	-0.91340637E 00	-0.14973869E 01

```
IVAKHNENKO POLYNOMIAL

Y =  A + B*U + C*V + D*U*U + E*V*V + F*U*V
A =    0.2091E 03
B =    0.3605E 03
C =   -0.2636E 02
D =   -0.1263E 04
E =    0.1118E 01
F =    0.2769E 02


U = X( 2 )      V = X( 3 )
```

Fig. 3 Output from GMDH and COMP.

After this information was read by the main program, it was printed as shown in Fig. 2.

The problem here is to find the model that predicts the variable Y. These data were run both by the author under subroutines GMDH and COMP and by a commercial system developed by Scientific Systems Inc. of Cambridge, Massachusetts. SSI is one of the few consulting companies that do advanced modeling with GMDH. Another is Adaptronics, Inc. of McLean, Virginia. The systems developed by SSI and Adaptronics are more general than the author's basic package and have more flexibility. However, to help the reader to get GMDH and COMP running on one's own system, it is the author's output that is listed here. The output of the main program after the input data of Fig. 2 have been run through GMDH and COMP is shown in Fig. 3.

In this simple example the GMDH algorithm converged in one step, the RMIN curve reaching a minimum of 0.00525. The square of the multiple correlation coefficient (summed over the 20 observations in the training set) was found to be $r^2 = 0.82$, which means that 82% of the variation of the dependent variable y is explained by the simple quadratic Ivakhnenko polynomial

$$y = 209 + 360x_2 - 26x_3 - 1263x_2^2 + 1.1x_3^2 + 28x_2x_3$$

Draper and Smith obtained the regression polynomial

$$y = 120 + 490x_2 - 5.7x_3 - 1107x_2^2$$

using a stepwise regression technique stopping when $r^2 = 0.97$. The program developed by Scientific Systems Inc. uses a variation of the basic GMDH algorithm which yielded the Ivakhnenko polynomial

$$y = 120 + 516x_2 - 5.7x_3 - 1167x_2^2$$

with a choice of NT = 20. SSI found that $r^2 = 0.90$.

The model that the SSI program chose is essentially the same as that obtained by Draper and Smith. The reason the author's program and SSI's did not yield the same Ivakhnenko polynomial was that different selection criteria were used (SSI's program also saves linear terms at each level in addition to the quadratic equations).

III. FORTRAN LISTINGS

A few notes on the following programs are first presented.

1. Subroutine GMDH uses the parameter PI ≥ 0 to determine how many new variables are to be kept at each level. The number of new variables NV2 at level k + 1 is NV2 = NV1 + PI*NV1, where NV1 is the number of variables at the preceding level, k. (If PI = 0.5, the number of variables grows by 50% at each level.) If the number of variables reaches 75, the program will keep the number fixed at that value (due to dimension statements).

There are, of course, other heuristics for determining how many of the new variables one should save, but this strategy works well.

2. The subroutine STAT(N, Y, X, RMS) computes the statistic

$$RMS = \frac{\Sigma (y_i - x_i)^2}{\Sigma y_i^2}$$

where the y_i and x_i, i = 1, 2, ..., N are the observed and computed values in the checking set. It is this statistic that orders the new variables, and the smallest one of these values is called RMIN. Users can change this sub-routine to other criteria if they wish.

3. Users can make the GMDH algorithm stop after a prescribed number of levels by selection of the input variable NITER. If NITER is nonzero, the number of levels performed by GMDH is NITER. If NITER = 0, then GMDH stops when the RMIN curve is a minimum.

4. The following programs are listed:

Main program (reads all data)
Subroutine GMDH (called by main program)
Subroutine COMP (called by main program)
Subroutine SORT (called by GMDH) sorts the values of r_j
Subroutine SYS (called by GMDH) solves a system of linear equations
Subroutine INTER (called by SYS) interchanges rows of a matrix
Subroutine STAT (called by GMDH) computes the statistic r_j (can be
     changed by the user)

The input to the main program is as follows:

M = number of independent variables

N = number of observations

NT = number of observations in the training set

NITER = 0; GMDH algorithm stops when the RMIN curve reaches a minimum
      ≠ 0; number of levels performed by GMDH algorithm

NPRNT = 0 if input observations are not printed
      ≠ 0 if input observations are printed

PI = fractional increase in the number of variables at each level (if
     PI = 0, the number of variables remains the same at each level;
     if PI = 0.5, the number of variables will increase by 50% each
     level). Subroutine GMDH will change the number of variables
     according to this parameter with the exception that the number
     of new variables will always be less than $M(M - 1)/2$, where M
     is the number of variables at the preceding level. Also, the total
     number of variables will not be allowed to be greater than 75 as
     a result of dimension statements.

Y(I) = dependent variable, I = 1, 2, ..., N

X(I, J) = independent variables, I = 1, 2, ..., N   J = 1, 2, ..., M

5. The main program will print the input data plus the value of the Ivakhnenko polynomial at the values of the input variables. If users wish, they may modify the main program by reading in new independent variables $z_1, \ldots, z_m$ and adding another CALL COMP(Z, DEP, ITREE, TREE, ITER,

ITR). The variables ITREE, TREE, ITER, and ITR have already been
computed. The value of DEP will be the evaluated value of the Ivakhnenko
polynomial at $z_1, \ldots, z_m$.

The programs listed here print the Ivakhnenko polynomial only when it
is a simple quadratic. That is, this sequence of programs is meant to be
used for prediction purposes and not for identification of relevant variables
in the model.

```
00010 C
00020 C
00030 C    THIS MAIN PROGRAM READS IN THE DATA ARRAYS NECESSARY TO FIND THE
00040 C    IVAKHNENKO POLYNOMIAL (IP).  SUBROUTINE GMDH FINDS THE IP AND STORES
00050 C    IT IN THE ARRAYS TREE, ITREE, AND ITER.  THESE ARRAYS ARE THEN USED
00060 C    BY SUBROUTINE COMP TO EVALUATE THE IP AT THE ORIGINAL DATA POINTS.
00070 C    THE USER CAN EASILY MODIFY THIS PROGRAM BY EITHER ADDING A
00080 C    PREPROCESSOR SUBROUTINE TO TRANSFORM THE DATA BEFORE GMDH OR BY
00090 C    ADDING ANOTHER READ STATEMENT TO EVALUATE THE IP AT OTHER POINTS.
00100 C
00110 C
00120 C    INPUT DATA IS  M = NUMBER OF VARIABLES
00130 C                   N = NUMBER OF DATA POINTS
00140 C                  NT = NUMBER OF DATA POINTS IN THE TRAINING SET (THE FIRST
00150 C                       NT DATA POINTS ARE ASSUMED TO BE IN THE TRAINING SET)
00160 C               NITER = NUMBER OF LEVELS GMDH PERFORMS BEFORE STOPPING
00170 C                       (IF NITER = 0 GMDH DECIDES ITSELF)
00180 C               NPRNT = 0  IF INPUT IS NOT PRINTED
00190 C                     = 1  IF INPUT IS PRINTED
00200 C                  PI = FRACTIONAL INCREASE IN THE NUMBER OF VARIABLES
00210 C                       AT EACH ITERATION (SHOULD BE BETWEEN 0 AND 1)
00220 C                   X = N BY M ARRAY OF INDEPENDENT VARIABLES
00230 C                   Y = N ARRAY OF DEP VARIABLES
00240 C
00250 C    ORDER OF INPUT: CARDS 1-10 FOLLOW INFO TO BE PRINTED
00260 C    CARD 11   M,N,NT,PI    FORMAT(3I5,F10.2)
00270 C    CARDS 12-  THE I TH CARD CONTAINS  Y(I), (X(I,J),J=1,M)   I=1,N
00280 C
00290        DIMENSION X(100,30),Y(100),ITREE(10,75),TREE(10,75,6)
00300       &,ZZ(30),EV(100,30),YSAVE(100),ITR(256),DATA(80)
00310 C
00320 C      READ NO. OF VARIABLES,NO. OF OBSERVATIONS,NO. OF OBSERVATIONS
00330 C          IN THE TRAINING SET, RMS VALUE.
00340 C
00350        WRITE(3,4)
00360        DO 60 I = 1,10
00370        READ(1,42) DATA
00380     42 FORMAT(80A1)
00390        WRITE(3,45) DATA
00400     45 FORMAT(10X,80A1)
00410     60 CONTINUE
00420        READ (1,1) M,N,NT,NITER,NPRNT,PI
00430      1 FORMAT(5I5,F10.2)
00440        DO 2 I = 1,N
00450 C
00460 C      READ DEP. VARIABLE, M INDEPENDENT VARIABLES.
00470 C
00480        READ(1,3) Y(I),(X(I,J),J=1,M)
00490      2 YSAVE(I) = Y(I)
00500      3 FORMAT(8F10.2)
00510 C
00520      4 FORMAT(1H1)
00530        WRITE(3,43)
00540     43 FORMAT('  -------------------------------------------------------
00550       &-------------------')
00560        WRITE(3,5) M,N
00570      5 FORMAT(' NUMBER OF INDEP VAR = ',I4/' NUMBER OF OBS = ',I4)
00580        WRITE(3,6) NT
00590      6 FORMAT(' NUMBER OF VAR IN TRAINING SET = ',I4)
00600        WRITE(3,7) PI
00610      7 FORMAT(' FRACTIONAL INCREASE IN VARIABLES = ',F8.3//)
00620        WRITE(3,43)
00630        IF(NPRNT.EQ.0.) GO TO 32
00640        WRITE(3,30)
00650     30 FORMAT(' PRINT INPUT DATA'//)
00660        WRITE(3,8)
00670      8 FORMAT(3X,' OBS       Y        (X(I),I=1,M)'/)
00680        DO 9 I = 1,N
00690      9 WRITE(3,10) I, Y(I),(X(I,J),J=1,M)
00700     10 FORMAT(I6,10F7.2)
00710        WRITE(3,4)
00720 C      SAVE THE DATA ARRAY FOR LATER
```

```
00730      32 DO 11 I = 1,N
00740         DO 11 J = 1,M
00750      11 EV(I,J) = X(I,J)
00760    C
00770    C      CALL GMDH SUBROUTINE.
00780    C
00790         CALL GMDH(N,M,NT,PI,X,Y,ITREE,TREE,ITER,NITER)
00800         WRITE(3,40)
00810      40 FORMAT(' CASE NC.    CBSERVED VALUE      ESTIMATE        ERROR
00820        C       PERCENT ERROR'/)
00830         DO 12 I = 1,N
00840         DO 13 J = 1,M
00850      13 ZZ(J) = EV(I,J)
00860    C
00870    C      CALL SUBROUTINE TC EVALUATE THE IVAKHNENKO POLYNOMIAL
00880    C
00890         CALL COMP(ZZ,YY,ITREE,TREE,ITER,ITR)
00900         ER = YSAVE(I) - YY
00910         PERER = 100.*ER/YSAVE(I)
00920      12 WRITE(3,14) I,YSAVE(I),YY,ER,PERER
00930      14 FORMAT(I5,3X,4E18.8)
00940         WRITE(3,43)
00950         WRITE(3,4)
00960    C THE IVAKHNENKO POLYNOMIAL IS PRINTED ONLY IF IT IS A SIMPLE
00970    C CUADRATIC
00980         IF(ITER.GT.1) STCP
00990         WRITE(3,70) (TREE(1,1,J), J = 1,6)
01000      70 FORMAT(//' IVAKHNENKO POLYNOMIAL '//
01010        *'  Y = A + B*L + C*V + D*L*U + E*V*V + F*U*V '/
01020        *'  A = ',E12.4/' B = ',E12.4/' C = ',E12.4/
01030        *'  D = ',E12.4/' E = ',E12.4/' F = ',E12.4//)
01040         WRITE (3,71) ITR(2),ITR(3)
01050      71 FORMAT(' U = X(',I2,' )      V = X(',I2,' )')
01060         WRITE(3,4)
01070         STOP
01080         END

02010         SUBROUTINE GMDH(N,M,NT,PI,X,Y,ITREE,TREE,ITER,NITER)
02020    C    -------------------------------------------------------------
02030    C
00040    C GMDH COMPUTES THE IVAKHNENKO POLYNOMIAL FROM THE N BY M DATA ARRAY
02050    C X (INDEP VAR) AND THE N ARRAY Y (DEP VAR).   THE IP IS STORED IN
00060    C THE ARRAY TREE UPON EXIT AND THE ARRAY ITREE PROVIDES THE
00070    C POINTERS FOR TREE.   ITER IS THE NUMBER OF ITERATIONS IT TOOK FOR
00080    C GMDH TO CONVERGE.   TREE, ITREE, AND ITER ARE OUTPUT FROM GMDH
00090    C AND ARE NEEDED FOR THE SUBROUTINE COMP.
00100    C    -------------------------------------------------------------
00110    C
00120    C      DEFINITION OF VARIABLES IN THE CALLING STATEMENT.
00130    C
00140    C      N = NUMBER OF OBSERVATIONS
00150    C      M = NUMBER OF VARIABLES
00160    C      NT = NUMBER OF OBSERVATIONS IN THE TRAINING SET
00170    C      PI = FRACTICNAL INCREASE IN THE NUMBER OF VARIABLES AT
00180    C          EACH ITERATION
00190    C
00200    C      X = N BY M ARRAY: THE I TH ROW CONTAINS THE I TH OBSERVATION
00210    C      OF THE M DEPENDENT VARIABLES
00220    C
00230    C      Y = N VECTOR: THE I TH ELEMENT CONTAINS THE I TH DEP VARIABLE
00240    C
00250    C
00260    C      ITREE = A TWO DIMENSIONAL ARRAY WHOSE ROW DIMENSION MUST BE
00270    C      AT LEAST AS LARGE AS THE NUMBER OF ITERATIONS PERFORMED AND
00280    C      WHOSE COLUMN DIMENSION MUST BE AT LEAST AS LARGE AS THE MAX
00290    C      NUMBER OF VARIABLES SAVED AT ANY ITERATION.
00300    C
00310    C      TREE = A THREE DIM ARRAY WHOSE FIRST TWO DIMENSIONS ARE THE
00320    C      SAME AS ITREE AND WHOSE THIRD DIMENSION IS ALWAYS SIX.
00330    C
00340    C      ITER = OUTPUT VARIABLE EQUAL TO THE NO. OF ITERATIONS PERFORMED
00350    C      NITER = NUMBER OF LEVELS GMDH PERFORMS BEFORE STOPPING
00360    C          (IF NITER = 0 GMDH DECIDES ITSELF)
00370    C
00380         DIMENSION X(100,30),Y(100),TREE(10,75,6),ITREE(10,75)
00390        &,POLY(6,75),XTX(6,7),XTY(6),INDEX(435),WORK(100,75)
00400        &,D(435),IND(435),ZZZ(6),MA(20),WRK(100)
00410         DIMENSION XWORK(100),YWORK(100)
00420         RMS = PI
00430         NTP1 = NT + 1
00440         NC = N - NT
00450         NITER = NITER + 1
00460         MM = M
00470         ITER = 1
00480         DMIN = 1.0 E20
00490      28 L = 1
00500         MM1 = M-1
00510    C
00520    C      BEGIN LOOP TO COMPUTE NEW VARIABLES
00530    C
00540         DO 4I = 1,MM1
00550         IP1 = I+1
```

```
00560              DO 4 J = IPL,M
00570              DO 5 II = 1,6
00580              XTY(II) = 0.0
00590              DO 5 JJ = 1,6
00600       5      XTX(II,JJ) = 0.
00610              XTX(1,1) = NT
00620              DO 7 K = 1,NT
00630              XTX(1,2) = XTX(1,2) + X(K,I)
00640              XTX(1,3) = XTX(1,3) + X(K,J)
00650              XTX(1,4) = XTX(1,4) + X(K,I)**2
00660              XTX(1,5) = XTX(1,5) + X(K,J)**2
00670              XTX(1,6) = XTX(1,6) + X(K,I)*X(K,J)
00680              XTX(2,2) = XTX(2,2) + X(K,I)**2
00690              XTX(2,3) = XTX(2,3) + X(K,I)*X(K,J)
00700              XTX(2,4) = XTX(2,4) + X(K,I)**3
00710              XTX(2,5) = XTX(2,5) + X(K,I)*X(K,J)**2
00720              XTX(2,6) = XTX(2,6) + X(K,I)*X(K,I)*X(K,J)
00730              XTX(3,3) = XTX(3,3) + X(K,J)**2
00740              XTX(3,4) = XTX(3,4) + X(K,I)*X(K,I)*X(K,J)
00750              XTX(3,5) = XTX(3,5) + X(K,J)**3
00760              XTX(3,6) = XTX(3,6) + X(K,J)**2
00770              XTX(4,4) = XTX(4,4) + X(K,I)**4
00780              XTX(4,5) = XTX(4,5) + (X(K,I)*X(K,J))**2
00790              XTX(4,6) = XTX(4,6) + X(K,I)**3
00800              XTX(5,5) = XTX(5,5) + X(K,J)**4
00810              XTX(5,6) = XTX(5,6) + X(K,I)*X(K,J)**3
00820              XTX(6,6) = XTX(6,6) + (X(K,I)*X(K,J))**2
00830              XTY(1) = XTY(1) + Y(K)
00840              XTY(2) = XTY(2) + X(K,I)*Y(K)
00850              XTY(3) = XTY(3) + X(K,J)*Y(K)
00860              XTY(4) = XTY(4) + X(K,I)*X(K,I)*Y(K)
00870              XTY(5) = XTY(5) + X(K,J)*X(K,J)*Y(K)
00880       7      XTY(6) = XTY(6) + X(K,I)*X(K,J)*Y(K)
00890              DO 45 II = 2,6
00900              IM1 = II-1
00910              DO 45 JJ = 1,IM1
00920       45     XTX(II,JJ) = XTX(JJ,II)
00930  C
00940  C          SOLVE REGRESSION EQUATIONS FOR VARIABLES I AND J.
00950  C
00960              CALL SYS(XTX,ZZZ,XTY,6,IFLAG)
00970              IF(IFLAG.EQ.1) GO TO 41
00980              DO 40 III = 1,6
00990       40     POLY(III,L) = ZZZ(III)
01000              DO 15 K = 1,N
01010              WW = POLY(1,L) + POLY(2,L)*X(K,I) + POLY(3,L)*X(K,J)
01020              WW = WW + POLY(4,L)*X(K,I)**2 + POLY(5,L)*X(K,J)**2
01030              WW = WW + POLY(6,L)*X(K,I)*X(K,J)
01040       15     WORK(K,L) = WW
01050              IND(L) = 100*(I+10) + (J+10)
01060              IF(L.EQ.75) GO TO 385
01070              L = L+1
01080       41     CONTINUE
01090       4      CONTINUE
01100  C
01110  C          COMPLETED CONSTRUCTION OF M*(M-1)/2 NEW VARIABLES
01120  C
01130              L = L-1
01140       385    DO 120 I = 1,NC
01150       120    YWORK(I) = Y(NT+I)
01160              DO 122 J = 1,L
01170              DO 124 I = 1,NC
01180       124    XWORK(I) = WORK(NT+I,J)
01190  C
01200  C          COMPUTE THE GOODNESS OF FIT STATISTIC
01210  C
01220              CALL STAT(NC,YWORK,XWORK,ST)
01230              D(J) = ST
01240       122    INDEX(J) = J
01250  C
01260  C          SORT THE VALUES OF THE STATISTICS FROM LOW TO HIGH
01270  C
01280              CALL SORT(D,L,INDEX)
01290              M = M + RMS*M
01300              IF(M.GT.L) M=L
01310       53     CONTINUE
01320  C     THE LARGEST NUMBER OF VAR IS SET TO 75.
01330              IF(M.GT.75) M = 75
01340              IF(M.LT.MM) M = MM
01350              DO 22 J = 1,M
01360              ITREE(ITER,J) = IND(INDEX(J))
01370              DO 23 K = 1,6
01380       23     TREE(ITER,J,K) = POLY(K,INDEX(J))
01390       22     CONTINUE
01400  C
01410  C          TEST FOR CONVERGENCE OF GMDH ALGORITHM
01420  C
01430              IF(NITER.EQ.1) GO TO 55
01440              IF(ITER.EQ.NITER) GO TO 60
01450              GO TO 56
```

```
01460     55 TEST = D(1) - DMIN + .0005
01470        IF(TEST.GE.0.) GO TO 60
01480     56 WRITE(3,300)
01490    300 FORMAT(' --------------------------------------------------
01500       &---------------------')
01510        WRITE(3,200) ITER
01520    200 FORMAT(' LEVEL NUMBER = ',I3)
01530        WRITE(3,203) M,D(1)
01540    203 FORMAT(' NO. VARIABLES SAVED = ',I3/' RMIN VALUE (SUMMED ',
01550       &'OVER CHECKING SET) = ',E12.5)
01560        DMIN = D(1)
01570        MA(ITER) = M
01580        ITER = ITER + 1
01590  C MAX NUMBER OF ITERATIONS IS 11.
01600        DO 26 I = 1,N
01610        DO 26 J = 1,M
01620     26 X(I,J) = WORK(I,INDEX(J))
01630        SUM = 0.
01640        DO 190 I = 1,NT
01650    190 SUM = SUM + Y(I)
01660        SUM = SUM/NT
01670        SUM1 = 0.
01680        SUM2 = 0.
01690        DO 191 I = 1,NT
01700        SUM1 = SUM1 + (SUM - X(I,1))**2
01710    191 SUM2 = SUM2 + (Y(I) - SUM)**2
01720        SUM = SUM1/SUM2
01730        WRITE(3,92) SUM
01740        WRITE(3,300)
01750        GO TO 28
01760     60 IF(ITER.EQ.1) GO TO 65
01770        ITER = ITER - 1
01780     65 M = MM
01790        WRITE(3,130) ITER
01800    130 FORMAT( /' GMDH CONVERGED AFTER ',I3,' GENERATION(S)')
01810        SUM = 0.
01820        DO 90 I = 1,NT
01830     90 SUM = SUM + Y(I)
01840        SUM = SUM/NT
01850        SUM1 = 0.
01860        SUM2 = 0.
01870        DO 91 I = 1,NT
01880        SUM1 = SUM1 + (SUM-X(I,1))**2
01890     91 SUM2 = SUM2 + (Y(I) - SUM)**2
01900        SUM = SUM1/SUM2
01910        WRITE(3,92) SUM
01920     92 FORMAT(' MULTIPLE CORRELATION (SUMMED OVER TRAINING SET) = '
01930       &,F6.2)
01940        WRITE(3,300)
01950        RETURN
01960        END
01970        SUBROUTINE SORT(A,LA,IR)
01980  C
01990  C ----------------------------------------------------------------
02000  C THIS IS A SORT PROGRAM USED BY GMDH (TAKEN FROM THE IBM LIBRARY)
02010  C ----------------------------------------------------------------
02020  C
02030        INTEGER        LA,IR(LA)
02040        REAL           A(LA)
02050  C
02060        INTEGER        IU(21),IL(21),I,M,J,K,IJ,IT,L,ITT  SPECIFICATIONS FOR LOCAL VARIABLES
02070        REAL           T,TT,R
02080  C                                    FIRST EXECUTABLE STATEMENT
02090  C                                    FIND ABSOLUTE VALUES OF ARRAY A
02100        IF (LA.LE.0) RETURN
02110        DO 5 I=1,LA
02120        IF(A(I).LT.0.) A(I)=-A(I)
02130      5 CONTINUE
02140        M=1
02150        I=1
02160        J=LA
02170        R=.375
02180     10 IF (I .EQ. J) GO TO 55
02190     15 IF (R .GT. .5898437) GO TO 20
02200        R=R+3.90625E-2
02210        GO TO 25
02220     20 R=R-.21875
02230     25 K=I
02240  C                                    SELECT A CENTRAL ELEMENT OF THE
02250  C                                    ARRAY AND SAVE IT IN LOCATION T
02260        IJ=I+(J-I)*R
02270        T=A(IJ)
02280        IT=IR(IJ)
02290  C                                    IF FIRST ELEMENT OF ARRAY IS GREATER
02300  C                                    THAN T, INTERCHANGE WITH T
02310        IF (A(I) .LE. T) GO TO 30
02320        A(IJ)=A(I)
02330        A(I)=T
02340        T=A(IJ)
02350        IR(IJ)=IR(I)
02360        IR(I)=IT
```

```
02370          IT=IR(IJ)
02380    30  L=J
02390  C                                    IF LAST ELEMENT OF ARRAY IS LESS THAN
02400  C                                    T, INTERCHANGE WITH T
02410          IF (A(J) .GE. T) GO TO 40
02420          A(IJ)=A(J)
02430          A(J)=T
02440          T=A(IJ)
02450          IR(IJ)=IR(J)
02460          IR(J)=IT
02470          IT=IR(IJ)
02480  C                                    IF FIRST ELEMENT OF ARRAY IS GREATER
02490  C                                    THAN T, INTERCHANGE WITH T
02500          IF (A(I) .LE. T) GO TO 40
02510          A(IJ)=A(I)
02520          A(I)=T
02530          T=A(IJ)
02540          IR(IJ)=IR(I)
02550          IR(I)=IT
02560          IT=IR(IJ)
02570          GO TO 40
02580    35  IF (A(L).EQ.A(K)) GO TO 40
02590          TT=A(L)
02600          A(L)=A(K)
02610          A(K)=TT
02620          ITT=IR(L)
02630          IR(L)=IR(K)
02640          IR(K)=ITT
02650  C                                    FIND AN ELEMENT IN THE SECOND HALF OF
02660  C                                    THE ARRAY WHICH IS SMALLER THAN T
02670    40  L=L-1
02680          IF (A(L) .GT. T) GO TO 40
02690  C                                    FIND AN ELEMENT IN THE FIRST HALF OF
02700  C                                    THE ARRAY WHICH IS GREATER THAN T
02710    45  K=K+1
02720          IF (A(K) .LT. T) GO TO 45
02730  C                                    INTERCHANGE THESE ELEMENTS
02740          IF (K .LE. L) GO TO 35
02750  C                                    SAVE UPPER AND LOWER SUBSCRIPTS OF
02760  C                                    THE ARRAY YET TO BE SORTED
02770          IF (L-I .LE. J-K) GO TO 50
02780          IL(M)=I
02790          IU(M)=L
02800          I=K
02810          M=M+1
02820          GO TO 60
02830    50  IL(M)=K
02840          IU(M)=J
02850          J=L
02860          M=M+1
02870          GO TO 60
02880  C                                    BEGIN AGAIN ON ANOTHER PORTION OF
02890  C                                    THE UNSORTED ARRAY
02900    55  M=M-1
02910          IF (M .EQ. 0) RETURN
02920          I=IL(M)
02930          J=IU(M)
02940    60  IF (J-I .GE. 11) GO TO 25
02950          IF (I .EQ. I) GO TO 10
02960          I=I-1
02970    65  I=I+1
02980          IF (I .EQ. J) GO TO 55
02990          T=A(I+1)
03000          IT=IR(I+1)
03010          IF (A(I) .LE. T) GO TO 65
03020          K=I
03030    70  A(K+1)=A(K)
03040          IR(K+1)=IR(K)
03050          K=K-1
03060          IF (T .LT. A(K)) GO TO 70
03070          A(K+1)=T
03080          IR(K+1)=IT
03090          GO TO 65
03100          END
03110          SUBROUTINE SYS(A,X,B,N,IFLAG)
03120  C
03130  C  ----------------------------------------------------------------
03140  C  SUBROUTINE SYS AND INTER FIND THE SOLUTION OF A 6 BY 6 LINEAR SYSTEM.
03150  C  SYS IS CALLED BY GMDH TO FIND THE LEAST SQUARES SOLUTION.
03160  C  ----------------------------------------------------------------
03170  C
03180          DIMENSION A(6,7),X(6),INDEX(6),B(6)
03190          NP1 = N + 1
03200          DO 11 I = 1,N
03210          INDEX(I) = I
03220    1  A(I,NP1) = B(I)
03230          DO 5 K = 1,N
03240          IF(K.EQ.N) GO TO 2
03250          CALL INTER(A,N,K,INDEX)
```

```
03260        2    KP1 = K + 1
03270             DO 4J = KP1,NP1
03280             IF(ABS(A(K,K)).LT..000001) GO TO 8
03290             A(K,J) = A(K,J)/A(K,K)
03300             DO 4I = 1,N
03310             IF(K-I)3,4,3
03320        3    A(I,J) = A(I,J)-A(I,K)*A(K,J)
03330        4    CONTINUE
03340        5    CONTINUE
03350             DO 7I = 1,N
03360             DO 6J = 1,N
03370             IF(INDEX(J).NE.I) GO TO 6
03380             X(I) = A(J,NP1)
03390             GO TO 7
03400        6    CONTINUE
03410        7    CONTINUE
03420             IFLAG = 0
03430             RETURN
03440        8    IFLAG = 1
03450             RETURN
03460             END
03470             SUBROUTINE INTER(A,N,K,INDEX)
03480    C
03490    C     -------------------------------------------------------------
03500    C     SUBROUTINE INTER INTERCHANGES ROWS AND COLUMNS OF A MATRIX
03510    C     AND IS CALLED BY SUBROUTINE SYS
03520    C     -------------------------------------------------------------
03530    C
03540             DIMENSION A(6,7),INDEX(6)
03550             NP1 = N + 1
03560             NR = K
03570             NC = K
03580             AB = ABS(A(K,K))
03590             DO 2I = K,N
03600             DO 2J = K,N
03610             IF(ABS(A(I,J))-AB) 2,2,1
03620        1    AB = ABS(A(I,J))
03630             NR = I
03640             NC = J
03650        2    CONTINUE
03660             DO 3J = K,NP1
03670             DE = A(NR,J)
03680             A(NR,J) = A(K,J)
03690        3    A(K,J) = DE
03700             DO 4I = 1,N
03710             DE = A(I,NC)
03720             A(I,NC) = A(I,K)
03730        4    A(I,K) = DE
03740             IS = INDEX(NC)
03750             INDEX(NC) = INDEX(K)
03760             INDEX(K) = IS
03770             RETURN
03780             END
00010            SUBROUTINE CCMP(X,Y,ITREE,TREE,ITER,ITR)
00020    C
00030    C     -------------------------------------------------------------
00040    C     SUBROUTINE CCMP EVALUATES THE IF AT THE POINT X(1),...,X(M)
00050    C     CCMP NEEDS THE INPUT ARRAYS, ITREE, TREE, AND ITER THAT WERE
00060    C     COMPUTED BY GMDH.  THE ONLY OUTPUT IS THE VALUE OF THE IP Y.
00070    C     THE USER COULD CALL ON CCMP ANY TIME AFTER GMDH IF HE/SHE
00080    C     WISHED TO EVALUATE THE IF AT SOME POINT.
00090    C     -------------------------------------------------------------
00100    C
00110    C     X = INPUT VECTOR THAT CONTAINS THE VECTOR THAT WILL BE
00120    C     EVALUATED (AN M VECTOR WHERE M IS THE SAME AS IN GMDH).
00130    C
00140    C     Y = OUTPUT VALUE OF THE IVAKHNENKO POLYNOMIAL.
00150    C
00160    C     ITREE, TREE AND ITER ARE ARRAYS WHOSE INPUT HAS BEEN COMPUTED
00170    C     BY GMDH BUT ARE NEEDED BY CCMP.
00180    C
00190             DIMENSION X(30),ITREE(10,75),TREE(10,75,6),WCRK(256)
00200            &,ITR(256)
00210             IT = ITER
00220             ITR(1) = 1
00230             I = 1
00240        6    L = 0
00250             NN = 2 **(I-1)
00260             N1 = 2**I
00270             NZ = 2**(I+1)-1
00280             J = N1
00290        4    JJ = ITR(NN+L)
00300             XX = ITREE(ITER,JJ)
00310             ITR(J) = ITREE(ITER,JJ)/100-10
00320             IZ = ITREE(ITER,JJ)/100
00330             ITR(J+1) = XX - 100*IZ - 10
00340             J = J+2
00350             IF(J.GT.NZ) GO TO 3
00360             L = L+1
00370             GO TO 4
00380        3    IF(ITER.EQ.1) GO TO 5
```

```
00390          ITER = ITER - 1
00400          I = I + 1
00410          GO TO 6
00420     5    ITER = IT
00430          NZ = 2 **(ITER-1)
00440          NZZ = NZ
00450          N1 = 2 **ITER
00460          DO 8 J = 1,NZZ
00470          JJ1 = ITR(NZ)
00480          JJ2 = ITR(N1)
00490          JJ3 = ITR(N1+1)
00500          WK = TREE(1,JJ1,1) + TREE(1,JJ1,2)*X(JJ2) + TREE(1,JJ1,3)*X(JJ3)
00510          WK = WK + TREE(1,JJ1,4)*X(JJ2)**2 + TREE(1,JJ1,5)*X(JJ3)**2
00520          WK = WK + TREE(1,JJ1,6)*X(JJ2)*X(JJ3)
00530          WORK(J) = WK
00540          NZ = NZ + 1
00550     8    N1 = N1 + 2
00560          ITER = ITER -1
00570          IF(ITER.NE.C) GC TO 12
00580          Y = WORK(1)
00590          ITER = IT
00600          RETURN
00610    12    I = 2
00620    18    NZ = 2**(ITER-1)
00630          N1 = 2**ITER
00640          NZZ = NZ
00650          N11 = N1
00660          DO 13 J = 1,NZZ
00670          JJ = 2*J-1
00680          JJ1 = ITR(NZ)
00690          JJ2 = ITR(N1)
00700          JJ3 = ITR(N1+1)
00710          WK = TREE(I,JJ1,1) + TREE(I,JJ1,2)*WORK(JJ)
00720          WK = WK + TREE(I,JJ1,3)*WCRK(JJ+1) + TREE(I,JJ1,4)*WCRK(JJ)**2
00730          WK = WK + TREE(I,JJ1,5)*WCRK(JJ+1)**2
00740          WK = WK + TREE(I,JJ1,6)*WCRK(JJ)*WCRK(JJ+1)
00750          WORK(N11+J) = WK
00760          NZ = NZ + 1
00770    13    N1 = N1 + 2
00780          ITER = ITER -1
00790          IF(ITER.EQ.C) GC TO 15
00800          DO 14 L = 1,NZZ
00810    14    WORK(L) = WCRK(N11+L)
00820          I = I+1
00830          GO TC 18
00840    15    Y = WORK(3)
00850          ITER = IT
00860          RETURN
00870          END
00010          SUBROUTINE STAT(N,Y,X,RMS)
00020    C
00030    C     ----------------------------------------------------------
00040    C     STAT IS A SUBRCUTINE THAT CCMPUTES THE STATISTIC THE USER
00050    C     WISHES TC USE FCR COMPARING THE COMPUTED VARIABLES AGAINST
00060    C     THE CHECKING SET.  STAT IS CALLEC EY CMCH.
00070    C     ----------------------------------------------------------
00080    C
00090          DIMENSION Y(50),X(50)
00100          SS = 0.
00110          SSS = C.
00120          DO 5 I = 1,N
00130          SS = SS + (Y(I) - X(I))**2
00140     5    SSS = SSS + Y(I)**2
00150          RMS = SS/SSS
00160          RETURN
00170          END
```

REFERENCE

Draper, N. R., and Smith, H. (1966). Applied Regression Analysis. Wiley, New York.

16

An SAS Program for Simplified GMDH Models

MICHAEL H. PRAGER / Graduate School of Oceanography, University of
Rhode Island, Narragansett, Rhode Island

I. INTRODUCTION

A. SAS

The Statistical Analysis System (SAS) is an integrated set of statistical and
data-management programs. Also included in SAS are a high-level programming language and a powerful matrix language, both of which are used in
this program. Although SAS originally ran only on IBM mainframe computers, at the time of this writing (October 1982) the SAS Institute had
announced plans to adapt it to run on minicomputers as well. Any data can
be put into a SAS data set, usually with just a few statements, and thus can
be analyzed by all SAS procedures, as well as the program described herein.
Readers wishing more information on SAS can consult the user's guide (SAS,
1979) or the SAS Institute, Box 8000, Cary, NC 27511-8000. SAS is a
registered trademark of the SAS Institute, Inc.

B. Program Scope

The Group Method of Data Handling Program—referred to hereafter as
GMDH(PGM)—is a set of SAS statements that performs GMDH analyses on
SAS data sets and reports on the models retained after each iteration (layer).
Printer plots of observed and predicted values can also be included in the
output, as desired.

  Although GMDH(PGM) provides all the information necessary for determining the model, it does not print the model equation itself. The user
must do some algebra to determine, from the program output, the variables
that comprise the final model. The parameters are then estimated by an
ordinary least-squares (OLS) routine such as PROC GLM. Because the

program develops a GMDH model of greatly reduced complexity [compared to the multilayer selection algorithm of Ivakhnenko (1976)], the algebra is simple and the model (in my experience) is of few enough terms so that parameter estimation by OLS is possible. Thus one obtains the advantages of an OLS model, such as the ability to calculate confidence intervals for estimates. Of course, the validity of confidence intervals is conditional on the validity of the model, which may not meet the assumptions of classical linear regression (CLR). In particular, it is unlikely that GMDH and other empirical modeling techniques always select properly specified models. In addition, the data may fail to meet other assumptions of CLR, such as homoskedasticity. For an informal (but thorough) discussion of these assumptions, see Kennedy (1979). Although GMDH models are self-validating through the use of an external complement (the checking set) in their construction, the responsible user may want to test their validity further; methods of model validation are described in Draper and Smith (1981, pp. 419–422).

C. Computational Method

GMDH(PGM) follows for the most part the multilayer selection algorithm of Ivakhnenko (1976). [This method was also called the multilayer threshold algorithm by Ivakhnenko et al. (1979), and has recently been reviewed by Farlow (1981).] For ranking of the partial (quadratic) models at each level, the program uses the regularity criterion of Ivakhnenko (1976), which is a root-mean-squared (rms) error. However, the program does depart from the multilayer GMDH selection algorithm (as documented) in several ways, as explained in the next two sections.

1. Selection of Variables for Partial Functions

To reduce the complexity of the final prediction equation, the program subjects the terms of each partial model to best-subset selection. [This is a practice similar to that of the combinatorial algorithm GMDH of Ivakhnenko (1976).] The selection criterion is Mallows's C_p (Mallows, 1973), whose minimum indicates the subset with the lowest prediction error. Duffy (1974) used stepwise regression for the same purpose, and its use is also reported by Yurachkovskiy (1977). Using C_p on all possible regressions is a further improvement over stepwise procedures, as it always finds the best subset.

2. Final Model Determination

As mentioned in Sec. I.B, the user selects the best subset of the final GMDH model by again using the C_p statistic, and then estimates the model's parameters by OLS. (A stepwise procedure is used—instead of C_p—if a large number of potential terms in the final model makes evaluating all possible regressions impractical.) Both of these steps may be performed using all

the data (i.e., the training and checking sets combined). There is some
evidence, however (see Chap. 9), that using only the checking set for best-
subset selection results in a model with lower prediction error. Use of all
the data for parameter estimation is suggested by Duffy (1974), Ivakhnenko
(1976), and Ivakhnenko et al. (1979).

3. Computer Resources Needed

GMDH(PGM) was designed for research use, and was written in SAS for
ease of programming and data management. Computational cost was not a
consideration. A problem with 20 data points and 20 input variables took
about 3 minutes of CPU time and 1000K of memory to execute on URI's
NAS-7 mainframe computer.

II. USE OF THE PROGRAM

Program use and limitations are discussed below. A complete example is
given in Sec. III; the actual program listings are given in Sec. IV.
      One caveat: GMDH(PGM) was developed under the IBM OS/MVS oper-
ating system, and the examples below reflect that. Users of different
operating systems (such as VM for CMS) will have to modify the JCL statements (Sec. II.C.1) to suit their systems; modification of the program's
references to external (non-SAS) files may also be necessary.

A. Loading the Program Statements

GMDH(PGM) consists of three parts: GMDH(INPUT), GMDH(SETUP), and
GMDH(RUN), which must be used in sequence. The easiest way of using
GMDH(PGM) is to load its three parts into members of a partitioned data
set (PDS), naming the members INPUT, SETUP, and RUN. The program
can then be invoked by SAS %INCLUDE statements as illustrated in Sec. 3.
[For documentation of the SAS %INCLUDE statement, see SAS (1981, p. 7)
or (1982, p. 316).] The explanation and examples below assume that
GMDH(PGM) will be used in that way, and that a DDname of GMDH will be
used for the program PDS. [See Sec. II.C.1, statement (1), and Sec. III.A,
statement (1), for examples of a correct DDname.]
      Users not familiar with loading program statements into a PDS should
consult their computer center personnel for assistance in this routine
operation.

B. Usage Notes

1. Number of Input Variables

The maximum number of input variables is 20; the minimum is 3. The
program terminates automatically if, after an iteration, fewer than three
variables are left for the next iteration.

2. Missing Data

GMDH(PGM) is not designed for analysis of data sets with missing values. Observations with missing values should not be included, even at the beginning or end of the data set. Unpredictable results will occur if this program is used with incomplete data.

3. Abends

The 1979 version of SAS provides no way to end an iterative program other than by executing an ABORT statement. Therefore, the program will always appear to abend (i.e., end "abnormally" as far as the computer is concerned), even when it runs perfectly. For this reason GMDH(PGM) must be the last program run in any job.

4. Error Messages

Error messages issued by the program begin with three asterisks, distinguishing them from error messages issued by the SAS supervisor.

5. Singularity Problems

Occasionally, in higher iterations, singularity (caused by extreme collinearity between two of the partial functions) prevents the program from calculating the C_p statistic. In this case an error message will advise which variable is causing the problem. The solution is either to remove one of the original input variables and rerun the job, or to use the results of the preceding iteration, which may fit the data well enough so that rerunning is unnecessary.

6. Input Variable Names

As the program adds an "S" to variable names to indicate the square of a variable, the input variable names must be limited to seven characters, not the usual eight. For the same reason, variable names ending in "S" should be avoided.

The program also creates new variables of the forms Ennnn and CPnnnn, which represent estimates and cross-products. Input variable names of this form should be changed to avoid conflict.

7. Internal Data Set and Macro Names

The program generates several data sets and macros (named program segments) for internal use. To avoid conflict with user data sets and macros, all internal macro names begin with underscores, and all internal data set names begin and end with underscores.

8. Output Format

The output of GMDH(PGM) is formatted for line printers with standard 132-position line length. It may be confusing when sent to printers or terminals of shorter line lengths.

9. Titles

As the program uses titles on its output to pass information to the user, user TITLE statements are overridden.

C. Statements Needed to Run GMDH(PGM)

The statements needed to run GMDH(PGM) after it has been loaded into a PDS are explained in the next two sections and are illustrated in Sec. III.A. In the statements shown below, underscoring indicates material supplied by the user. Statement numbers (in parentheses) refer to the example (Sec. III.A), and are not part of the statements.

1. JCL Statements

(1) //GMDH DD DSN=data.set.name, DISP=SHR
   This statement defines the partitioned data set in whose members the GMDH(PGM) statements reside.

(2) //GSOURCE DD UNIT=SYSDA, SPACE=(TRK, (10, 5, 1))
   The GSOURCE DD statement allocates a temporary PDS used by the program to store intermediate SAS code; it must have the DDname GSOURCE.

(3) //FT10F001 DD UNIT=SYSDA, SPACE=(TRK, (2, 2))
   Statement (3) also allocates temporary storage. Its DDname (of the form FTnnF001) refers to FORTRAN logical unit nn, in this case unit 10, which in most computer systems is free of any preassigned system function. If unit 10 is assigned to a system function at your installation, you must change the nn portion of statement (3) to a free unit number. You must also change to the same unit number the SAS INFILE statement (statement 31250) and the PROC PRINTTO UNIT statement (statement 31170) of GMDH(RUN) as listed in Sec. IV.

2. SAS Statements

Many of the statements in this section are macros that pass the names of SAS variables and data sets to the program and set the values of program options. Note that SAS macro statements end with percent signs, not semicolons. [For more information on SAS macro statements, see SAS (1979, p. 12).] Except where indicated, each macro statement below must be included.
   Underscoring indicates material to be supplied by the user.

(4)   MACRO _TRAIN data set name%

This macro identifies the training data set, which may bear any valid
SAS name. The data set may be permanent (i.e., with a two-part
name) or temporary.

(5)   MACRO _CHECK data set name%

This statement, coded like the preceding one, identifies the checking
data set.

(6)   MACRO _DEPVAR variable name%

The _DEPVAR macro identifies the dependent variable (output variable). The variable may have any SAS name, and must, of course, be
a variable in the checking and training sets.

(7)   MACRO _METHOD n%

The value of n fixes the method of computing how many variables to
retain at each iteration. The possible values of n are the integers
from 0 to 9. Table 1 shows the effect of each value of n after an
iteration starting with k variables.

(8)   MACRO _PLOT n%

The value of n specifies the first iteration at which printer plots are
produced. If plots are not desired, omit this macro or code $n = 0$.

(9)   MACRO _PLOTMAX n%

When plots are produced, the number made is the lesser of the
_PLOTMAX macro and the number of models retained at that iteration.
If no plots are desired, omit this macro or code $n = 0$.

(10) MACRO _ID variable name%

This statement defines a variable for the x-axis of printer plots. It
may be any variable (except the output variable) in the checking and
training sets. If no plots are desired, leave two blanks before the
percent sign.

(11) OPTIONS NONOTES NOSOURCE2;

This statement suppresses unnecessary information from the SAS log.

(12) %INCLUDE GMDH(INPUT);

This statement reads and checks the input variable names. It must
be followed immediately by a CARDS statement (13), and then by the
variable names (14) one per line in columns 1 to 7.

(15) %INCLUDE GMDH(SETUP);

This statement copies the training and checking sets for internal use.

Table 1 Effect of Value of _METHOD Macro:
Number of Models Retained for k Variables to Start

Value of METHOD macro (n)	Number of models retained after iteration	Number rounded up or down?
0	k	—
1	k - 1	—
2	k - 2	—
3	k - 3	—
4	k - 4	—
5	0.5k	Up
6	0.6k	Up
7	0.7k	Up
8	0.5k	Down
9	(0.4 + iteration number) k (but not more than k)	

(16) %INCLUDE GMDH(RUN);

This statement invokes the main program to perform GMDH. Include as many of these statements as the maximum number of iterations contemplated. The program will terminate automatically if a solution is reached earlier.

III. EXAMPLE

A sample GMDH job was run on the test data listed in Tables 2 and 3. The input and output of the job are explained in the next three sections.

A. Source Statements

The statements below were used to run the test job at the University of Rhode Island. Statement numbers (in parentheses) are for reference only and are not part of the statements. Be aware that the JOB card varies from system to system.

Table 2 Training Data Set for Example of Sec. III

Year	Y	X1	X2	X3	X4	X5
1950	22.1850	1.4324	21.4820	1.3957	1.4309	23.5353
1951	15.2626	0.3434	21.6878	-0.0306	-4.7577	22.1819
1952	15.1236	0.2527	22.2133	1.8564	-4.0028	24.1853
1953	16.1317	-1.0725	22.8959	-0.9865	-1.0060	23.6404
1954	15.0561	-0.6404	23.0648	-1.9811	-0.5350	21.2181
1955	18.4930	-1.0965	23.3373	0.8944	2.8190	23.4471
1958	15.4711	1.0645	24.8361	-1.4909	3.4950	22.1102
1962	15.4483	0.6644	26.5820	-1.0806	0.7533	25.8578
1963	15.0973	-0.6407	26.9486	-1.6113	-7.0433	25.5455
1964	16.2165	0.6480	27.2197	0.4729	-0.5110	28.3249
1965	15.6560	-0.8836	27.7938	-1.0194	2.3419	25.5810
1967	17.2672	0.7592	28.8897	1.8300	-4.8300	31.9060
1969	19.0760	1.2589	29.3235	0.6369	-4.5846	27.9323
1970	15.1402	-0.2705	30.1815	0.1823	-1.4904	29.8858
1971	15.7351	-1.0699	30.4115	-1.3615	0.1228	29.6108
1972	15.6332	-0.6103	30.9182	-0.2790	-5.1598	31.7440
1973	15.0727	-0.0540	31.2739	1.4213	-1.8428	33.0698
1974	14.9354	-0.0472	31.7831	-0.3899	0.0973	30.9779

Table 3 Checking Data Set for Example of Sec. III

Year	Y	X1	X2	X3	X4	X5
1956	15.0456	0.2489	23.8578	-1.4783	0.0080	22.4089
1957	15.1338	0.0113	24.4009	-0.8482	1.0198	23.7643
1959	21.4214	-1.2599	25.3547	1.9878	1.4693	26.5690
1960	18.1862	-0.9801	25.8602	1.4438	1.2403	25.5077
1961	17.2739	0.8444	26.0204	1.0789	2.3604	27.3899
1966	18.4304	0.9876	28.3329	1.3969	-2.4026	30.8196
1968	18.6360	-1.1676	28.9977	0.5527	-4.3260	30.2351
1975	15.7772	-0.6454	32.2099	-0.2656	-2.0461	31.0517

```
//GMDHJOB JOB (LNS101),'MIKE PRAGER',TIME=(1,0),MSGLEVEL=(2,0)
// EXEC SAS79,REGION.GO=600K
//TEST DD DSN=URI.LNS1.SAS.TEST.DATA,DISP=SHR
//GMDH DD DSN=URI.LNS1.GMDH.PGM,DISP=SHR          (1)
//GSOURCE   DD UNIT=SYSDA,SPACE=(TRK,(10,5,1))    (2)
//FT10F001 DD UNIT=SYSDA,SPACE=(TRK,(2,2))        (3)
//SYSIN DD *
MACRO  TRAIN TEST.TRAIN%                          (4)
MACRO  CHECK TEST.CHECK%                          (5)
MACRO  DEPVAR Y%                                  (6)
MACRO  METHOD 0%                                  (7)
MACRO  PLOT 1%                                    (8)
MACRO  PLOTMAX 2%                                 (9)
MACRO  ID YEAR%                                   (10)
OPTIONS NONOTES NOSOURCE2;                        (11)
%INCLUDE GMDH(INPUT);                             (12)
CARDS;                                            (13)
X1
X2
X3                                               (14)
X4
X5
%INCLUDE GMDH (SETUP);                            (15)
%INCLUDE GMDH (RUN);
%INCLUDE GMDH (RUN);
%INCLUDE GMDH (RUN);                              (16)
%INCLUDE GMDH (RUN);
%INCLUDE GMDH (RUN);
/*
//
```

B. GMDH(PGM) Output

The output from the job shown above is reproduced in Fig. 1. The report,
generated after each iteration, includes (see the circled letters on Fig. 1)

(A) Titles listing the dependent variable, data set names, and iteration
number.
(B) A listing of the number of variables input to the iteration, the number
of partial models evaluated, and the number retained.
(C) A table listing, for each model retained, (1) the program-generated
variable name given to the estimate, (2) the number of variables included in it, (3) its r^2 on the training set, (4) its rms error on the
checking set, (5) a list of the variables in the model, and (6) the variables that make up the cross-product term of the model.
(D) An explanation of the notation used in the report.
(E) If applicable, a note that duplicate model(s) have been dropped. (This
occurs when, during best-subset selection, two partial models end up
with the same terms.) Any dropping takes place before the best models
are selected.

G M D H   A L G O R I T H M

REPORT ON MODEL EVALUATION FOR THIS ITERATION

(B) 5 VARIABLES WERE INPUT FOR POSSIBLE USE.
10 QUADRATIC PARTIAL MODELS WERE EVALUATED.
5 MODELS WERE RETAINED ON THE BASIS OF LOWEST CHECKING-SET RMS ERROR.

(C)

NAME OF ESTIMATE VARIABLE	NUMBER OF VARS IN MODEL	TRAINING SET RSQUARE	CHECKING SET RMS ERROR	LIST OF VARIABLES INCLUDED IN MODEL	FACTORS IN THE CROSSPRODUCT VARIABLE ("CPNNNN")	
E1002	2	0.883	0.0440	X3 X1S	X3	* X1
E1008	3	0.416	0.0791	X3 X4 CP1008	X4	* X3
E1001	2	0.744	0.0889	X1 X1S	X2	* X1
E1005	1	0.213	0.0892	X3	X3	* X2
E1009	2	0.323	0.0909	X3 X5S	X5	* X3

(D) NOTES: ALTHOUGH ITS FACTORS ARE ALWAYS LISTED, A CROSSPRODUCT VARIABLE MAY OR MAY NOT BE INCLUDED IN EACH MODEL.
A VARIABLE ENDING IN "S" REPRESENTS THE SQUARE OF THE ORIGINAL VARIABLE.
EACH MODEL ALSO INCLUDES AN INTERCEPT.

(E) NOTE: AT LEAST 1 DUPLICATE MODEL WAS DROPPED, WHICH MAY CAUSE FEWER MODELS TO BE RETAINED THAN EXPECTED.

(F) THE 5 VARIABLES LISTED ABOVE WILL BE USED AS INPUT FOR ITERATION 2 .

G M D H   A L G O R I T H M

DEPENDENT VARIABLE: Y
TRAINING DATA SET: TEST.TRAIN
CHECKING DATA SET: TEST.CHECK
ITERATION NO. 2

(A)

REPORT ON MODEL EVALUATION FOR THIS ITERATION

(B) 5 VARIABLES WERE INPUT FOR POSSIBLE USE.
10 QUADRATIC PARTIAL MODELS WERE EVALUATED.
5 MODELS WERE RETAINED ON THE BASIS OF LOWEST CHECKING-SET RMS ERROR.

(C)

NAME OF ESTIMATE VARIABLE	NUMBER OF VARS IN MODEL	TRAINING SET RSQUARE	CHECKING SET RMS ERROR	LIST OF VARIABLES INCLUDED IN MODEL	FACTORS IN THE CROSSPRODUCT VARIABLE ("CPNNNN")		
E2003	4	0.998	0.0056	E1002 E1005 E1005S CP2003	E1005	*	E1002
E2002	4	0.983	0.0263	E1002 E1001 E1002S E1001S	E1001	*	E1002
E2004	4	0.979	0.0264	E1002 E1002S E1009S CP2004	E1009	*	E1002
E2001	2	0.967	0.0299	E1002 E1002S	E1008	*	E1002
E2009	4	0.957	0.0539	E1001 E1009 E1001S CP2009	E1009	*	E1001

(D) NOTES: ALTHOUGH ITS FACTORS ARE ALWAYS LISTED, A CROSSPRODUCT VARIABLE MAY OR MAY NOT BE INCLUDED IN EACH MODEL.
A VARIABLE ENDING IN "S" REPRESENTS THE SQUARE OF THE ORIGINAL VARIABLE.
EACH MODEL ALSO INCLUDES AN INTERCEPT.

(E) NOTE: AT LEAST 1 DUPLICATE MODEL WAS DROPPED, WHICH MAY CAUSE FEWER MODELS TO BE RETAINED THAN EXPECTED.

(F) THE 5 VARIABLES LISTED ABOVE WILL BE USED AS INPUT FOR ITERATION 3 .

G M D H   A L G O R I T H M

DEPENDENT VARIABLE: Y
TRAINING DATA SET: TEST.TRAIN
CHECKING DATA SET: TEST.CHECK
ITERATION NO. 3

(A)

REPORT ON MODEL EVALUATION FOR THIS ITERATION

(B) 5 VARIABLES WERE INPUT FOR POSSIBLE USE.
     10 QUADRATIC PARTIAL MODELS WERE EVALUATED.
     5 MODELS WERE RETAINED ON THE BASIS OF LOWEST CHECKING-SET RMS ERROR.

(C)

NAME OF ESTIMATE VARIABLE	NUMBER OF VARS IN MODEL	TRAINING SET RSQUARE	CHECKING SET RMS ERROR	LIST OF VARIABLES INCLUDED IN MODEL	FACTORS IN THE CROSSPRODUCT VARIABLE ("CPNNNN")	
E3001	1	0.998	0.0056	E2003	E2002	* E2003
E3004	4	0.999	0.0142	E2003 E2009 E2003S E2009S	E2009	* E2003
E3007	2	0.986	0.0193	E2002 E2009	E2009	* E2002
E3005	1	0.983	0.0263	E2002	E2004	* E2002
E3008	1	0.979	0.0264	E2004	E2001	* E2004

(D) NOTES: ALTHOUGH ITS FACTORS ARE ALWAYS LISTED, A CROSSPRODUCT VARIABLE MAY OR MAY NOT BE INCLUDED IN EACH MODEL.
     A VARIABLE ENDING IN "S" REPRESENTS THE SQUARE OF THE ORIGINAL VARIABLE.
     EACH MODEL ALSO INCLUDES AN INTERCEPT.

(E) NOTE: AT LEAST 1 DUPLICATE MODEL WAS DROPPED, WHICH MAY CAUSE FEWER MODELS TO BE RETAINED THAN EXPECTED.

(F) THE CHECKING-SET RMS ERROR HAS NOT IMPROVED SINCE THE LAST ITERATION, WHICH INDICATES
    THAT THE SOLUTION IS AT THE PREVIOUS ITERATION.

Fig. 1 Output of GMDH(PGM) for the example of Sec. III.A.

(F) An indication as to whether the program will continue to the next iteration or not.

Also included in the report, but not reproduced here, are a listing of the input variables and printer plots.

C. Calculating the Final Model

From Fig. 1, the model E2003 is seen to be the best GMDH estimate of the output variable Y. The expanded terms of the final model are calculated (by the user) as follows, omitting all coefficients (as they will be estimated later) and letting I represent any intercept.

First, after substituting for CP2003 and E1005S, we have

$$\hat{Y} = E1002 + E1005 + (E1005)(E1005) + (E1005)(E1002) + I \tag{18}$$

We next take definitions of the first-iteration estimates from the report (Fig. 1). Substituting the original variables, we have

$$\hat{Y} = (X_3 + X_1^2 + I) + (X_3 + I) + (X_3 + I)(X_3 + I) + (X_3 + X_1^2 + I)(X_3 + I) \tag{19}$$

Multiplying terms, grouping, and lumping intercepts again as I gives us

$$\hat{Y} = X_3 + X_1^2 + X_3^2 + X_1^2 X_3 \tag{20}$$

The following program segment was used to select the best subset of (20):

```
COMMENT  Make a new data set with the vars of interest;
    DATA TEST.MODEL;
        SET TEST.TRAIN TEST.CHECK;
        KEEP U1-U4  Y  YEAR;
        U1 = X3;
        U2 = X1 ** 2;
        U3 = X3 ** 2;
        U4 = (X1 ** 2) * X3;
COMMENT  Next calculate the C(p) for each possible model;
    PROC RSQUARE CP;
        MODEL Y = U1-U4 / PRINT=1;
```

The output of PROC RSQUARE is shown in Table 4. Mallows's C_p reached its minimum for the model with two variables, U2 and U4, indicating that those two comprised the best subset. The following program segment was then used to estimate the final model parameters with PROC GLM:

Table 4 Output of Best–Subset Selection (PROC RSQUARE)
for Example of Sec. III.C

NUMBER IN MODEL	R–SQUARE	C(P)	VARIABLES IN MODEL
1	0.74326324	3177.99037	U4
2	0.99819620	2.48270341	U2 U4
3	0.99827586	3.48986956	U1 U2 U4
4	0.99831516	5.00000000	U1 U2 U3 U4

Table 5 Output of OLS Estimation (PROC GLM) for Example of Sec. III.C

GENERAL LINEAR MODELS PROCEDURE

DEPENDENT VARIABLE: Y

SOURCE	DF	SUM OF SQUARES	MEAN SQUARE
MODEL	2	101.44273199	50.72136600
ERROR	23	0.18331286	0.00797012
CORRECTED TOTAL	25	101.62604486	

MODEL F =	6363.94		PR > F = 0.0001

R–SQUARE	C.V.	STD DEV	Y MEAN
0.998196	0.5362	0.08927555	16.65020775

SOURCE	DF	TYPE I SS	F VALUE	PR > F
U2	1	73.22716273	9187.71	0.0001
U4	1	28.21556927	3540.17	0.0001

SOURCE	DF	TYPE IV SS	F VALUE	PR > F
U2	1	25.90782856	3250.62	0.0001
U4	1	28.21556927	3540.17	0.0001

PARAMETER	ESTIMATE	T FOR HO: PARAMETER=0	PR > T	STD ERROR OF ESTIMATE
INTERCEPT	15.01751369	526.16	0.0001	0.02854156
U2	1.99164183	57.01	0.0001	0.03493239
U4	1.03827662	59.50	0.0001	0.01745023

PROC GLM DATA = TEST.MODEL;
MODEL Y = U2 U4;

The output from PROC GLM is shown in Table 5. From the parameter estimates at the bottom of Table 5, the predictive equation is

$$\hat{Y} = 15.02 + 1.99X_1^2 + 1.04X_1^2X_3 \qquad (21)$$

IV. PROGRAM STATEMENTS

Users will probably find it expeditious to obtain a card deck for GMDH(PGM) rather than keypunch the statement listings below. A card deck or magnetic tape can be obtained from the author for a nominal fee to cover duplication costs.

A. GMDH(INPUT) Listing

```
*----GMDH(INPUT)----;                                              00010010
OPTIONS NOSOURCE2 NONOTES  ERRORABEND GEN=0 NOSPOOL;               00010020
*----------------------------MODULE 1--------------------------;   00010030
*------------------CHECK FOR MISSING MACROS-------------------;    00010040
DATA _NULL_;                                                       00010050
   IF_ METHOD = . THEN DO;                                         00010060
      PUT '***ERROR:  _METHOD MACRO OMITTED.'/OVERPRINT           00010070
          '***ERROR:  ‾METHOD MACRO OMITTED.';                     00010080
      ABORT ABEND 101 ‾    /*TERMINATE PROGRAM*/;                  00010090
      END              /*END OF DO-GROUP*/;                        00010100
*NOTE:  _PLOT AND _PLOTMAX MACROS DEFAULT IN MODULE 24;           00010110
*-----------------------------------------------------------;      00010120
*---------------------------MODULE 2-------------------------;     00010130
*----------------INPUT THE PREDICTOR VARIABLES----------------;    00010140
DATA _LIST_;                                                       00010150
   ‾TITLE‾ ;                                                       00010160
   FILE PRINT;                                                     00010170
   IF _N_ = 1 THEN          /*PRINT MESSAGE IN OUTPUT*/            00010180
   PUT‾ ///                                                        00010190
      @4 'SIMPLIFIED GMDH ALGORITHM'/ @4 25 * '-' //               00010200
      @4 'AUTHOR: MICHAEL H. PRAGER'/                              00010210
      @4 'GRADUATE SCHOOL OF OCEANOGRAPHY'/                        00010220
      @4 'UNIVERSITY OF RHODE ISLAND'/                             00010230
      @4 'NARRAGANSETT, R.I., U.S.A.  02882-1197'//                00010240
      'LIST OF VARIABLES INPUT AS PREDICTORS:'/                    00010270
      37 * '-' /;                                                  00010280
   INPUT VARNAME $  1-7 CHECK $ 8 /*READ VARIABLE NAMES*/;         00010290
   DROP CHECK;                                                     00010300
   IF CHECK NE ' ' THEN DO    /*CHECK FOR VARNAME LONGER THAN 7*/; 00010310
      FILE LOG                  /*ROUTE MESSAGE TO LOG*/;          00010320
      PUT '***ERROR: VAR NAME LONGER THAN 7 CHARACTERS '          00010330
          'OR DOES NOT START IN COLUMN 1.'  /OVERPRINT             00010340
          '***ERROR: VAR NAME LONGER THAN 7 CHARACTERS '          00010350
```

```
      'OR DOES NOT START IN COLUMN 1.' /                              00010360
      @8 'VARIABLE NAME CONCERNED IS 'VARNAME $7. CHECK '.'/          00010370
      '        THIS IS NOT ALLOWED.  PROGRAM TERMINATED.';            00010380
   ABORT ABEND 201                      /*STOP EXECUTION*/;           00010390
   END                                  /*END OF DO-GROUP*/;          00010400
 PUT _N_ 15-16 '. ' VARNAME             /*ECHO PRINT VARIABLE NAME*/; 00010410
*---------------------------------------------------------------------; 00010420
```

B. GMDH(SETUP) Listing

```
*----GMDH(SETUP)----;                                                 00020005
OPTIONS NOSOURCE2 NONOTES ERRORABEND GEN=0 NOSPOOL;                   00020010
*------------------------MODULE 2.5---------------------------------; 00020020
*PRESET THE MACRO "_ITER" TO 1--;                                    00020030
  MACRO _ITER 1 %                                                    00020040
* GENERATE A PROGRAM TO COPY THE CHECKING DATA SET FOR INTERNAL USE; 00020050
DATA _NULL_ ;                                                        00020060
  FILE GSOURCE (COPY1)     /*ROUTE SOURCE OUTPUT*/;                  00020070
  IF _N_ = 1 THEN PUT                                                00020080
  '*GENERATED PGM FROM MODULE 2.5;' / '*;' /                         00020090
  'DATA _CHECK_ ;'/                                                  00020100
  '   SET _CHECK (KEEP = _DEPVAR _ID '@;                             00020110
  SET _LIST_ END=LAST;                                              00020120
  PUT +2 VARNAME ;                                                   00020130
  IF LAST THEN PUT ');'    /*FINISH WRITING "SET" STATEMENT*/;       00020140
* GENERATE A PROGRAM TO COPY THE TRAINING DATA SET FOR INTERNAL USE; 00020150
DATA _NULL_ ;                                                        00020160
  FILE GSOURCE (COPY2)     /*ROUTE SOURCE OUTPUT*/;                  00020170
  IF _N_ = 1 THEN PUT                                                00020180
  '*GENERATED PGM FROM MODULE 2.5;' / '*;' /                         00020190
  'DATA _TRAIN_ ;'/        /*WRITE DATA STATEMENT*/                  00020200
      'SET _TRAIN (KEEP = _DEPVAR _ID '@ /*START WRITING "SET" STMT*/;00020210
  SET _LIST_ END=LAST;                                              00020220
  PUT +2 VARNAME ;                                                   00020230
  IF LAST THEN PUT ');'    /*FINISH WRITING "SET" STATEMENT*/;       00020240
RUN                          /*RUN THE ABOVE STEPS*/;                00020250
%INCLUDE GSOURCE(COPY1)      /*INCLUDE THE PGM TO COPY CHECKING SET*/;00020260
%INCLUDE GSOURCE(COPY2)      /*INCLUDE THE PGM TO COPY TRAINING SET*/;00020270
RUN;                                                                 00020280
```

C. GMDH(RUN) Listing

```
*----GMDH(RUN)----;                                                  00030010
*;                                                                   00030020
*;                                                                   00030030
*-----MODULE 3:  PREPARE A LIST OF ALL PREDICTOR--------------------; 00030040
*----------VARIABLES TAKEN 2 AT A TIME, PLUS THEIR SQUARES---------; 00030050
*----------AND CROSSPRODUCTS.  DEFINE MACROS.---------------------; 00030060
*----------CHECK FOR TOO MANY OR TOO FEW VARIABLES.--------------; 00030070
DATA _VARSET_;                                                       00030080
   IF _N_ = 2 THEN STOP     /*THIS DATA STEP ONLY EXECUTES ONCE*/;   00030090
   N = _ITER * 1000         /*INITIALIZE N*/;                        00030100
   RETAIN N M VAR1;                                                  00030110
   FORMAT N 4.;                                                      00030120
```

```
LENGTH VAR1-VAR3 $ 8 VAR4 $ 8                                        00030130
   VAR5 $ 19                  /*SET LENGTHS*/;                       00030140
FILE GSOURCE (MACROS) OLD     /*DEFINE THE "PUT" FILE*/;             00030150
L = 1                         /* L = DUMMY VAR TO SET POINTER=1*/;   00030160
SET _LIST_ POINT=L NOBS=M     /*GET THE NO. OF VARIABLES (= M) */;   00030170
LINK SETUP                    /*EXECUTE SETUP SUBROUTINE BELOW*/;    00030180
DO I=1 TO (M-1)               /*START MAKING VARIABLE PAIRS*/;       00030190
   SET _LIST_ POINT=I         /*TAKE AN OBS TO GET VAR1*/;           00030200
   VAR1 = VARNAME             /*ASSIGN VALUE OF VAR1*/;              00030210
   DO J = (I+1) TO M          /*MAKE PAIRS WITH VAR1*/;              00030220
      N=N+1                   /*INCREMENT*/;                         00030230
      SET _LIST_ POINT = J    /*TAKE A VARIABLE NAME*/;              00030240
      VAR2 = VARNAME          /*ASSIGN IT TO VAR2*/;                 00030250
      VAR3 = TRIM(VAR1)||'S'     /*MAKE A NAME FOR 3RD VAR*/;        00030260
      VAR4 = TRIM(VAR2)||'S'     /*VARS 3 & 4 = 1 & 2 **2 */;        00030270
      VAR5 = VAR2||' * '||VAR1   /*VAR5 IS THE CROSSPRODUCT*/;       00030280
      OUTPUT;                                                        00030290
   END;                                                              00030300
END;                                                                00030310
KEEP N VAR1-VAR5;                                                    00030320
RETURN;                                                             00030330
SETUP:    *A SUBROUTINE TO DEFINE MACROS FOR LATER IN THE PROGRAM;   00030340
          *ALSO CHECKS FOR TOO MANY OR TOO FEW VARIABLES INPUT;     00030350
PUT 'MACRO _NVARS ' M '%'     /*A MACRO TO HOLD # OF VARIABLES*/;    00030360
MM = M*(M-1)/2                /*MM = # OF COMBINATIONS OF VARS*/;    00030370
PUT 'MACRO _MODELS 'MM'%'     /*A MACRO TO HOLD # OF COMBINATIONS*/; 00030380
IF M LE 2 THEN DO             /*CHECK FOR TOO FEW VARS*/;            00030390
   FILE LOG                   /*WRITE MESSAGE TO LOG*/;              00030400
   PUT '***ERROR: 'M ' VARIABLES INPUT.'/ OVERPRINT                 00030410
       '***ERROR: 'M ' VARIABLES INPUT.'/                           00030420
       'THIS IS LESS THAN THE MINIMUM OF 3 NEEDED.'/OVERPRINT       00030430
       'THIS IS LESS THAN THE MINIMUM OF 3 NEEDED.'/                00030440
       'PROGRAM TERMINATED BECAUSE OF THIS CONDITON.' / OVERPRINT   00030450
       'PROGRAM TERMINATED BECAUSE OF THIS CONDITON.';             00030460
   ABORT ABEND  301           /*END THE PROGRAM*/;                  00030470
   END                        /*END OF DO-GROUP*/;                  00030480
IF M > 20 THEN DO             /*CHECK FOR TOO MANY VARS*/;           00030490
   FILE LOG                   /*WRITE MESSAGE TO LOG*/;              00030500
   PUT '***ERROR: 'M 'VARIABLES INPUT.'/ OVERPRINT                  00030510
       '***ERROR: 'M 'VARIABLES INPUT.'/                            00030520
       'THIS IS MORE THAN THE MAXIMUM OF 20 ALLOWED.'/OVERPRINT     00030530
       'THIS IS MORE THAN THE MAXIMUM OF 20 ALLOWED.'/              00030540
       'PROGRAM TERMINATED BECAUSE OF THIS ERROR.' / OVERPRINT      00030550
       'PROGRAM TERMINATED BECAUSE OF THIS ERROR.';                00030560
   ABORT ABEND  302           /*END THE PROGRAM*/;                  00030570
   END                        /*END OF DO-GROUP*/;                  00030580
* THE NEXT GROUP OF STATEMENTS DETERMINES;                          00030590
* HOW MANY VARIABLES TO USE IN THE NEXT ITERATION;                  00030600
IF _METHOD = 0 THEN KEEP = M;                                       00030610
ELSE IF _METHOD = 1 THEN KEEP = M - 1;                              00030620
ELSE IF _METHOD = 2 THEN KEEP = M - 2;                              00030630
ELSE IF _METHOD = 3 THEN KEEP = M - 3;                              00030640
ELSE IF _METHOD = 4 THEN KEEP = M - 4;                              00030650
ELSE IF _METHOD = 5 THEN KEEP = CEIL(M * 0.5);                      00030660
ELSE IF _METHOD = 6 THEN KEEP = CEIL(M * 0.6);                      00030670
ELSE IF _METHOD = 7 THEN KEEP = CEIL(M * 0.7);                      00030680
```

```
      ELSE IF _METHOD = 8 THEN KEEP = FLOOR(M*0.5) + 2;            00030690
      ELSE IF _METHOD=9 THEN KEEP=CEIL(M*(MIN((0.4 + .1*_ITER),1)));  00030700
      ELSE IF _METHOD = 99 THEN KEEP = MM;                         00030710
      IF KEEP LT 3 THEN KEEP=2;                                    00030720
      PUT 'MACRO _KEEP ' KEEP '%';                                 00030730
      * "KEEP" IS THE # OF PREDICTORS TO USE NEXT ITERATION.;      00030740
      RETURN                       /*END OF SUBROUTINE*/;          00030750
*-------------------------MODULE 4-------------------------------------;  00030760
RUN                              /*RUN THE PROGRAM SO FAR*/;       00030770
%INCLUDE GSOURCE (MACROS);                                         00030780
*---------------------------------------------------------------------;  00030790
*----------MODULE 6:  GENERATE A PROGRAM TO ADD THE CROSSPRODUCTS-----;  00030800
*----------OF EACH VARIABLE PAIR & THE SQUARE OF EACH VARIABLE -------;  00030810
*----------TO THE CHECKING & TRAINING SETS---------------------------;  00030820
DATA _NULL_;                                                       00030830
      FILE GSOURCE (XPRODS) OLD   /*ROUTE OUTPUT*/;                00030840
      LENGTH VS $ 8               /*VS IS A DUMMY VARIABLE      */;  00030850
      DROP VS                     /*TO HOLD SQUARED VARIABLE NAME*/;  00030860
      SET _VARSET_ END=LAST       /*TAKE A VAR NAME*/;             00030870
      BY VAR1 NOTSORTED;                                           00030880
      IF _N_=1 THEN LINK FIRSTOBS /*EXECUTE SUBROUTINE*/;          00030890
      PUT '   CP'N ' = ' VAR5';';                                  00030900
      IF FIRST.VAR1 THEN LINK SQUARES;                             00030910
      IF LAST THEN DO;                                             00030920
        VS=TRIM(VAR2)||'S';                                        00030930
        PUT @5 VS ' = ' VAR2 '**2;'/                               00030940
        @5 'IF INCHECK THEN OUTPUT _CHECK_;'/                      00030950
        @5 'ELSE OUTPUT _TRAIN_;';                                 00030960
        END;                                                       00030970
      RETURN;                                                      00030980
    SQUARES: VS = TRIM(VAR1)||'S';                                 00030990
      PUT @5 VS ' = ' VAR1 '**2;';                                 00031000
      RETURN;                                                      00031010
    FIRSTOBS:                                                      00031020
        PUT '*GENERATED PROGRAM FROM MODULE 6;'/ '*;'/             00031030
           'DATA _TRAIN_ _CHECK_;'/                                00031040
        '     SET _TRAIN_ _CHECK_(IN=INCHECK);' ;                  00031050
        RETURN;                                                    00031060
*---------------------------------------------------------------------;  00031070
*---------------MODULE 8:  GENERATE A PROGRAM TO RUN------------------;  00031080
*----------PROC RSQUARE ON EACH SET OF VARIABLES, AND UPDATE ---------;  00031090
*----------DATASET _VARSET_ WITH THE SUBSET HAVING LOWEST C_SUB_P------; 00031100
*;                                                                 00031110
DATA _NULL_;                                                       00031120
      IF _N_ = 2 THEN STOP         /*THE STEP ONLY EXECUTES ONCE*/;  00031130
      FILE GSOURCE (RSQUARE) OLD   /*ROUTE PUT STATEMENT OUTPUT*/;  00031140
      M = _MODELS                  /*GET THE NUMBER OF MODELS*/;   00031150
      PUT '*GENERATED PROGRAM FROM MODULE 8;'/                     00031160
         'PROC PRINTTO NEW UNIT = 10 /*ROUTE PROC RSQARE OUTPUT*/;'/  00031170
         'PROC RSQUARE CP DATA=_TRAIN_;';                          00031180
      DO I = 1 TO M               /*WRITE MODEL STATEMENTS*/;      00031190
        SET _VARSET_               /*GET VARIABLE NAMES*/;         00031200
        PUT 'MODEL _DEPVAR=' VAR1-VAR4'CP'N'/PRINT=1 ;';           00031210
        END                       /* END OF DO-LOOP */;           00031220
      PUT 'PROC PRINTTO           /*RESET SAS OUTPUT STREAM*/;'/   00031230
           'DATA _VARS2_;'/        /*READ PROC RSQUARE OUTPUT*/    00031240
```

```
'   INFILE FT10F001 MISSOVER END=_E_ ;'/                              00031250
'   INPUT @6 NUM ??1. @14 RSQ ??10.8 @25 C_SUB_P ??14.8 '/            00031260
'   @42 VAR1 $ VAR2 $ VAR3 $ VAR4 $ VAR5 $ ;'/                        00031270
'   IF 1 =< NUM <=5;'  /   /*DELETE GARBAGE LINES*/                   00031280
'   IF C_SUB_P =. THEN LINK SING;'/                                   00031290
'   IF NUM=1 THEN MODEL + 1;'//*INCREMENT MODEL # */                  00031300
'   RETURN;'/                                                         00031310
'SING:  *A SUBROUTINE TO HANDLE SINGULARITIES;'/                      00031320
'   FILE LOG;'/                                                       00031330
'   PUT ''***ERROR:  SINGULARITY ENCOUNTERED.'''/'/                   00031340
'   OVERPRINT ''***ERROR:  SINGULARITY ENCOUNTERED.'';'/              00031350
'   I = 1;'                /* INITIALIZE INDEX VARIABLE */            00031360
'   DO UNTIL (_E_>0 OR I=25);'/ /* E>0 ONLY @ LAST OBS */             00031370
'     INPUT;'/                  /* READ A LINE */                     00031380
'     PUT _INFILE_;'/           /* ECHO PRINT TO LOG */               00031390
'     I = I + 1;'/              /* INCREMENT INDEX */                 00031400
'     END;'/                                                          00031410
'   ABORT ABEND   801;'/       /* TERMINATE PROGRAM */                00031420
'   RETURN;'/                  /* END OF "SING" ROUTINE */            00031430
'PROC SORTT;'/                                                        00031440
'   BY MODEL C_SUB_P;'/                                               00031450
'DATA _VARS2_;'/              /*THIS STEP SELECTS THE     */          00031460
'   SET _VARS2_;'/            /*SUBSET WITH LOWEST C-SUB-P*/          00031470
'   BY MODEL;'/               /*TO BE IN EACH MODEL       */          00031480
'   IF FIRST.MODEL;';                                                 00031490
*;                                                                    00031500
*----------------------MODULE 10----------------------------------;   00031510
RUN                     /*RUN THE PROGRAM SO FAR        */;           00031520
%INCLUDE GSOURCE (XPRODS)      /*RUN THE PGM TO GENERATE XPRODS */;   00031530
%INCLUDE GSOURCE (RSQUARE)     /*RUN THE PGM FOR PROC RSQUARE   */;   00031540
*;                                                                    00031550
*-------MODULE 12:  GENERATE A PROGRAM TO RUN PROC SYSREG----------;  00031560
*-------ON EACH SUBSET OF VARIABLES.------------------------------;   00031570
*;                                                                    00031580
DATA _NULL_;                                                          00031590
   SET _VARS2_                 /* GET VARIABLE NAMES*/;               00031600
   FILE GSOURCE (SYSREG) OLD   /* ROUTE OUTPUT STMNTS */;             00031610
   IF _N_=1 THEN PUT                                                  00031620
   '* GENERATED PROGRAM FROM MODULE 12;'/'*;'/                        00031630
   'PROC SYSREG NOPRINT DATA=_TRAIN_ OUTEST=_EST_'/                   00031640
   '  (DROP=_TYPE_ _SIGMA_ _DEPVAR);';                               00031650
   PUT '    MODEL _DEPVAR_ = ' VAR1-VAR5 ';';                         00031660
*;                                                                    00031670
*-----------------------MODULE 13---------------------------------;   00031680
RUN                     /* RUN THE PROGRAM SO FAR */;                 00031690
%INCLUDE GSOURCE (SYSREG)      /*INCL THE PGM FOR PROC SYSREG*/;      00031700
*;                                                                    00031710
*-----------------------MODULE 14---------------------------------;   00031720
*---------TRANSPOSE THE ESTIMATE DATASET (DS "EST") TO PUT THE-----;  00031730
*---------VARIABLE NAMES INTO A CHARACTER VARIABLE (_NAME_).-------;  00031740
*---------GENERATE A PROGRAM TO MULTIPLY VAR NAMES BY-------------;   00031750
*-----------------------COEFFICIENTS.-----------------------------;   00031760
*;                                                                    00031770
PROC TRANSPOSE DATA=_EST_ OUT=_TRANSP_ (DROP=_LABEL_) PREFIX=MODEL;   00031780
DATA _NULL_;                                                          00031790
```

```
        FILE GSOURCE (ESTS1) OLD;                                        00031800
        IF _N_=2 THEN STOP;                                              00031810
        M = LEFT (_MODELS)          /*GET NUMBER OF MODELS*/;            00031820
        PUT                                                             00031830
          '*GENERATED PROGRAM FROM MODULE 14;'/'*;'/                     00031840
          'DATA _TERMS_;' /                                              00031850
          'KEEP MODL1-MODL' M ';' /                                      00031860
          'LENGTH MODL1-MODL' M '$ 25.;' /                              00031870
          'ARRAY EXP1 (J)  MODEL1-MODEL' M ';' /                         00031880
          'ARRAY EXP2 (J) MODL1-MODL'M';'/                               00031890
          'SET _TRANSP_;' /                                              00031900
          'IF _NAME_ = ''INTERCEP'' THEN _NAME_ = ''1'';' /             00031910
          'DO OVER EXP1;' /                                              00031920
          '   IF EXP1=. THEN EXP2 = ''0'' ;' /                          00031930
          '   ELSE EXP2 = TRIM(_NAME_)|| ''*'' || LEFT(EXP1);'/         00031940
          '   END;'/                                                     00031950
          'PROC TRANSPOSE DATA=_TERMS_ OUT=_TERMS2_(DROP=_NAME_)' /      00031960
          '   PREFIX=TERM;' /                                            00031970
          '   VAR _ALL_;' ;                                              00031980
*-----------------------------------------------------------------;     00031990
RUN                           /*RUN THE PROGRAM SO FAR*/;                00032000
%INCLUDE GSOURCE (ESTS1);                                               00032010
*-----------------------------------------------------------------;     00032020
*----------------------------MODULE 16----------------------------;     00032030
*---------GENERATE   A MACRO CONTAINING THE NAMES-----------------;     00032040
*---------OF THE VARIABLES IN DATASET TERMS2.  THESE VARIABLES----;     00032050
*---------CONTAIN THE EXPRESSIONS WHICH, SUMMED, GIVE THE---------;     00032060
*---------ESTIMATES OF THE DEPENDENT VARIABLE.-------------------;      00032070
*;                                                                      00032080
DATA _NULL_;                                                            00032090
        FILE GSOURCE (MACRO2) OLD;                                       00032100
        IF _N_=2 THEN STOP;                                             00032110
        L=1 /* L IS A DUMMY VARIABLE*/;                                 00032120
        SET _TERMS_ POINT=L NOBS=T;                                      00032130
        * T IS THE # OF TERMS TO ADD UP;                                00032140
        PUT '* GENERATED MACRO FROM MODULE 16;';                         00032150
        PUT 'MACRO _TERMS ' @;                                          00032160
        DO I = 1 TO T;                                                  00032170
          II = LEFT (I);                                                 00032180
          PUT ' '' + '' TERM' II @;                                     00032190
          END;                                                           00032200
        PUT '%';                                                         00032210
*----------------------------MODULE 17----------------------------;     00032220
RUN                           /* RUN THE PROGRAM SO FAR */;              00032230
%INCLUDE GSOURCE (MACRO2)     /* INCLUDE THE _TERMS MACRO */;            00032240
*-----------------------------------------------------------------;     00032250
*----------------------------MODULE 18----------------------------;     00032260
*---------GENERATE A PROGRAM TO ADD ESTIMATED VALUES (ONE PER MODEL---;  00032270
*----------PER OBSERVATION) TO THE TRAINING & CHECKING SETS---------;    00032280
DATA _TERMS2_;                                                          00032290
        LENGTH ESTNAME $8;                                               00032300
        FILE GSOURCE (EST2) OLD;                                         00032310
        IF _N_ = 1 THEN  LINK SETUP;                                    00032320
        SET _TERMS2_ END=FINAL;                                          00032330
        EST = _N_ + KC;                                                  00032340
        ESTNAME = 'E'||LEFT(EST)    /*MAKE NAMES OF VARS FOR ESTIMATES*/; 00032350
```

```
        PUT ESTNAME ' = ' _TERMS ';' ;                              00032360
        IF FINAL THEN PUT                                           00032370
          'IF INCHECK THEN OUTPUT _CHECK_;'/                        00032380
          'ELSE OUTPUT _TRAIN_;'/                                   00032390
          'KEEP _ID _DEPVAR E' FIRST '-E' LAST';';                 00032400
        RETURN;                                                     00032410
  SETUP: CYCLE = _ITER              /*GET VALUES FROM MACROS*/;     00032420
        MODELS = _MODELS;                                           00032430
        KC = CYCLE * 1000;                                          00032440
        FIRST = LEFT (KC + 1);                                      00032450
        LAST = LEFT (KC + MODELS);                                  00032460
        DROP CYCLE MODELS KC FIRST LAST EST;                        00032470
        RETAIN KC FIRST LAST;                                       00032480
        PUT '*GENERATED PGM FROM MODULE 18;'/'*;'/                  00032490
          'DATA _TRAIN_ _CHECK_;'/   /*WRITE DATASET NAMES*/        00032500
          '   SET _TRAIN_ _CHECK_ (IN=INCHECK);';                   00032510
        RETURN;                                                     00032520
*----------------------MODULE 19-------------------------------; 00032530
RUN;                                                               00032540
%INCLUDE GSOURCE (EST2)         /*RUN THE PGM FOR ESTIMATED VALUES*/; 00032550
RUN;                                                               00032560
*----------------------------MODULE 20-------------------------; 00032570
*---------GENERATE A PROGRAM TO FIGURE THE TOTAL SQUARED ERROR-------; 00032580
*---------OF EACH MODEL AND OUTPUT THE RMSE'S TO A SAS DATASET-------; 00032590
DATA _NULL_;                                                       00032600
        IF _N_ =2 THEN STOP;                                       00032610
        FILE GSOURCE (MATRIX) OLD /*ROUTE GENERATED STATEMENTS*/;  00032620
        LINK SETUP              /*EXECUTE SUBROUTINE BELOW*/;      00032630
        PUT                     /*GENERATE PROGRAM*/              00032640
          'PROC MATRIX ;'/                                         00032650
          'FETCH OBSD DATA=_CHECK_  /*FETCH CHECKING DATASET*/'/   00032660
          '   (KEEP = _DEPVAR)   /*OBSERVED VALUES ONLY*/;'/       00032670
          'FETCH ESTS DATA=_CHECK_'/                               00032680
          '   (KEEP = E' FIRST '-E' LAST ') /*NOW GET ESTIMATES*/;'/ 00032690
          'ERR = ESTS           /* DIMENSION ERROR MATRIX*/;'/     00032700
          'DO I = 1 TO ' MODELS ';'/                               00032710
          '   ERR(,I) = OBSD(,1) - ESTS(,I) /*MAKE ERRORS*/;'/     00032720
          '   END               /*END OF DO-GROUP*/;'/            00032730
          'OBSDSQ = OBSD  # OBSD  /* SQUARE OBSERVED  */;'/        00032740
          'ERRSQ = ERR # ERR    /* SQUARE ERRORS */;'/            00032750
          'OBSDSSQ = OBSDSQ(+,)  /* SUM OBSD SQUARES  */;'/        00032760
          'ERRSSQ = ERRSQ(+,)   /* SUM ERRS SQUARES  */;'/        00032770
          'RMSE = SQRT(ERRSSQ #/ OBSDSSQ) /*FIND RMSE*/;'/         00032780
          'RMSE = RMSE''       /*TRANSPOSE RMSE MATRIX*/;'/        00032790
          'FETCH MODEL DATA=_VARS2_(KEEP=MODEL) /*GET ID # OF EA MODEL*/;'/ 00032800
          'RMSE = MODEL || RMSE /* CONCATENATE */;'/               00032810
          'OUTPUT RMSE OUT=_RMSE_ /*OUTPUT RMSE MATRIX TO SAS DATASET*/'/ 00032820
          '  (RENAME=(COL1=MODEL COL2=RMSE));'/                    00032830
          'STOP;';                                                 00032840
        RETURN;                                                   00032850
  SETUP: CYCLE = _ITER           /*GET VALUES FROM MACROS*/;       00032860
        CYC = LEFT (CYCLE);                                        00032870
        MODELS = _MODELS;                                          00032880
        KC = CYCLE * 1000;                                         00032890
        FIRST = LEFT (KC + 1);                                     00032900
        LAST = LEFT (KC + MODELS);                                 00032910
        RETURN;                                                   00032920
*------------------------------------------------------------------; 00032930
```

```
RUN;                                                                  00032940
%INCLUDE GSOURCE (MATRIX);                                            00032950
*------------------------MODULE 22----------------------------------; 00032960
*------REPORT ON THE MODELS THIS ITERATION.  UPDATE DATASET----------;00032970
*-------------------- _LIST_ AND THE _ITER_ MACRO.-----------------;  00032980
DATA _REPORT_                        /*MAKE A DATASET FOR REPORTING USE*/; 00032990
  MERGE _RMSE_ _TERMS2_(KEEP=ESTNAME) _VARS2_                         00033000
  (KEEP=NUM RSQ VAR1-VAR5) _VARSET_ (KEEP=VAR5 RENAME=(VAR5=XPROD));  00033010
PROC SORTT;                                                           00033020
  BY RMSE                            /* RANK BY RMSE */;              00033030
DATA _REPORT_                        /*THIS STEP FLAGS DUPLICATE MODELS*/; 00033040
  RETAIN MODELS;                                                      00033050
  KILL = 0                           /* INITIALIZE KILL VARIABLE*/;   00033060
  DROP D MODELS;                                                      00033070
  SET _REPORT_ END=LAST              /* TAKE AN OBSERVATION */;       00033080
  FILE GSOURCE(MACRO22)              /* WRITE TO OS FILE*/;           00033090
  IF _N_ = 1 THEN MODELS = _MODELS /* INITIALIZE MODELS VARIABLE */;  00033100
  D = DIF(RMSE)                      /* DEFINE DIFFERENCE VARIABLE */; 00033110
  IF ABS (D) LE .0005*RMSE AND D NE . THEN DO;                       00033120
    MODELS = MODELS - 1              /* DECREMENT # MODELS */;        00033130
    KILL = 1                         /* FLAG THE OBSERVATION*/;       00033140
    PUT 'MACRO _KILL ' KILL 1. '%';                                  00033150
    END                             /* END OF DO-LOOP */;            00033160
  IF LAST THEN DO;                                                    00033170
    IF MODELS < _KEEP THEN PUT 'MACRO _KEEP ' MODELS 3. '%';         00033180
    END;                                                             00033190
RUN;                                                                  00033200
%INCLUDE GSOURCE(MACRO22);                                            00033210
DATA _REPORT_                        /* THIS STEP DROPS DUPLICATE MODELS*/;00033220
  SET _REPORT_;                                                       00033230
  IF KILL = 0;                                                        00033240
  DROP KILL;                                                          00033250
RUN;                                                                  00033260
*;                                                                    00033270
* NEXT UPDATE THE INPUT VARIABLE LIST WHILE WRITING THE REPORT;       00033280
*;                                                                    00033290
DATA _LIST_ (KEEP=ESTNAME RENAME=(ESTNAME=VARNAME));                  00033300
  TITLE G M D H   A L G O R I T H M;                                 00033310
  TITLE2  ;                                                           00033320
  TITLE3 DEPENDENT VARIABLE: _DEPVAR;                                 00033330
  TITLE4 TRAINING DATA SET: _TRAIN;                                  00033340
  TITLE5 CHECKING DATA SET: _CHECK;                                  00033350
  TITLE6 ITERATION NO. _ITER;                                        00033360
  FILE PRINT                         /* ROUTE OUTPUT TO PRINTER */;  00033370
  SET _REPORT_ END= _LAST_          /* WRITE REPORT NEXT */;         00033380
  IF _N_ =1 THEN DO;                                                 00033390
    LINK SETUP                       /* EXECUTE "SETUP" SUBROUTINE */; 00033400
    LINK HEADER                      /* EXECUTE "HEADER" SUBROUTINE*/; 00033410
    LINK RMSECHK                     /* COMPARE RMSE TO LAST TIME */; 00033420
    END                             /* END OF DO-LOOP */;            00033430
  IF _N_ GT KEEP THEN STOP;                                          00033440
  PUT ESTNAME @16 NUM @24 RSQ 5.3 /* PRINT A LINE OF OUTPUT */       00033450
    @36 RMSE 6.4 @48 VAR1-VAR5 @103 XPROD/;                          00033460
  IF _N_ = KEEP OR _LAST_ GT 0 THEN GO TO FINISH;                   00033470
  RETURN;                                                            00033480
```

```
HEADER: PUT              //          /* PRINT HEADER ON OUTPUT*/        00033490
'REPORT ON MODEL EVALUATION FOR THIS ITERATION' / 46*'-'//              00033500
@7 NVARS 'VARIABLES WERE INPUT FOR POSSIBLE USE.'/                      00033510
@7 MODELS 'QUADRATIC PARTIAL MODELS WERE EVALUATED.'/                   00033520
@7 KEEP 'MODELS WERE RETAINED ON THE BASIS OF LOWEST CHECKING-SET '     00033530
'RMS ERROR.' ///                                                        00033540
'NAME OF' @13 'NUMBER' @ 23 'TRAINING' @35 'CHECKING'                   00033550
@103 'FACTORS IN THE'/                                                  00033560
'ESTIMATE' @13 'OF VARS' @23 'SET' @ 35 'SET'                           00033570
@48 'LIST OF' @103 'CROSSPRODUCT'/                                      00033580
'VARIABLE' @13 'IN MODEL' @23 'RSQUARE' @ 35 'RMS ERROR'                00033590
@48 'VARIABLES INCLUDED IN MODEL' @103                                 00033600
'VARIABLE ("CPNNNN")'/                                                  00033610
9*'-' @13 8*'-' @35 9*'-' @48 48*'-' @103 27*'-'                        00033620
       @23 9*'-'/;                                                      00033630
RETURN;                                                                 00033640
SETUP: *GET VALUES OF MACROS;                                          00033650
  RETAIN KEEP NEXT STOP KILL;                                          00033660
  NEXT = _ITER + 1;                                                    00033670
  KILL = _KILL;                                                        00033680
  MODELS = _MODELS;                                                    00033690
  NVARS = _NVARS;                                                      00033700
  KEEP = _KEEP;                                                        00033710
RETURN;                                                                00033720
RMSECHK: * CHECK RMSE VS. OLD RMSE, RECORD "_RMSE" AND "_STOP" MACROS;  00033730
  FILE GSOURCE(MACRO3) OLD        /*WRITE TO OS FILE*/;                 00033740
  PUT 'MACRO _RMSE ' RMSE BEST15. '%' /* WRITE NEW _RMSE MACRO*/;       00033750
  IF KEEP < 3 THEN STOP = 2       /*SET STOP IF TOO FEW VARS LEFT*/;    00033760
  IF NEXT GT 2 THEN DO            /*NEXT IS NEXT ITERATN # */;          00033770
    IF RMSE GE _RMSE * .995       /*COMPARE RMSE TO OLD RMSE*/          00033780
      THEN STOP = 1               /* STOP PGM IF NO IMPRVMNT */;        00033790
    END                           /*END OF DO-GROUP*/;                 00033800
  PUT 'MACRO _STOP ' STOP 1. '%'  /*WRITE MACRO */;                     00033810
  FILE PRINT MOD                  /*RESET "PUT" STREAM*/;               00033820
  RETURN                          /*END OF SUBROUTINE*/;                00033830
FINISH: *PRINT NOTE ON REPORT ;                                        00033840
  PUT // 'NOTES: ALTHOUGH ITS FACTORS ARE ALWAYS LISTED,'              00033850
  ' A CROSSPRODUCT VARIABLE MAY OR MAY NOT'                            00033855
  ' BE INCLUDED IN EACH MODEL.'/                                      00033860
  '        A VARIABLE ENDING IN "S" REPRESENTS THE SQUARE '           00033870
  'OF THE ORIGINAL VARIABLE.'/                                        00033880
  '        EACH MODEL ALSO INCLUDES AN INTERCEPT.'/;                  00033890
  IF KILL = 1 THEN PUT                                                00033900
    'NOTE:  AT LEAST 1 DUPLICATE MODEL WAS DROPPED, WHICH MAY '       00033910
    'CAUSE FEWER MODELS TO BE RETAINED THAN EXPECTED.'/;              00033920
  IF STOP = 1 THEN PUT                                                00033930
    'THE CHECKING-SET RMS ERROR HAS NOT IMPROVED '                    00033940
    'SINCE THE LAST ITERATION, WHICH '                                00033950
    'INDICATES ' / 'THAT THE SOLUTION '                               00033960
    'IS AT THE PREVIOUS ITERATION.';                                  00033970
  ELSE IF STOP = 2 THEN PUT                                           00033980
    'NOT ENOUGH VARIABLES ARE AVAILABLE TO RUN THE NEXT '             00033990
    'ITERATION.  (THE MINIMUM IS 3.)' /                               00034000
    'THE PROGRAM WILL END '                                           00034010
    'DUE TO THIS CONDITION.';                                         00034020
```

```
      ELSE PUT                                                                 00034030
        'THE 'KEEP 'VARIABLES LISTED ABOVE WILL BE USED AS'                    00034040
        ' INPUT FOR ITERATION ' NEXT '.';                                      00034050
 *-------------------------------MODULE 23-----------------------------;       00034060
 RUN;                                                                          00034070
 %INCLUDE GSOURCE (MACRO3)            /* INCLUDE MACROS FROM MOD 22 */;         00034080
 *-----------------------MODULE 24---------------------------------;           00034090
 *-------------------GENERATE A PROGRAM TO---------------------------;         00034100
 *-------------------MAKE PLOTS IF DESIRED---------------------------;         00034110
 TITLE;                                                                        00034120
 DATA _PLOTS_                         /*MAKE A DATASET FOR PLOTTING*/;          00034130
 FILE GSOURCE (PLOTS) OLD             /* ROUTE GENERATED STATEMENTS */;        00034140
   IF _PLOT LE 0 THEN LINK NOPLOTS    /* NO PLOTS IF MACRO = 0 OR . */;        00034150
   IF _PLOTMAX LE 0 THEN LINK NOPLOTS /* NO PLOTS IF MACRO = 0 OR . */;        00034160
   IF _STOP = 1 THEN LINK NOPLOTS     /*NO PLOTS IF SOLN @ LAST ITER*/;        00034170
   IF _ITER - _PLOT LT 0 THEN LINK NOPLOTS;                                    00034180
   SET_ CHECK_ (IN=INCHECK_) _TRAIN_ END=EOF_;                                 00034190
   IF INCHECK_ THEN _SET_ = 'C';                                              00034200
   ELSE _SET_ = 'T';                                                           00034210
   OUTPUT;                                                                     00034220
   IF EOF_ THEN DO     /* GENERATE PLOTTING PROGRAM */;                        00034230
     KEEP_ = MIN ( _KEEP, _PLOTMAX)   /*KEEP = # OF PLOTS TO MAKE*/;           00034240
     PUT 'PROC PLOT DATA=_PLOTS_;'    /*INVOKE PROC PLOT*/;                    00034250
     DO I = 1 TO KEEP_                /*GENERATE PLOT STMNTS*/;                 00034260
       SET _LIST_                     /*GET ESTIMATE NAME*/;                    00034270
       PUT 'PLOT _DEPVAR * _ID = _SET_ '/                                      00034280
           VARNAME_ * _ID = '*'' / OVERLAY;';                                 00034290
       END                            /*END OF INNER DO-GROUP*/;              00034300
     END                              /*END OF OUTER DO-GROUP*/;              00034310
   RETURN;                                                                     00034320
   NOPLOTS: PUT ' ';                                                           00034330
   STOP;                                                                       00034340
 *-------------------------MODULE 25-------------------------------;           00034350
 RUN                                  /* RUN THE PROGRAM SO FAR */;            00034360
 %INCLUDE GSOURCE (PLOTS)             /*INCLUDE THE PLOTTING PROGRAM*/;        00034370
 RUN;                                                                          00034380
 *-----------------------MODULE 26---------------------------------;           00034390
 *--------------INCREMENT _ITER MACRO-------------------------------;          00034400
 DATA _NULL_                          /* CHECK WHETHER TO CONTINUE */;         00034410
   IF _STOP = 1 THEN ABORT ABEND 2601 /* END THE PROGRAM */;                   00034420
   ELSE IF _STOP = 2 THEN ABORT ABEND 2602;                                   00034430
   ELSE IF _STOP > 0 THEN ABORT ABEND 2603;                                   00034440
   ELSE FILE GSOURCE (MACRO4)         /* ROUTE OUTPUT */;                     00034450
   ITER = _ITER + 1                   /* DEFINE ITER VARIABLE */;             00034460
   PUT 'MACRO _ITER ' ITER '%'        /* INCREMENT _ITER MACRO */;            00034470
 *-------------------------------------------------------------------;        00034480
 RUN                                  /* RUN THE PROGRAM SO FAR */;            00034490
 %INCLUDE GSOURCE (MACRO4)            /* INCLUDE NEW _ITER MACRO */;          00034500
```

ACKNOWLEDGMENTS

Dr. Saul Saila, my major professor, first suggested that I work on GMDH, and has encouraged me to continue. Henry Walker aided me with computer and statistical problems, gave many helpful suggestions, and provided much encouragement.
The use of the SAS language and several SAS procedures simplified the writing of this program considerably. I would also like to acknowledge the help of the programming consultants at SAS Institute and of Craig Dickstein at the University of Rhode Island.
The manuscript was reviewed by John Hoenig, Cynthia Jones, Dr. Saul Saila, Dr. Sybil Seitzinger, and Henry Walker, all of whom suggested improvements. Any remaining errors are entirely my own.
The bulk of this work was funded by a National Science Foundation graduate fellowship to the author; however, the University of Rhode Island and its Graduate School of Oceanography, which generously contributed computer time, also deserve acknowledgment.

REFERENCES

Draper, N. R., and Smith, H. (1981). Applied Regression Analysis, 2nd ed. Wiley, New York.
Duffy, J. J. (1974). Identification of Soil Nitrogen Transformations in an Agricultural Ecosystem. Doctoral dissertation, Sever Institute of Technology, Washington University, pp. 12-26.
Farlow, S. J. (1981). The GMDH Algorithm of Ivakhnenko. Am. Stat. 35(4):210-215.
Ivakhnenko, A. G. (1976). The Group Method of Data Handling in Prediction Problems. Sov. Autom. Control 9(6):21-30.
Ivakhnenko, A. G., Krotov, G. I., and Visotsky, V. N. (1979). Identification of the Mathematical Model of a Complex System by the Self-Organization Method. In Theoretical Systems Ecology, E. Halfon (Ed.). Academic Press, New York, pp. 325-352.
Kennedy, P. (1979). A Guide to Econometrics. M.I.T. Press, Cambridge, Mass.
Mallows, C. L. (1973). Some Comments on C_p. Technometrics 15:661-675.
SAS (1979). SAS User's Guide, 1979 Edition. SAS Institute Press, Raleigh, N.C.
SAS (1981). SAS System Enhancements. SAS Commun. 7(4):6-7.
SAS (1982). SAS User's Guide: Basics, 1982 Edition. SAS Institute Press, Cary, N.C.
Yurachkovskiy, Yu. P. (1977). Improved GMDH Algorithms for Process Prediction. Sov. Autom. Control 10(5):61-71.

Bibliography

Reprinted from the INSPEC data base by permission of the Institution of Electrical Engineers

The following bibliography provides a comprehensive list of articles on the theory and application of the GMDH algorithm from 1970 to 1982. The articles are ordered according to date of publication (with the exception of the first 21, which were not "date" indexed). A few of the very early papers (1968-1970) are not included in this list. Three of the important ones are

AU Ivakhnenko, A. G.
TI The Group Method of Data Handling—A Rival of the Method of Stochastic Approximation
TR Soviet Automatic Control, Vol. 3, 1968.

AU Ivakhnenko, A. G., Koppa, Yu. U., Wu Suan Min
TI Polynomial and Logical Theory of Dynamic Systems (Part 1)
TR Soviet Automatic Control, Vol. 3, No. 3 (1970)

AU Ivakhnenko, A. G., Koppa, Yu. U., Wu Suan Min
TI Polynomial and Logical Theory of Dynamic Systems (Part 2)
TR Soviet Automatic Control, Vol. 3, No. 4 (1970)

The bibliography lists

AU = author of article
TI = title of article
SO = name of journal where the article was originally published (most of Ivakhnenko's articles were originally published in the Soviet journal Avtomatika
TR = name of translated journal in the case the original source is translated into English (Avtomatika is translated into the English language journal Soviet Automatic Control)

317

***************************** INSB *********************************
AU IVAKHNENKO, A. G. TODUA, M. M. CHUKIN, YU. V.
TI GMDH ALGORITHM WITH SUCCESSIVE DETERMINATION OF POLYNOMIAL TRENDS
   USING TEE MCST ESSENTIAL VARIABLES.
SO AVTOMATIKA (USSR). 60-73. 8.
TR SOV. AUTOM. CONTR. (USA). SAUCBZ. VOL. 5, NO. 2. 44-54. MARCH-APRIL
   1972.

***************************** INSB *********************************
AU IVAKHNENKO, A. G. TODUA, M. M. CHUKIN, YU. V.
TI GMDH ALGORITHM WITH SUCCESSIVE DETERMINATION OF POLYNOMIAL TRENDS
   USING TEE MOST ESSENTIAL VARIABLES.
SO AVTOMATIKA (USSR). 60-73. 8.
TR SOV. AUTOM. CONTR. (USA). SAUCBZ. VOL. 5, NO. 2. 44-54. MARCH-APRIL
   1972.

***************************** INSB *********************************
AU IVAKHNENKO, A. G.
TI STRUCTURAL IDENTIFICATION OF DIFFERENTIAL EQUATIONS BY A SMALL NUMBER
   OF OBSERVATIONS USING SELF-ORGANIZATION METHODS.
SO AVTOMATIKA (USSR). 43-59. 8.
TR SOV. AUTOM. CONTR. (USA). SAUCBZ. VOL. 5, NO. 2. 31-44. MARCH-APRIL
   1972.

***************************** INSB *********************************
AU IVAKHNENKO, A. G. KOPPA, YU. V. TODUA, M. M. PETRACHE, G.
TI MATHEMATICAL SIMULATION OF COMPLEX ECOLOGICAL SYSTEMS.
SO AVTOMATIKA (USSR). 20-34. 10.
TR SOV. AUTOM. CCNTR. (USA). SAUCBZ. VOL.4, NO. 4. 15-26. JULY-AUG.
   1971.

***************************** INSB *********************************
AU STEPANENKO, V. N. STEPASHKO, V. S.
TI VARIANCE ANALYSIS OF REGULARITY (ADEQUACY) OF MULTIPLE REGRESSION
   EQUATIONS FOR GMDH.
SO AVTOMATIKA (USSR). 35-41. 4.
TR SOV. AUTOM. CCNTR. (USA). SAUCBZ. VOL.4, NO. 4. 27-32. JULY-AUG.
   1971.

***************************** INSB *********************************
AU KOPPA, YU. V.
TI AN ALGOL-BESM-6 PROGRAM FOR INTERPOLATION USING THE GMDH ALGORITHM
   WITH POLYNOMIALS OF FIRST CR SECOND DEGREE.
SO AVTOMATIKA (USSR).
TR SOV. AUTCM. CONTR. (USA). SAUCBZ. VOL. 4, NO. 1. 69-71. JAN. 1971.

***************************** INSB *********************************
AU IVAKHNENKO, N. O.
TI A POLYNCMIAL GMDH ALGORITHM WITH SUCCESSIVE INTRODUCTION OF
   VARIABLES.
SO AVTOMATIKA (USSR). 64-6.
TR SOV. AUTOM. CONTR. (USA). SAUCBZ. VOL.3, NO.5. SEPT.-OCT. 1970.

***************************** INSB *********************************

AU DIMITROV, V. D.
TI A STOCHASTIC GMDH ALGORITHM WITH SUCCESSIVE INTRODUCTION OF PAIRS OF
   VARIABLES.
SO AVTOMATIKA (USSR). 61-3.
TR SOV. AUTOM. CONTR. (USA). SAUCBZ. VOL.3, NO.5. SEPT.-OCT. 1970.

****************************** INSB ********************************
AU IVAKHNENKO, A. G. SPYNU, YE. I. DIMITROV, V. D. PATEREU, S. G.
   PATRATIY, L. Z.
TI PREDICTING THE DURATION OF THE DESTRUCTIVE EFFECT AND PESTICIDE
   CONTENT IN PLANTS BY A STOCHASTIC GMDH ALGORITHM.
SO AVTOMATIKA (USSR). 31-40.
TR SOV. AUTOM. CONTR. (USA). SAUCBZ. VOL.3, NO.5. SEPT.-OCT. 1970.

****************************** INSB ********************************
AU IVAKHNENKO, A. G. KOPPA, YU. V.
TI REGULARIZATION OF DECISION FUNCTIONS IN THE GROUP METHOD OF DATA
   HANDLING.
SO AUTOMATIKA (USSR). 10.
TR SOU AUTOMAT. CCNTR. (USA). SAUCBZ. VOL.3, NO.2. 28-37. MARCH-APRIL
   1970.

****************************** INSB ********************************
AU IMAS, L. N.
TI AN ALGOL-60 PROGRAM FOR GMDH ALGORITHMS WITH QUADRATIC POLYNOMIALS.
SO AUTOMATIKA (USSR). 56-9.
TR SOV. AUTOMAT. CONTR. (USA). SAUCBZ. VOL. 3, NO. 4. JULY-AUG. 1970.

****************************** INSB ********************************
AU IVAKHNENKO, A. G. KOPPA, YU. V. WU SUAN MIN.
TI POLYNOMIAL AND LOGICAL THEORY OF DYNAMIC SYSTEMS. I.
SO AVTOMATIKA (USSR).
TR SOV. AUTOM. CCNTR. (USA). SAUCBZ. VOL.3, NO.3. 1-13. MAY-JUNE 1970.

****************************** INSB ********************************
AU IVAKHNENKO, A. G.
TI REMARKS ON SOME PAPERS OF PARKS AND PYATT CONCERNING CONTROL OF
   DYNAMIC MODELS IN ECONOMICS.
SO AUTOMATIKA (USSR). 31-3.
TR SOV. AUTOMAT. CONTR. (USA). SAUCBZ. VOL. 3, NO. 4. JULY-AUG. 1970.

****************************** INSB ********************************
AU LEBEDEV, S. D. SHCHERBININ, YU. H.
TI SOLUTION OF A COMBINATORIAL PROBLEM IN THE DESIGN OF SELF-CHECKING
   EQUIPMENT.
SO AVTOMATIKA (USSR). 40-4.
TR SCV. AUTOM. CONTR. (USA). SAUCBZ. VOL.3, NO.5. SEPT.-OCT. 1970.

****************************** INSB ********************************
AU DIMITROV, V. D.
TI GMDH ALGORITHMS ON FUZZY SETS OF ZADEH.
SO AUTOMATIKA (USSR). 40-5.
TR SOV. AUTOMAT. CCNTR. (USA). SAUCBZ. VOL. 3, NO. 4. JULY-AUG. 1970.

****************************** INSB ********************************
AU IVAKHNENKO, A. G. KOPPA, YU. V. WU SUAN MIN.
TI POLYNOMIAL AND LOGICAL THEORY OF DYNAMIC SYSTEMS. II.

SO AUTOMATIKA (USSR). 11-30.
TR SOV. AUTOMAT. CONTR. (USA). SAUCBZ. VOL. 3, NO. 4. JULY-AUG. 1970.

******************************* INSB *******************************
AU IVAKHNENKO, A. G.
TI THE GROUP METHOD OF DATA HANDLING IN OPTIMAL CONTROL WITH
   ACCUMULATION OF INFORMATION.
SO AVTOMATIKA (USSR).
TR SOV. AUTOMAT. CONTR. (USA). NO.5. 57-60. 1969.

******************************* INSB *******************************
AU DIMITROV, V. D.
TI USE OF GENETIC THEORY FOR SELECTING PARTIAL DESCRIPTIONS IN THE GROUP
   METHOD OF DATA HANDLING.
SO AUTOMATIKA (USSR). 57-63.
TR SOV. AUTOMAT. CONTR. (USA). NO.6. 1969.

******************************* INSP *******************************
AU DYLBOKOV, I. S. LEFTEROV, A. G. DYLBOKOVA, D. L.
TI INSTRUMENTATION OF POLYNOMIAL GMDH ALGORITHMS.
SO AVTOMATIKA (USSR). JAN-FEB 1976. 19-28. 9.
TR SOV. AUTOM. CONTROL (USA). SAUCBZ. VOL.9, NO.1. 13-22. JAN.-FEB.
   1976..

******************************* INSP *******************************
AU KAZANSKIY, A. V. PERCHUK, V. L. GOLENKOV, YE. A.
TI GRAPH DIALOGUE FOR HEURISTIC CONTROL OF SELF-SAMPLING OF USEFUL
   INFORMATION ON THE BASIS OF ADONIS SYSTEM.
SO AVTOMATIKA (USSR). JAN-FEB 1976. 39-47. 13.
TR SOV. AUTOM. CONTROL (USA). SAUCBZ. VOL.9, NO.1. 31-7. JAN.-FEB.
   1976.

******************************* INSP *******************************
AU IHARA, C. IVAKHNENKO, A. G. IVAKHNENKO, N. A.
TI UNIQUE SELECTION OF MODEL BY BALANCE-OF-VARIABLES CRITERION.
SO AVTOMATIKA (USSR). JAN-FEB 1976. 86-8. 2.
TR SOV. AUTOM. CONTROL (USA). SAUCBZ. VOL.9, NO.1. 70-2. JAN.-FEB.
   1976..

******************************* INSB *******************************
AU IVAKHNENKO, A. G. DIMITROV, V. D.
TI STOCHASTIC ALGORITHMS OF THE METHOD OF DATA HANDLING BY GROUPS IN THE
   PROBLEM OF PREDICTING THE RANDOM EVENTS.
SO AVTOMATIKA (USSR). NO.3. 49.
YR 69.
TR SOVIET AUTOMATIC CONTROL (USA).

******************************* INSB *******************************
AU IVAKHNENKO, A. G. KOPPA, YU. V. TIMCHENKO, I. K. IVAKHNENKO, N.
   A.
TI METHOD OF GROUP DATA HANDLING IN THE IDENTIFICATION OF STATISTICAL
   CHARACTERISTICS IN MULTI-EXTREMAL PLANT AND ANIMAL SELECTION.
SO AVTOMATIKA (USSR). NO.2. 43-53.
YR 69.

******************************* INSB *******************************

AU IVAKHNENKO, L. M. DIMITRCV, V. D. GULYAN, N. V.
TI THE GROUP METHOD OF DATA HANDLING USED IN THE CONSTRUCTION OF AN
   'ASSOCIATIVE' MODEL OF HUMAN DECISION MAKING.
SO AVTOMATIKA (USSR). VOL.3, NO.1. 23-33.
YR 70.
TR SOV. AUTCMAT. CONTR. (USA). SAUCBZ.

***************************** IN SB *********************************
AU IVAKHNENKO, A. G.
TI HEURISTIC SELF-ORGANIZATION IN PROBLEMS OF ENGINEERING CYBERNETICS.
SO AUTOMATICA (GB). VCL.6, NO.2. 207-19. 13.
YR 70.

***************************** INSB *********************************
AU IVAKHNENKO, A. G. SPYNU, E. I. DIMITROV, V. D. PATEREU, S. G.
   PATRATII, I. Z.
TI MULTISERIES ALGORITHMS OF THE THEORY OF STATISTICAL SOLUTIONS (WITH
   REFERENCE TC PROBLEMS OF CLASS RECOGNITION).
SO PROCEEDINGS OF THE 5TH ALL-UNION CONFERENCE ON CONTROL PROBLEMS. II.
   MOSCOW, USSR. MOSCOW, USSR. 236. 110-12. NAT. COMMITTEE OF AUTOMATIC
   CONTRCL OF THE USSR. NAT. COMMITTEE OF AUTOMATIC CONTROL OF THE USSR.
   4-8 OCT 1971.
YR 71.

***************************** INSB *********************************
AU IVAKHNENKO, A. G.
TI POLYNOMIAL THEORY OF CCMPLEX SYSTEMS.
SO IEEE TRANS. SYST. MAN AND CYBERN. (USA). VOL. SMC-1, NO. 4. 364-78.
YR 71.

***************************** IN SB *********************************
AU PISKUNOV, V. O.
TI CONSTRUCTICN OF AN ANALYTICAL MODEL OF A GAME SITUATION BY THE GROUP
   METHOD OF DATA HANDLING.
SO AVIOMATIKA (USSR). VOL.5, NO.6. 21-6. 3.
YR 72.
TR SOV. AUTCM. CONTROL (USA). SAUCBZ. VOL.5, NO.6. NOV.-DEC. 1972.

***************************** IN SB *********************************
AU IVAKHNENKO, A. G. STETSENKO, M. D. SVETALSKIY, B. K.
TI STRUCTURAL IDENTIFICATION OF PHOTOSYNTHESIS PROCESS BY
   SELF-ORGANIZATICN METHODS.
SO AVTOMATIKA (USSR). VOL.5, NO.6. 13-21. 11.
YR 72.
TR SOV. AUTCM. CONTROL (USA). SAUCBZ. VOL.5, NO.6. NOV.-DEC. 1972.

***************************** IN SB *********************************
AU IVAKHNENKO, A. G. KOVALCHUK, P. I.
TI UNIQUE CCNSTRUCTION OF REGRESSION CURVE USING A SMALL NUMBER OF
   POINTS.
SO AVTOMATIKA (USSR). VOL.5, NC.5. 34-41. 15.
YR 72.
TR SOV. AUTCM. CCNTROL (USA). SAUCBZ. VOL.5, NO.5. 26-32. SEPT.-OCT.
   1972.

***************************** IN SB *********************************

AU IVAKHNENKO, N. A.   KVASKO, M. Z.
TI COMBINATION-GENERATING GMDH ALGORITHMS IN WHICH THE REGULARITY OF
   BOTH 'SYMMETRICAL' AND 'NONSYMMETRICAL' POLYNOMIALS IS CHECKED.
SC AVTOMATIKA (USSR). VOL.5, NO.5. 42-8. 6.
YR 72.
TR SOV. AUTOM. CONTROL (USA). SAUCBZ. VOL.5, NO.5. 33-8. SEPT.-OCT.
   1972.

********************************** INSB **********************************
AU IVAKHNENKO, A. G.   TODUA, M. M.
TI PREDICTION OF RANDOM PROCESSES USING SELF-ORGANIZATION OF THE
   PREDICTION EQUATIONS. I. PROBLEMS OF SIMPLE MEDIUM-TERM PREDICTION.
SO AVTOMATIKA (USSR). VOL.5, NO.3. 46-68. 15.
YR 72.
TR SOV. AUTOM. CONTROL (USA). SAUCBZ. VOL.5, NO.3. 35-52. MAY-JUNE 1972.

********************************** INSB **********************************
AU PATRATIY, I. Z.   SHELUDKO, O. I.   DROBOTYA, A. L.
TI PERFORMANCE OF STOCHASTIC GMDH ALGORITHMS IN ACTUAL EXAMPLES.
SO AVTCMATIKA (USSR). VOL.5, NO.5. 49-58. 10.
YR 72.
TR SOV. AUTOM. CONTROL (USA). SAUCBZ. VOL.5, NO.5. 39-46. SEPT.-OCT.
   1972.

********************************** INSB **********************************
AU IVAKHNENKO, A. G.   PETRACHE, G.   KRASYTSKYY, M. S.
TI A GMDH ALGORITHM WITH RANDOM SELECTION OF PAIRS.
SO AVTCMATIKA (USSR). VOL.5, NO.4. 33-41. 4.
YR 72.
TR SOV. AUTOM. CONTROL (USA). SAUCBZ. VOL.5, NO.4. 23-30. JULY-AUG.
   1972.

********************************** INSB **********************************
AU SVETALSKYY, D. K.
TI A GRADIENT METHOD FOR EXTENDING THE GMDH TO OBJECTS THAT DID NOT TAKE
   PART IN EXPERIMENTS.
SO AVTCMATIKA (USSR). VOL.5, NO.3. 68-76. 11.
YR 72.
TR SOV. AUTOM. CONTROL (USA). SAUCBZ. VOL.5, NO.3. 52-8. MAY-JUNE 1972.

********************************** INSB **********************************
AU IVAKHNENKO, A. G.   DIMITROV, V. D.   GULYAN, N. V.   IVAKHNENKO, L. N.
TI PROBLEMS IN THE SIMULATION OF COMPLEX OBJECTS ON THE BASIS OF
   HEURISTIC SELF-ORGANIZATION.
SO KIBERN. AND VYCHISL. TEKH. (USSR). NO.13. 18-38. 21.
YR 72.

********************************** INSB **********************************
AU IVAKHNENKO, A. G.   SPINU, E. I.   DIMITROV, V. D.   PATEREU, S. G.
   PATRATY, I. S.
TI MULTILAYER ALGORITHMS OF DECISION THEORY.
SO ABSTRACTS OF THE IFAC 5TH WORLD CONGRESS. PARIS, FRANCE. PITTSBURGH,
   PA. USA. XIII&179. 119. INSTRUMENT SOC. AMERICA. IFAC UNION OF
   INTERNAT. ENGNG. ORGANIZATIONS. 12-17 JUN 1972.
YR 72.

********************************** INSB **********************************

AU IVAKHNENKO, A. G. DELEUR, YE. G. YAREMENKO, A. G. PATEREU, S. G.
   TODUA, M. M.
TI GMDH ALGORITHM FOR SYNTHESIS OF MULTIPLICATIVE MODELS OF COMPLEX
   SYSTEMS.
SO AVTOMATIKA (USSR). VOL.6, NO.6. 72-6. 11.
YR 73.
TR SOV. AUTOM. CONTROL (USA). SAUCBZ. VOL.6, NO.6. 52-7. NOV.-DEC.
   1973.

******************************* INSB ********************************
AU IVAKHNENKO, A. G. KOVALCHUK, P. I. TODUA, M. M. SHELUDKO, O. I.
   DUBROVIN, O. F.
TI UNIQUE CONSTRUCTION OF REGRESSION CURVE USING A SMALL NUMBER OF
   POINTS. II.
SO AVTOMATIKA (USSR). VOL.6, NO. 5. 35-49. 16.
YR 73.
TR SOV. AUTOM. CONTROL (USA). SAUCBZ. VOL6, NO. 5. 29-41. SEPT.-OCT.
   1973.

******************************* INSB ********************************
AU IVAKHNENKO, A. G. OVCHINNIKOV, V. A. PETRACHE, G. GULYAN, N. V.
TI CONTROL OF COMPLEX PLANTS WITH PREDICTION OPTIMIZATION BASED ON
   SELF-ORGANIZATION.
SO AVTOMATIKA (USSR). VOL. 6, NO. 4. 39-52. 9.
YR 73.
TR SOV. AUTOM. CONTROL (USA). SAUCBZ. VOL. 6, NO. 4. 29-40. JULY-AUG.
   1973.

******************************* INSB ********************************
AU PISKUNOV, V. A.
TI DETERMINATION OF SAMPLE SIZE IN APPROXIMATION PROBLEMS.
SO AVTOMATIKA (USSR). VOL.6, NO. 5. 50-4.
YR 73.
TR SOV. AUTOM. CONTROL (USA). SAUCBZ. VOL6, NO. 5. 42-6. SEPT.-OCT.
   1973.

******************************* INSB ********************************
AU PATEREU, S. G. SHELUDKO, O. I.
TI A PROBABILISTIC GMDH ALGORITHM WITH SEQUENTIAL DISCRIMINATION OF
   INPUT FEATURES.
SO AVTOMATIKA (USSR). VOL.5, NO.3. 37-43. 6.
YR 73.
TR SOV. AUTOM. CONTROL (USA). SAUCBZ. VOL.6, NO.3. 29-33. MAY-JUNE 1973.

******************************* INSB ********************************
AU IVAKHNENKO, A. G. TODUA, M. M. GULYAN, N. V. SVETALSKIY, E. K.
TI STATISTICAL PREDICTION OF RANDOM PROCESSES USING SELF-ORGANIZATION OF
   THE PREDICTION EQUATIONS. III. EXAMPLES OF SYSTEMATIC MULTIPLE
   DIFFERENTIAL PREDICTION OF RANDOM PROCESSES.
SO AVTOMATIKA (USSR). VOL.5, NO.3. 15-36. 9.
YR 73.
TR SOV. AUTOM. CONTROL (USA). SAUCBZ. VOL.6, NO.3. 11-28. MAY-JUNE 1973.

******************************* INSB ********************************
AU IVAKHNENKO, A. G. EYKOFF, P.
TI SELF-ORGANIZATION THEORY AS THE BASIS OF DIRECT COMPLEX SYSTEM
   MODELLING AFTER EXPERIMENTAL DATA.

SO PROCEEDINGS OF THE 3RD IFAC SYMPOSIUM ON IDENTIFICATION AND SYSTEM
   PARAMETER ESTIMATION. II. THE HAGUE, DELFT, NETHERLANDS. AMSTERDAM,
   NETHERLANDS. XII&645PP. 1060-3. 4. NORTH HOLLAND. IFAC. 12-15 JUNE
   1973.
RN 0 7204 20830.
YR 73.

******************************* INSB ********************************
AU IVAKHNENKO, A. G. TOLOKHNYANENKO, V. A.
TI CONTROL WITH PREDICTION OPTIMIZATION IN SIMPLE CONTINUOUS PLANT
   MODELS.
SO AVTOMATIKA (USSR). VOL.6, NO.2. 64-72. 7.
YR 73.
TR SOV. AUTOM. CONTROL (USA). SAUCBZ. VOL.6, NO.2. 48-5. MARCH-APRIL
   1973.

******************************* INSB ********************************
AU IVAKHNENKO, A. G. TODUA, M. M. CHUKIN, YU. V.
TI STATISTICAL PREDICTION OF RANDOM PROCESSES ON THE BASIS OF
   SELF-ORGANIZATION OF PREDICTION EQUATIONS. II. SYSTEMATIC MULTIPLE
   DIFFERENTIAL PREDICTION WITH SEPARATION OF TWO TRENDS.
SO AVTOMATIKA (USSR). VOL.6, NO.2. 44-63. 9.
YR 73.
TR SOV. AUTOM. CONTROL (USA). SAUCBZ. VOL.6, NO.2. 32-47. MARCH-APRIL
   1973.

******************************* INSB ********************************
AU IVAKHNENKO, A. G. TODUA, M. M. FOMICHEV, A. A.
TI STRUCTURAL SYNTHESIS OF FILTERS FOR INDIRECT MEASUREMENT OF VARIABLES
   ON THE GENERALIZED GMDH (GROUP METHOD OF DATA HANDLING) ALGORITHM.
SO AVTOMATIKA (USSR). VOL.6, NO.1. 49-63. 4.
YR 73.
TR SOV. AUTOM. CONTROL (USA). SAUCBZ. VOL.6, NO.1. 37-47. JAN.-FEB.
   1973.

******************************* INSB ********************************
AU DUFFY, J. FRANKLIN, M.
TI A CASE STUDY OF ENVIRONMENTAL SYSTEM MODELING WITH THE GROUP METHOD
   OF DATA HANDLING.
SC 14TH JOINT AUTOMATIC CONTROL CONFERENCE OF THE AMERICAN CONTROL
   COUNCIL. (PREPRINTS). COLUMBUS, OHIO, USA. NEW YORK, USA. XIX&963 PP.
   101-11. 2. IEEE. AMERICAN AUTOMATIC CONTROL COUNCIL ASME IEEE ISA ET
   AL. 20-22 JUNE 1973.
YR 73.

******************************* INSB ********************************
AU IVAKHNENKO, A. G.
TI PROBLEMS OF COMPLEX SYSTEM MODELING AND APPLIED MATHEMATICAL
   STATISTICS.
SO INT. J. COMPUT. AND INF. SCI. (GB). VOL.2, NO.1. 49-60. 14.
YR 73.

******************************* INSB ********************************
AU IVAKHNENKO, A. G. TOLOKHNYANENKO, V. A. YAREMENKO, A. G.
TI CONTROL WITH PREDICTION OPTIMISATION USING DISCRETE LINEAR PREDICTION
   MODELS OF THE PLANT.
SO AVTOMATIKA (USSR). VOL.7, NO.2. 38-44. 6.

YR 74.
TR SOV. AUTOM. CONTROL (USA). SAUCBZ. VOL. 7 NO. 1. 22-31. JAN.-FEB.
   1974.

***************************** INSB *********************************
AU SHERVASHIDZE, V. V. SHARGOVSKIY, V. P.
TI STUDY OF ABERRATION OF OPTICAL SYSTEMS BY HEURISTIC SELF-ORGANISATION
   METHODS.
SO AVTOMATIKA (USSR). VOL.7, NO.2. 45-7. 6.
YR 74.
TR SCV. AUTOM. CONTROL (USA). SAUCBZ. VOL. 7 NO. 1. 32-7. JAN.-FEB.
   1974.

***************************** INSB *********************************
AU IVAKHNENKO, A. G. KOPPA, YU. V.
TI SELECTION OF FEATURE ENSEMBLE AND SYNTHESIS OF MULTILAYER PERCEPTRON
   USING SELF-ORGANIZATION.
SO AVTOMATIKA (USSR). VOL.7 NO.2. 41-53. 5.
YR 74.
TR SOV. AUTCM. CONTROL (USA). SAUCBZ. VOL. 7 NO.2.. 33-43. MARCH-APRIL
   1974.

***************************** INSB *********************************
AU BRAVERMAN, V. YA. POMYCHOV, O. O.
TI SOME ASPECTS OF THE USE OF THE GROUP METHOD OF DATA HANDLING IN THE
   CASE OF SMALL SAMPLES.
SO AVTOMATIKA (USSR). VOL.7 NO.2. 34-40. 4.
YR 74.
TR SOV. AUTOM. CCNTROL (USA). SAUCBZ. VOL. 7 NO.2.. 26-32. MARCH-APRIL
   1974.

***************************** INSB *********************************
AU SHEIUDKC, O. I.
TI GMDH ALGORITHM WITY ORTHOGONALISED COMPLETE DESCRIPTION FOR SYNTHESIS
   OF MODELS BY THE RESULTS CF A PLANNED EXPERIMENT.
SO AVTOMATIKA (USSR). VOL.7, NO.5. 32-42. 5.
YR 74.
TR SOV. AUTOM. CCNTROL (USA). SAUCBZ. VO.7, NO.5. 24-33. SEPT.-OCT.
   1974.

***************************** INSB *********************************
AU IVAKHNENKO, A. G. TOLOKHNYANENKO, V. A. YAREMENKO, A. G.
TI MODEL SELF-ORGANISATION AND CONTROL WITH PREDICTION OPTIMISATION AS A
   MEANS OF TRANSITION FROM AUTOMATED TO AUTOMATIC CONTROL SYSTEMS.
SO AVTOMATIKA (USSR). VOL.7, NO.5. 22-31. 4.
YR 74.
TR SOV. AUTOM. CONTROL (USA). SAUCBZ. VO.7, NO.5. 16-21. SEPT.-CCT.
   1974.

***************************** INSB *********************************
AU DUBROVIN, C. F. STEPASHKO, V. S.
TI AN ADAPTIVE GMDH FILTER IN A CLOSED CONTROL LOOP.
SO AVTOMATIKA (USSR). VOL.7, NO.5. 11-16. 6.
YR 74.
TR SOV. AUTOM. CONTROL (USA). SAUCBZ. VO.7, NO.5. 7-11. SEPT.-OCT.
   1974.

**************************** INSB ********************************
AU PARKS, P.   IVAKHNENKO, A. G.   BOYCHUK, L. M.   SVETALSKIY, V. K.
TI SELF-ORGANIZATION OF A MODEL OF THE BRITISH ECONOMY BY
   BALANCE-OF-VARIABLES CRITERION FOR CONTROL WITH PREDICTION
   OPTIMIZATION.
SO AVTOMATIKA (USSR). VOL.7, NO.6. 30-52. 9.
YR 74.
TR SOV. AUTOM. CONTROL (USA). SAUCBZ. VOL.7, NO.6. 25-43. NOV.-DEC.
   1974.

**************************** INSB ********************************
AU YAREMENKO, A. G.
TI SYNTHESIS OF REGRESSION EQUATION FOR GROSS PRODUCT OF SOUTH CAROLINA
   USING GMDH ALGORITHMS.
SO AVTOMATIKA (USSR). VOL.7, NO.4. 83-6. 5.
YR 74.
TR SOV. AUTOM. CONTROL (USA). SAUCBZ. VOL.7, NO.4. 70-3. JULY-AUG.
   1974.

**************************** INSB ********************************
AU STEPANOV, V. A.
TI RESULTS OF MASS CHECKING OF GMDH EFFICIENCY IN LONG-TERM PREDICTION
   OF DEMAND FOR CONSUMER GOODS.
SO AVTOMATIKA (USSR). VOL.7, NO.4. 60-6. 1.
YR 74.
TR SOV. AUTOM. CONTROL (USA). SAUCBZ. VOL.7, NO.4. 46-51. JULY-AUG.
   1974.

**************************** INSB ********************************
AU IVAKHNENKO, A. G.   IVAKHNENKO, N. A.
TI LONG-TERM PREDICTION OF RANDOM PROCESSES BY GMDH ALGORITHMS USING THE
   UNBIASEDNESS CRITERION AND BALANCE-OF-VARIABLES CRITERION.
SO AVTOMATIKA (USSR). VOL.7, NO.4. 52-9. 4.
YR 74.
TR SOV. AUTOM. CONTROL (USA). SAUCBZ. VOL.7, NO.4. 40-5. JULY-AUG.
   1974.

**************************** INSB ********************************
AU IVAKHNENKO, A. G.   DYLBOKOV, I. S.   KARAN, L. P.   DYLBOKOVA, D. L.
TI RECOVERY OF OMITTED DIGITS OF BINARY CODES USING STOCHASTIC GMDH
   ALGORITHMS.
SO AVTOMATIKA (USSR). VOL.7, NO.3. 20-9. 2.
YR 74.
TR SOV. AUTOM. CONTROL (USA). SAUCBZ. VOL.7, NO.3. 15-23. MAY-JUNE 1974.

**************************** INSB ********************************
AU BUSHUYEV, YE. I.   YEGORTSEV, YE. YA.   KONYUKHOV, S. M.   KOPPA, YU.
   V.   MIMRIKOV, C. K.
TI IDENTIFICATION OF PARAMETERS OF A HIGH-PRESSURE TANK FROM PRODUCTION
   TEST RESULTS.
SO AVTOMATIKA (USSR). VOL.7, NO.3. 13-19. 4.
YR 74.
TR SOV. AUTOM. CONTROL (USA). SAUCBZ. VOL.7, NO.3. 10-14. MAY-JUNE 1974.

**************************** INSB ********************************
AU IVAKHNENKO, A. G.   KOVALCHUK, P. I.   VYSOTSKIY, V. M.
TI SIMULATION OF RANDOM-FIELD DYNAMICS OF WATER POLLUTION USING GMDH

ALGORITHMS.
SO AVTOMATIKA (USSR). VOL.7, NO.3. 30-41. 17.
YR 74.
TR SOV. AUTOM. CONTROL (USA). SAUCBZ. VOL.7, NJ.3. 24-33. MAY-JUNE 1974.

******************************** INSB ********************************
AU IVAKHNENKO, A. G. TOLOKHNYANENKO, V. A. YAREMENKO, A. G.
TI ON THE STABILITY OF DISCRETE CONTROL SYSTEMS WITH PREDICTICN
   OPTIMIZATION (ANSWER TO LETTER TO THE EDITOR).
SO AVTOMATIKA (USSR). VOL.7, NO.3. 90. 2.
YR 74.
TR SOV. AUTOM. CONTROL (USA). SAUCBZ. VOL.7, NO.3. 74. MAY-JUNE 1974.

******************************** INSB ********************************
AU SHERVASHIDZE, V. V. SHABGCVSKIY, V. P.
TI APPROXIMATE SOLUTION OF NONLINEAR SYSTEMS OF EQUATIONS BY OBJECTIVE
   STRUCTURAL IDENTIFICATION USING GMDH ALGORITHMS.
SO AVTOMATIKA (USSR). VCL.8, NO.3. 24-31. 6.
YR 75.
TR SCV. AUTOM. CONTROL (USA). SAUCBZ. VOL.8, NO.3. 18-24. MAY-JUNE 1975.

******************************** INSB ********************************
AU IVAKHNENKO, A. G. VYSOTSKIY, V. N. FEDOROV, V. D. MAKSIMCV, V. M.
   SCKOLOVA, S. A.
TI SIMULATICN OF THE DYNAMICS OF THE ENVIRONMENT-PLANKTON ECOLOGICAL
   SYSTEM OF THE WHITE SEA AND ANALYSIS OF ITS STABILITY.
SO AVTOMATIKA (USSR). VOL.8, NO.3. 9-18. 16.
YR 75.
TR SOV. AUTCM. CONTROL (USA). SAUCBZ. VOL.8, NO.3. 6-13. MAY-JUNE 1975.

******************************** INSB ********************************
AU PATEREU, S. G. RAVSKA, N. S. KULCHIY, V. M.
TI A MATHEMATICAL MODEL OF THE METAL CUTTING PROCESS.
SO AVTOMATIKA (USSR). VOL.8, NO.3. 19-23. 4.
YR 75.
TR SOV. AUTOM. CONTROL (USA). SAUCBZ. VOL.8, NO.3. 14-17. MAY-JUNE 1975.

******************************** INSB ********************************
AU DYLBOKOVA, D. L. DYIBCKCV, I. S.
TI PREDICTION OF TRENDS OF DEVELOPMENT OF DIGITAL COMPUTERS BY GMDH
   USING LINEAR-FRACTION PARTIAL DESCRIPTIONS AND BALANCE-OF-VARIABLES
   CRITERION.
SO AVTOMATIKA (USSR). VOL.8, NC.2. 30-7. 7.
YR 75.
TR SOV. AUTCM. CCNTROL (USA). SAUCBZ. VOL.8, NO.2. 24-30. MARCH-APRIL
   1975.

******************************** INSB ********************************
AU IVAKHNENKO, A. G. OVCHINNIKOV, V. A.
TI CONTRCL CF DNIEPR HYDROELECTRIC POWER STATION WATER RESERVOIRS WITH
   TWO OPTIMALITY CRITERIA BASED ON SELF-ORGANIZATION.
SC AVTOMATIKA (USSR). VOL.8, NO.2. 49-59. 8.
YR 75.
TR SOV. AUTOM. CCNTROL (USA). SAUCBZ. VOL.8, NO.2. 41-9. MARCH-APRIL
   1975.

******************************** INSB ********************************

AU PARKS, P. C. IVAKHNENKO, A. G. BOICHUK, L. M. SVETALSKY, B. K.
TI A SELF-ORGANIZING MODEL OF THE BRITISH ECONOMY FOR CONTROL WITH
   OPTIMAL PREDICTION USING THE BALANCE-OF-VARIABLES CRITERION.
SO INT. J. COMPUT. AND INF. SCI. (GB). VOL.4, NO.4. 349-79. 10.
YR 75.

******************************** INSB ********************************
AU VASILEV, V. I. ZAKHAROV, YU. YE. KANATOV, I. I. KLESHCHOV, V. V.
   KONOVALENKO, V. V.
TI ANALYSIS AND CLASSIFICATION OF SITUATIONS GENERATED BY RANDOM
   PROCESSES USING THE GROUP METHOD OF DATA HANDLING.
SO AVTOMATIKA (USSR). VOL.8, NO.1. 57-65. 2.
YR 75.
TR SOV. AUTOM. CONTROL (USA). SAUCBZ. VOL.8, NO.1. 46-53. JAN.-FEB.
   1975.

******************************** INSB ********************************
AU VYSOTSKIY, V. M. IVAKHNENKO, A. G. CHEBERKUS, V. I.
TI LONG-TERM PREDICTION OF OSCILLATORY PROCESSES BY FINDING A HARMONIC
   TREND OF OPTIMUM COMPLEXITY BY THE BALANCE-OF-VARIABLES CRITERION.
SO AVTOMATIKA (USSR). VOL.8, NO.1. 23-31. 5.
YR 75.
TR SOV. AUTOM. CONTROL (USA). SAUCBZ. VOL.8, NO.1. 18-24. JAN.-FEB.
   1975.

******************************** INSB ********************************
AU IVAKHNENKO, A. G. MIKRYUKOV, B. G. BOSHNYAKOV, A. N. SVETALSKIY,
   B. K.
TI OBJECTIVE IDENTIFICATION OF THERMAL STATE OF BLAST FURNACE BY
   SELF-ORGANIZATION METHODS.
SO AVTOMATIKA (USSR). VOL.8, NO.6. 44-9. 8.
YR 75.
TR SOV. AUTOM. CONTROL (USA). SAUCBZ. VOL.8, NO. 6. 35-9. NOV.-DEC.
   1975.

******************************** INSB ********************************
AU IVAKHNENKO, A. G. DYLEOKOVA, D. L. SIRBILADZE, D. D. DYLBOKOV, I.
   S.
TI DEPTH OF MINIMUM OF SELECTION CRITERION AS AN INDICATOR OF MODEL
   ADEQUACY.
SO AVTOMATIKA (USSR). VOL.8, NO.6. 13-18. 5.
YR 75.
TR SOV. AUTOM. CONTROL (USA). SAUCBZ. VOL.8, NO.6. 9-13. NOV.-DEC.
   1975.

******************************** INSB ********************************
AU IVAKHNENKO, A. G. STEPASHKO, V. S.
TI SELF-ORGANISATION OF MODELS AND LONG-TERM PREDICTION OF RIVER RUNOFF
   BY THE BALANCE CRITERION.
SO AVTOMATIKA (USSR). VOL.8, NC.5. 34-41. 5.
YR 75.
TR SOV. AUTCM. CONTROL (USA). SAUCBZ. VOL.8, NO.5. 27-33. SEPT.-OCT.
   1975.

******************************** INSB ********************************
AU IVAKHNENKO, A. G.
TI PRESENT STATE OF THE THEORY OF COMPUTER-AIDED SELF-ORGANISATION OF

MATHEMATICAL MODELS (SURVEY).
SO AVTOMATIKA (USSR). VOL.8, NO.5. 23-33. 17.
YR 75.
TR SOV. AUTOM. CONTROL (USA). SAUCBZ. VOL.8, NO.5. 18-26. SEPT.-OCT.
   1975.

****************************** INSB ********************************
AU IVAKHNENKO, A. G. IVAKHNENKO, N. A.
TI LONG-TERM PREDICTION OF RANDOM PROCESSES BY GMDH ALGORITHMS USING THE
   UNBIASEDNESS CRITERION AND THE BALANCE-OF-VARIANCE CRITERION. II.
SO AVTOMATIKA (USSR). VOL.8, NO.4. 31-47. 8.
YR 75.
TR SOV. AUTOM. CONTROL (USA). SAUCBZ. VOL.8, NO.4. 24-38. JULY-AUG.
   1975.

****************************** INSB ********************************
AU AKISHIN, B. A. IVAKHNENKO, A. G.
TI EXTRAPOLATION (PREDICTION) USING MONOTONICALLY VARYING NOISY DATA.
SO AVTOMATIKA (USSR). VOL.8, NO.4. 22-30. 6.
YR 75.
TR SOV. AUTOM. CONTROL (USA). SAUCBZ. VOL.8, NO.4. 17-23. JULY-AUG.
   1975.

****************************** INSB ********************************
AU IHARA, J.
TI IMPROVED GMDH-A CASE OF DYNAMICAL WORLD POPULATION MODELS.
SO SYST. AND CONTROL (JAPAN). VCL.19, NO.4. 201-10. 5.
YR 75.

****************************** INSB ********************************
AU IVAKHNENKO, A. G. SVETALSKIY, B. K.
TI SELF-ORGANIZATION OF WORLD DYNAMICS MODEL ACCORDING TO FORRESTER'S
   DATA AND CONTROL SYNTHESIS BY SELECTING THE VERTICES OF THE HYPERCUBE
   OF FEASIBLE CONTROLS.
SO AVTOMATIKA (USSR). VOL.8, NO.1. 32-50. 14.
YR 75.
TR SOV. AUTOM. CONTROL (USA). SAUCBZ. VOL.8, NO.1. 25-40. JAN.-FEB.
   1975.

****************************** INSB ********************************
AU IKEDA, S. SAWARAGI, Y.
TI GROUP METHOD OF DATA HANDLING AND MODELLING OF COMPLEX SYSTEMS.
SO J. SOC. INSTRUM. AND CONTROL ENG. (JAPAN). VOL.14, NO.2. 185-95. 34.
YR 75.

****************************** INSB ********************************
AU IVAKHNENKO, A. G. MIKRYUKOV, B. G. BOSHNYAKOV, A. N. SVETALSKIY,
   B. K.
TI OBJECTIVE IDENTIFICATION OF THERMAL STATE OF BLAST FURNACE BY
   SELF-ORGANIZATION METHODS.
SO AVTOMATIKA (USSR). VOL.8, NO.6. 44-9. 8.
YR 75.
TR SOV. AUTOM. CONTROL (USA). SAUCBZ. VOL.8, NO.6. 35-9. NOV.-DEC.
   1975.

****************************** INSB ********************************

AU IVAKHNENKO, A. G. DYLBOKOVA, D. L. SIRBILADZE, D. D. DYLBOKOV, I. S.
TI DEPTH OF MINIMUM OF SELECTION CRITERION AS AN INDICATOR OF MODEL ADEQUACY.
SC AVTOMATIKA (USSR). VOL.8, NO.6. 13-18. 5.
YR 75.
TR SOV. AUTOM. CONTROL (USA). SAUCBZ. VOL.8, NO.6. 9-13. NOV.-DEC. 1975.

******************************** INSB ********************************
AU IKEDA, S. OCHIAI, M. SAWARAGI, Y.
TI SEQUENTIAL GMDH ALGORITHM AND ITS APPLICATION TO RIVER FLOW PREDICTION.
SO TRANS. SOC. INSTRUM. AND CONTROL ENG. (JAPAN). VOL.12, NO.2. 209-14. 12.
YR 76.

******************************** INSB ********************************
AU TAMURA, H. AOTANI, T.
TI LARGE-SPATIAL AIR POLLUTION IDENTIFICATION BY COMBINED APPROACH OF SOURCE-RECEPTOR MATRIX AND GMDH.
SO TRANS. SOC. INSTRUM. AND CONTROL ENG. (JAPAN). VOL.12, NO.2. 121-6. 10.
YR 76.

******************************** INSB ********************************
AU IKEDA, S. SAWARAGI, Y. OCHIAI, M.
TI SEQUENTIAL GMDH ALGORITHM AND ITS APPLICATION TO RIVER FLOW PREDICTION.
SO IEEE TRANS. SYST. MAN AND CYBERN. (USA). VOL.SMC-6, NO.7. 473-9. 18.
YR 76.

******************************** INSB ********************************
AU TAMURA, H. AOTANI, T.
TI LARGE-SPATIAL IDENTIFICATION BY CONVERSATIONAL GMDH.
SO TRANS. SOC. INSTRUM. AND CONTROL ENG. (JAPAN). VOL.12, NO.1. 69-75. 8.
YR 76.

******************************** INSP ********************************
AU SCOTT, D. E. HUTCHINSON, C. E. VOGT, W. G. MICKLE, M. H.
TI THE GMDH ALGORITHM-A TECHNIQUE FOR ECONOMIC MODELING.
SO MODELING AND SIMULATION, VOL.7. II. PITTSBURGH, PA. USA. PITTSBURGH, PA. USA. XXIV8640. 729-33. 16. ISA. UNIV. PITTSBURGH ISA ET AL. 26-27 APRIL 1976.
BN 0 87664 323 3.
YR 76.

******************************** INSP ********************************
AU IVAKHNENKO, A. G.
TI THE GROUP METHOD OF DATA HANDLING IN PREDICTION PROBLEMS.
SO AVTOMATIKA (USSR). VOL.9, NO.6. 24-34. 5.
YR 76.
TR SOV. AUTOM. CONTROL (USA). SAUCBZ. VOL.9, NO.6. 21-30. NOV.-DEC. 1976.

******************************** INSP ********************************

AU IVAKHNENKO, A. G. IVAKHNENKO, N. A. VYSOTSKIY, V. M.
TI LONG-TERM PREDICTION OF RANDOM PROCESSES BY GMDH ALGORITHMS USING THE
   UNBIASEDNESS AND BALANCE-OF-VARIABLES CRITERIA. IV.
SO AVTOMATIKA (USSR). VOL.9, NO. 4. 21-35. 8.
YR 76.
TR SOV. AUTOM. CONTROL (USA). SAUCBZ. VOL.9, NO. 4. 16-27. JULY-AUG.
   1976..

***************************** INSP *******************************
AU MINATC, K. KUWAHARA, M.
TI SOME CONVERGENT PROPERTIES OF GMDH.
SO TRANS. INST. ELECTR. ENG. JAP. (JAPAN). VOL.96, NO. 2. 22.
YR 76.
TR ELECTR. ENG. JPN. (USA). EENJAU. VOL.96, NO.2. 106-10. MARCH-APRIL
   1976..

***************************** INSP *******************************
AU IVAKHNENKO, A. G. SIBBILADZE, D. D. KARPUKHINA, A. M.
TI OPTIMIZATION OF NONLINEAR TRANSFORMATION OF CONTINUOUS FEATURES INTO
   DISCRETE (BINARY) CODE BY SIFTING ACCORDING TO ANALYTIC FORMULAS.
SO AVTOMATIKA (USSR). VOL.9 NO. 5. 3-13. 5.
YR 76.
TR SOV. AUTOM. CONTROL (USA). SAUCBZ. VOL.9, NO. 5. 1-10. SEPT.-OCT.
   1976.

***************************** INSP *******************************
AU IVAKHNEBKO, A. G. PEKA, P. YU. KOSHULKO, A. I.
TI SIMULATION OF THE DYNAMICS OF THE MINERALIZATION FIELD OF AQUIFERS
   WITH OPTIMIZATION OF POROSITY ESTIMATE OF THE MEDIUM.
SO AVTOMATIKA (USSR). VOL.9, NO.4. 35-44. 8.
YR 76.
TR SOV. AUTOM. CONTROL (USA). SAUCBZ. VOL.9, NO.4. 28-35. JULY-AUG.
   1976.

***************************** INSP *******************************
AU VYSOTSKIY, V. N.
TI OPTIMUM PARTITIONING OF EXPERIMENTAL DATA ON GMDH ALGORITHMS.
SO AVTOMATIKA (USSR). VOL.9, NO. 3. 71-3. 3.
YR 76.
TR SOV. AUTOM. CONTROL (USA). SAUCBZ. VOL.9, NO. 3. 62-5. MAY-JUNE 1976.

***************************** INSP *******************************
AU IVAKHNENKO, A. G. IVAKHNENKO, N. A. VYSOTSKIY, V. M. CHEBERKUS, V.
   I.
TI LONG-TERM PREDICTION OF RANDOM PROCESSES BY GMDH ALGORITHMS USING THE
   UNBIASEDNESS CRITERION AND THE BALANCE-OF-VARIABLES CRITERION. III.
SO AVTOMATIKA (USSR). VOL.9, NO. 2. 36-53. 5.
YR 76.
TR SOV. AUTOM. CONTROL (USA). SAUCBZ. VOL.9, NO.2. 28-42. MARCH-APRIL
   1976..

***************************** INSP *******************************
AU SILIS, YA. YA. ROZENBLII, A. B.
TI ALGORITHM FOR CONSTRUCTION OF DECISION FUNCTION IN THE FORM OF A
   COMPLEX LOGIC PROPOSITION.
SO AVTOMATIKA (USSR). VOL.9, NO.2. 3-8. 9.
YR 76.

TR SOV. AUTOM. CONTROL (USA). SAUCBZ. VOL.9, NO.2. 1-5. MARCH-APRIL
1976.

******************************** INSP ********************************
AU SHERVASHIDZE, V. V.
TI AN INDUCTIVE METHOD OF OPTIMIZATION OF CONTROL SYSTEMS USING
INTERPOLATION BASED ON GMDH ALGORITHMS.
SO AVTOMATIKA (USSR). VOL.9, NO.3. 7-12. 7.
YR 76.
TR SOV. AUTCM. CONTROL (USA). SAUCBZ. VOL.9, NO.3. 5-9. MAY-JUNE 1976.

******************************** INSP ********************************
AU DYLBOKOVA, D. L.   DYLBOKOV, I. S.
TI RECOVERY OF OMITTED DATA BY GMDH ALGORITHMS USING THE UNBIASEDNESS
CRITERION AND REMAINDER FUNCTIONS.
SO AVTOMATIKA (USSR). VOL.9, NC.2. 27-35. 7.
YR 76.
TR SOV. AUTCM. CONTROL (USA). SAUCBZ. VOL.9, NO.2. 21-7. MARCH-APRIL
1976..

******************************** INSP ********************************
AU VYSOTSKIY, V. N.
TI OPTIMUM PARTITIONING OF EXPERIMENTAL DATA ON GMDH ALGORITHMS.
SO AVTOMATIKA (USSR). VOL.9, NO.3. 71-3. 3.
YR 76.
TR SOV. AUTOM. CONTROL (USA). SAUCBZ. VOL.9, NO.3. 62-5. MAY-JUNE 1976.

******************************** INSP ********************************
AU SATOH, T.
TI APPLICATION OF GMDH IN PREDICTION OF USED BOTTLE RECOVERY.
SO SYST. AND CONTROL (JAPAN). VOL.20, NO.7. 365-72. 9.
YR 76.

******************************** INSP ********************************
AU IVAKHNENKO, A. G.   IVAKHNENKO, N. A.
TI SELF-ORGANIZATION THEORY AS A BASIS FOR DIRECT MODELLING OF COMPLEX
SYSTEMS FOR EXPERIMENTAL DATA.
SO KIBERN. AND VYCHISL. TEKH. (USSR). NO.33. 31-43. 18.
YR 76.

******************************** INSP ********************************
AU IHARA, J.
TI STABILISATION OF A MULTILAYER GMDH ALGORITHM. I. STABILISATION METHOD
I USING A SIMPLE AVERAGE.
SO BULL. ELECTROTECH. LAB. (JAPAN). VOL.40, NO.8. 673-708. 15.
YR 76.

******************************** INSP ********************************
AU STONE, R.   IVAKHNENKO, A. G.   VYSOTSKYY, V. M.   SOMINA, L. P.
TI DISCOVERY OF LAWS GOVERNING THE BEHAVIOR OF COMPLEX OBJECTS WHOSE
OUTPUT VARIABLES ARE NOT INDICATED.
SO AVTOMATIKA (USSR). VOL.10, NO.6. 27-40. 11.
YR 77.
TR SOV. AUTOM. CONTROL (USA). SAUCBZ. VOL.10, NO.6. 21-32. NOV.-DEC.
1977.

******************************** INSP ********************************

AU VASYLYEV, V. I. KONOVALENKO, V. V. OVSYANNYKOVA, F. P.
TI SYNTHESIS OF DECISION RULE COLLECTIVES AND ITS APPLICATION TO PATTERN
   RECOGNITION.
SO AVTOMATIKA (USSR). VOL.10, NO.6. 3-12. 10.
YR 77.
TR SOV. AUTOM. CONTROL (USA). SAUCBZ. VOL.10, NO.6. 1-8. NOV.-DEC.
   1977.

******************************* INSP *********************************
AU IVAKHNENKO, A. G. STEPASHKO, V. S. KHOMOVNENKO, M. G. GALYAMIN,
   YE. P.
TI SELF-ORGANIZATION OF DYNAMIC MODELS OF GROWTH OF AGRICULTURAL CROPS
   FOR CONTROL OF IRRIGATED CROP ROTATION.
SO AVTOMATIKA (USSR). VOL.10, NO.5. 32-44. 6.
YR 77.
TR SOV. AUTOM. CONTROL (USA). SAUCBZ. VOL.10, NO.5. 23-33. SEPT.-OCT.
   1977.

******************************* INSP *********************************
AU IVAKHNENKO, A. G. KROTOV, G. I.
TI SIMULATION OF ENVIRONMENTAL POLLUTION IN THE ABSENCE OF INFORMATION
   ABOUT DISTURBANCES OF THE POLLUTION SOURCE.
SO AVTOMATIKA (USSR). VOL.10, NO.5. 14-31. 11.
YR 77.
TR SOV. AUTOM. CONTROL (USA). SAUCBZ. VOL.10, NO.5. 8-22. SEPT.-OCT.
   1977.

******************************* INSP *********************************
AU YURACHKOVSKIY, YU. P.
TI IMPROVED GMDH ALGORITHMS FOR PROCESS PREDICTION.
SO AVTOMATIKA (USSR). VOL.10, NO.5. 76-86. 8.
YR 77.
TR SOV. AUTOM. CONTROL (USA). SAUCBZ. VOL.10, NO.5. 61-71. SEPT.-OCT.
   1977.

******************************* INSP *********************************
AU VYSOTSKIY, V. N. YUNUSOV, N. I.
TI IMPROVING THE NOISE IMMUNITY OF A GMDH ALGORITHM USED FOR FINDING A
   HARMONIC TREND WITH NONMULTIPLE FREQUENCIES.
SO AVTOMATIKA (USSR). VOL.10, NO.5. 73-5. 5.
YR 77.
TR SOV. AUTOM. CONTROL (USA). SAUCBZ. VOL.10, NO.5. 57-60. SEPT.-OCT.
   1977.

******************************* INSP *********************************
AU IVAKHNENKO, A. G. VOSTROV, N. N. PYATKOV, V. I.
TI THREE-STAGE STRUCTURAL IDENTIFICATION OF COMPLEX SPATIAL OBJECTS.
SO AVTOMATIKA (USSR). VOL.10, NO.4. 16-22. 4.
YR 77.
TR SCV. AUTOM. CONTROL (USA). SAUCBZ. VOL.10, NO.4. 13-18. JULY-AUG.
   1977.

******************************* INSP *********************************
AU TRISEYEV, YU. P.
TI GMDH ALGORITHM WITH VARIABLE FREEDOM OF CHOICE IN SELECTION LAYERS
   BASED ON CRITERION OF DIVERSITY OF VARIABLES.
SO AVTOMATIKA (USSR). VOL.10, NO.4. 37-42. 2.

YR 77.
TR SOV. AUTOM. CONTROL (USA). SAUCBZ. VOL.10, NO.4. 30-3. JULY-AUG.
   1977..

**************************+++++++** INSP **********************+++++++++++++++
AU IVAKHNENKO, A. G. PEKA, P. YU. YAKOVENKO, P. I. KOSHULKO, A. I.
TI ESTIMATION OF LIQUID OUTFLOW AT THE BOUNDARY OF AN ARTESIAN AQUIFER
   REGION BY SELF-ORGANIZATION METHODS.
SO AVTOMATIKA (USSR). VOL. 10, NO. 4. 32-6. 4.
YR 77.
TR SOV. AUTCM. CONTROL (USA). SAUCBZ. VOL. 10, NO.4. 26-9. JULY-AUG.
   1977.

*******************************+** INSP **********************************+++*
AU PAK, V. G. KIRYAKOV, YU. M.
TI GMDH ALGORITHM FOR RECOGNITION OF BLACK-AND-WHITE LETTERS OR NUMBERS.
SO AVTOMATIKA (USSR). VOL. 10, NO. 4. 11-15. 2.
YR 77.
TR SOV. AUTCM. CONTROL (USA). SAUCBZ. VOL. 10, NO.4. 8-12. JULY-AUG.
   1977.

***************************+****** INSP **********************+++++++++++++++*
AU IVAKHNENKO, A. G. PEKA, P. YU. YAKOVENKO, P. I.
TI IDENTIFICATION OF DYNAMIC EQUATIONS OF A COMPLEX PLANT ON THE BASIS
   OF EXPERIMENTAL DATA BY USING SELF-ORGANIZATION OF MODELS. II.
   MULTIDIMENSIONAL PROBLEMS.
SO AVTOMATIKA (USSR). VOL.10, NO. 2. 38-46. 10.
YR 77.
TR SOV. AUTOM. CONTROL (USA). SAUCBZ. VOL.10, NO.2. 31-7. MARCH-APRIL
   1977..

*******************************+** INSP **********************************++++
AU IVAKHNENKO, A. G. KROTOV, G. I. CHEBERKUS, V. I. VYSOTSKIY, V. N.
TI IDENTIFICATION OF DYNAMIC EQUATIONS OF A COMPLEX PLANT ON THE BASIS
   OF EXPERIMENTAL DATA BY USING SELF-ORGANIZATION OF MODELS. I.
   ONE-DIMENSIONAL PROBLEMS.
SO AVTOMATIKA (USSR). VOL. 10, NO. 2. 30-7. 7.
YR 77.
TR SOV. AUTCM. CONTROL (USA). SAUCBZ. VOL. 10, NO.2. 24-30. MARCH-APRIL
   1977..

***************************+****** INSP **********************+++++++++++++++*
AU VASILYEV, V. I. KONOVALENKO, V. V. KOVAL, P. N.
TI USE OF SELF-ORGANIZATION (GMDH) ALGORITHMS FOR PREDICTING THE MELTING
   PROCESS IN OXYGEN-BLOWN CONVERTERS.
SC AVTOMATIKA (USSR). VOL. 10, NO. 3. 28-31. 2.
YR 77.
TR SOV. AUTOM. CONTROL (USA). SAUCBZ. VOL. 10, NO. 3. 21-3. MAY-JUNE 1977.

*********************+*+********** INSP **********************+++++++++++++++*
AU KOZUBOVSKIY, S. F.
TI USE OF THE GMDH IN MICROBIOLOGY.
SO AVTOMATIKA (USSR). VOL. 10, NO.1. 82-5. 10.
YR 77.
TR SOV. AUTCM. CONTROL (USA). SAUCBZ. VOL. 10, NO.1. 67-71. JAN.-FEB.
   1977.

***************************** INSP *******************************
AU TAMURA, H. KONDO, T.
TI REVISED GMDH ALGORITHM USING SELF-SELECTION OF OPTIMAL PARTIAL
   POLYNOMIALS AND ITS APPLICATION TO LARGE-SPATIAL AIR POLLUTION
   PATTERN IDENTIFICATION.
SO TRANS. SOC. INSTRUM. AND CONTROL ENG. (JAPAN). VOL.13, NO.4. 351-7.
   9.
YR 77.

***************************** INSP *******************************
AU MEHRA, R. K.
TI GROUP METHOD OF DATA HANDLING (GMDH): REVIEW AND EXPERIENCE.
SO PROCEEDINGS OF THE 1977 IEEE CONFERENCE ON DECISION AND CONTROL. I.
   NEW ORLEANS, LA, USA. NEW YORK, USA. XXI&1280. 29-34. 20. IEEE.
   IEEE. 7-9 DEC. 1977.
YR 77.

***************************** INSP *******************************
AU IVAKHNENKO, A. G.
TI PRINCIPAL STAGES IN THE DEVELOPMENT OF THE TECHNIQUE OF SIMULATION OF
   COMPLEX PROCESSES AND SYSTEMS.
SO AVTOMATIKA (USSR). VOL. 10, NO. 1. 86-91. 14.
YR 77.
TR SOV. AUTOM. CONTROL (USA). SAUCBZ. VOL. 10, NO. 1. 72-7. JAN.-FEB.
   1977..

***************************** INSP *******************************
AU IVAKHNENKO, A. G. VOSTROV, N. N.
TI IDENTIFICATION OF OIL HORIZONS BY COMPUTER-AIDED SELF-ORGANIZATION OF
   MATHEMATICAL MODELS.
SO AVTOMATIKA (USSR). VOL.10, NO. 1. 3-12. 11.
YR 77.
TR SOV. AUTOM. CONTROL (USA). SAJCBZ. VOL.10, NO.1. 1-8. JAN.-FEB.
   1977.

***************************** INSP *******************************
AU IHARA, J.
TI STABILIZATION OF MULTILAYER GMDH ALGORITHM. II. STABILIZATION METHOD
   II USING A SIMPLE AVERAGE.
SO BULL. ELECTROTECH. LAB. (JAPAN). VOL. 41, NO.4. 289-303. 12.
YR 77.

***************************** INSP *******************************
AU KOPYSITSKIY, T. I. YUSIFOV, R. YU.
TI IDENTIFICATION OF PARAMETERS OF COMPLEX SYSTEMS. APPLICATION OF GMDH
   TO IDENTIFICATION OF PETRO-CHEMICAL PLANTS AND PETROLEUM REFINERIES.
SO AVTOMATIKA (USSR). VOL. 11, NO. 6. 3-8. 8.
YR 78.
TR SOV. AUTOM. CONTROL (USA). SAUCBZ. VOL. 11, NO. 6. 1-5. NOV.-DEC.
   1978.

***************************** INSP *******************************
AU BRAVERMAN, V. YA.
TI PECULIARITIES OF GMDH APPLICATIONS TO ECONOMIC RESEARCH.
SO AVTOMATIKA (USSR). VOL. 11, NO.6. 58-61. 6.
YR 78.
TR SOV. AUTOM. CONTROL (USA). SAUCBZ. VOL. 11, NO.6. 49-52. NOV.-DEC.

1978.

******************************* INSP ********************************
AU IVAKHNENKO, A. G.   KOSTENKO, YU. V.
TI ON THE POSSIBLE AND IMPOSSIBLE IN MODELING OF AN INTELLECT.
SO AVTOMATIKA (USSR). VOL. 11, NO. 6. 75-82. 20.
YR 78.
TR SOV. AUTOM. CONTROL (USA). SAUCBZ. VOL. 11, NO. 6. 68-76. NOV.-DEC.
   1978.

******************************* INSP ********************************
AU IVAKHNENKO, A. G.   BELOZERSKII, E. A.   BURLAKOV, N. N.   KOZUBOVSKI,
   S. P.
TI SCHEMATIC REALIZATION OF THE GROUP METHOD OF DATA HANDLING (GMDH)
   USING AN ALGORITHM FOR SELECTING THE VALUES OF COEFFICIENTS OF THE
   PARTIAL DESCRIPTIONS.
SO AVTOMATIKA (USSR). VOL.11, NO.6. 41-8. 23.
YR 78.
TR SOV. AUTOM. CONTROL (USA). SAUCBZ. VOL.11, NO.6. 36-41. NOV.-DEC.
   1978.

******************************* INSP ********************************
AU IVAKHNENKO, A. G.   KROTOV, G. I.
TI COMPARATIVE STUDIES IN SELF-ORGANIZATION OF PHYSICAL FIELD MODELS.
SO AVTOMATIKA (USSR). VOL.11, NO.5. 51-65. 5.
YR 78.
TR SOV. AUTOM. CONTROL (USA). SAUCBZ. VOL.11, NO.5. 42-52. SEPT.-OCT.
   1978..

******************************* INSP ********************************
AU KROTOV, G. I.   STEPASHKO, V. S.   VYSOTSKIY, V. N.   PREOBRAZHENSKIY,
   V. YU.
TI METHODS FOR REDUCING OPERATIONAL COMPUTER MEMORY VOLUME WHEN USING
   GMDH ALGORITHMS.
SO AVTOMATIKA (USSR). VOL. 11, NO. 4. 89 ET SEQ. 7.
YR 78.
TR SOV. AUTOM. CONTROL (USA). SAUCBZ. VOL. 11, NO. 4. 74-8. JULY-AUG.
   1978..

******************************* INSP ********************************
AU PAVLENKO, V. D.
TI OPTIMUM LONG-RANGE PREDICTION OF STEEL PRODUCTION USING A MULTILEVEL
   GMDH ALGORITHM.
SO AVTOMATIKA (USSR). VOL. 11, NO. 3. 73-4. 4.
YR 78.
TR SOV. AUTOM. CONTROL (USA). SAUCBZ. 67-9..

******************************* INSP ********************************
AU IVAKHNENKO, N. A.   MARCHEV, A. A.
TI SELF-ORGANIZATION OF A MATHEMATICAL MODEL FOR LONG-RANGE PLANNING OF
   CONSTRUCTION AND INSTALLATION ACTIVITIES.
SO AVTOMATIKA (USSR). VOL. 11, NO. 3. 12-18. 5.
YR 78.
TR SOV. AUTOM. CONTROL (USA). SAUCBZ. 9-14..

******************************* INSP ********************************

AU IVAKHNEIKC, A. G.
TI INDUCTIVE METHCD FOR SELF-OBGANIZATION OF COMPLEX SYSTEM MODELS BY
   ELECTRONIC COMPUTERS.
SO AVTOMATIKA (USSR). VOL. 11, NO. 4. 11-26. 17.
YR 78.
TR SOV. AUTOM. CCNTROL (USA). SAUCBZ. VOL. 11, NO. 4. 7-19. JULY-AUG.
   1978.

****************************** INSP ******************************
AU TUMANOV, N. V.
TI A GMDH ALGORITHM WITH MUTUALLY ORTHOGONAL PARTIAL DESCRIPTIONS FOR
   SYNTHESIS OF POLYNOMIAI MODELS OF COMPLEX OBJECTS.
SO AVTOMATIKA (USSR). VOL. 11, NO. 3. 85-7. 3.
YR 78.
TR SOV. AUTOM. CCNTROL (USA). SAUCBZ. 82-4..

****************************** INSP ******************************
AU VYSOTSKIY, V. N. IHARA, J.
TI IMPROVEMENT OF NOISE IMMUNITY OF GMDH SELECTION CRITERIA BY USING
   VECTOR REPRESENTATIONS AND MINIMAX FORMS.
SO AVTOMATIKA (USSR). VOL. 11, NO. 3. 3-11. 8.
YR 78.
TR SOV. AUTOM. CONTRCL (USA). SAUCBZ. 1-8..

****************************** INSP ******************************
AU IVAKHNENKO, A. G. YURACHKOVSKIY, YU. P.
TI USE OF LONG-TERM PREDICTION IN SELECTION OF OPTIMUM CONTROL FROM THE
   DOMAIN OF PARETO-EFFICIENT DECISIONS.
SO AVTOMATIKA (USSR). VOL. 11, NO. 2. 39-43. 4.
YR 78.
TR SOV. AUTOM. CCNTROL (USA). SAUCBZ. VOL. 11, NO. 2. 31-4. MARCH-APRIL
   1978..

****************************** INSP ******************************
AU IVAKHNEIKO, A. G. PEKA, P. Y.
TI METEOD OF OPTIMAL CONTRCL SYNTHESIS FOR THE FIELD OF UNDERGRCUND
   WATEB MINERALIZATION.
SO AVTCMATIKA (USSR). VOL. 11, NO. 2. 31-8. 8.
YR 78.
TR SOV. AUTOM. CONTROL (USA). SAUCBZ. VOL. 11, NO. 2. 25-30. MARCH-APRIL
   1978..

****************************** INSP ******************************
AU RYOUBU, M. IKEDA, S. SAWARAGI, Y.
TI A PREDICTION MODEL FOR REGIONAL ECONOMIC SYSTEM BY SELF-ORGANIZATION
   METHOD.
SO PROCEEDINGS OF THE INTERNATIONAL CONFERENCE ON CYBERNETICS AND
   SOCIETY. I. TOKYO-KYOTO, JAFAN. NEW YORK, USA. 800. 619-22. 9. IEEE.
   IEEE. 3-7 NOV. 1978.
YR 78.

****************************** INSP ******************************
AU TAMURA, H. KONDO, T.
TI NONIINEAR MCDELING FOR SHORT-TERM PREDICTION OF AIR POLLUTION
   CONCENTRATICN BY A REVISED GMDH.
SO PROCEEDINGS OF THE INTERNATICNAL CONFERENCE ON CYBERNETICS AND
   SCCIETY. I. TOKYO-KYOTO, JAPAN. NEW YORK, USA. 800. 596-601. 9.

IEEE. IEEE. 3-7 NOV. 1978.
YR 78.

******************************* INSP ********************************
AU SIRBILADZE, D. D.
TI CHOICE OF PROTOTYPES FOR A BINARY PERCEPTRON WITH VOTING ALGORITHM
   USING THE GROUP METHOD OF DATA HANDLING.
SO AVTOMATIKA (USSR). VOL. 11, NO. 2. 3-12. 9.
YR 78.
TR SOV. AUTOM. CONTROL (USA). SAUCBZ. VOL. 11, NO. 2. 1-10. MARCH-APRIL
   1978..

******************************* INSP ********************************
AU MORITA, T. KONISHI, M. NOSE, K. ASADA, Y. YAMAMOTO, I. SAWARAGI,
   Y. AKASHI, H.
TI USE OF GMDH FOR ESTIMATION OF REGIONAL AIR QUALITY.
SO ENVIRONMENTAL SYSTEMS PLANNING, DESIGN AND CONTROL. I. KYOTO, JAPAN.
   OXFORD, ENGLAND. XVE441. 381-7. 4. PERGAMON. IFAC. 1-5 AUG. 1977.
IN 0 08 022016 9.
YR 78.

******************************* INSP ********************************
AU TAMURA, H. KONDO, T. SAWARAGI, Y. AKASHI, H.
TI LARGE-SPATIAL PATTERN IDENTIFICATION OF AIR POLLUTION BY A COMBINED
   MODEL OF SOURCE-RECEPTOR MATRIX AND REVISED GMDH.
SO ENVIRONMENTAL SYSTEMS PLANNING, DESIGN AND CONTROL. I. KYOTO, JAPAN.
   OXFORD, ENGLAND. XVE441. 373-80. 13. PERGAMON. IFAC. 1-5 AUG. 1977.
BN 0 C8 022016 9.
YR 78.

******************************* INSP ********************************
AU DAEAGYAN, A. V. KONONENKO, I. V.
TI AN ADAPTIVE MODEL FOR DIAGNOSIS AND PROGNOSIS OF THE FUNCTIONAL STATE
   OF THE BRAIN.
SO AVTOMATIKA (USSR). VOL. 11, NO. 1. 3-9. 6.
YR 78.
TR SOV. AUTOM. CONTROL (USA). SAUCBZ. VOL. 11, NO. 1. 1-6. JAN.-FEB.
   1978.

******************************* INSP ********************************
AU IVAKHNENKO, A. G. VYSOTSKIY, V. N. IVAKHNENKO, N. A.
TI PRINCIPAL VERSIONS OF THE MINIMUM BIAS CRITERION FOR A MODEL AND AN
   INVESTIGATION OF THEIR NOISE IMMUNITY.
SO AVTOMATIKA (USSR). VOL. 11, NO. 1. 32-53. 12.
YR 78.
TR SOV. AUTOM. CONTROL (USA). SAUCBZ. VOL. 11, NO. 1. 27-45. JAN.-FEB.
   1978.

******************************* INSP ********************************
AU TAMURA, H. KONDO, T.
TI REVISED GMDH ALGORITHM USING PREDICTION SUM OF SQUARES (PSS) AS A
   CRITERION FOR MODEL SELECTION.
SO TRANS. SOC. INSTRUM. AND CONTROL ENG. (JAPAN). VOL. 14, NO. 5. 519-24.
   7.
YR 78.

******************************* INSP ********************************

AU ICHIKAWA, A. VANSTEENKISTE, G. C.
TI PREDICTION AND SIMULATION OF RIVER WATER QUALITY BY USING G.M.D.H.
SO MODELING, IDENTIFICATION AND CONTROL IN ENVIRONMENTAL SYSTEMS. GHENT,
   BELGIUM. AMSTERDAM, NETHERLANDS. XVIII&1028. 79-87. 5.
   NORTH-HOLLAND. IFIP. 30 AUG. - 2 SEPT. 1977.
RN 0 444 85180 1.
YR 78.

******************************* INSP *********************************
AU IVAKHNENKO, A. G.
TI THE GROUP METHOD OF DATA HANDLING IN LONG-RANGE FORECASTING.
SO TECHNOL. FORECAST. AND SOC. CHANGE (USA). VOL.12, NO.2-3. 213-27. 5.
YR 78.

******************************* INSP *********************************
AU INOOKA, H. INOUE, A.
TI APPLICATION OF THE GMDH ALGORITHM TO A MANUAL CONTROL SYSTEM.
SO IEEE TRANS. SYST. MAN AND CYBERN. (USA). VOL.SMC-8, NO.11. 819-21.
   8.
YR 78.

******************************* INSP *********************************
AU ENDO, A.
TI IDENTIFICATION OF A NONLINEAR SYSTEM WITH THE MODIFIED GMDH.
SO TRANS. SOC. INSTRUM. AND CONTROL ENG. (JAPAN). VOL.14, NO.2. 130-5.
   9.
YR 78.

******************************* INSP *********************************
AU DUONG, N. WINN, C. B.
TI AN OPTIMAL CONTROL STRATEGY FOR PEAK LOAD REDUCTION APPLIED TO SOLAR
   STRUCTURES.
SO ENERGY '78. IEEE 1978 REGION V ANNUAL CONFERENCE. TULSA, OK. USA. NEW
   YORK, USA. 274. 163. 0. IEEE. IEEE. 16-18 APRIL 1978.
YR 78.

******************************* INSP *********************************
AU VAN ZYL, J. G. DE JONGH, D. C. J.
TI EXPERIMENTS IN SOCIO-ECONOMIC FORECASTING USING IVAKHNENKO'S
   APPROACH.
SO APPL. MATH. MODELLING (GB). VOL.2, NO.1. 49-56. 3.
YR 78.

******************************* INSP *********************************
AU IVAKHNENKO, A. G. RAO, M. G.
TI PREDICTION AND EXTRAPOLATION OF METEOROLOGICAL FIELDS BY MODEL
   SELF-ORGANIZATION.
SO AVTOMATIKA (USSR). VOL.12, NO.6. 18-35. 18.
YR 79.
TR SOV. AUTOM. CONTROL (USA). SAUCBZ. VOL.12, NO.6. 13-27. 1979..

******************************* INSP *********************************
AU BELYAEV, V. I. IVAKHNENKO, A. G. ELEISHMAN, B. S.
TI SIMULATION, SELF-ORGANIZATION, AND POTENTIAL EFFECTIVENESS.
SO AVTOMATIKA (USSR). VOL.12, NO.6. 9-17. 10.
YR 79.
TR SOV. AUTOM. CONTROL (USA). SAUCBZ. VOL.12, NO.6. 6-12. 1979.
******************************* INSP *********************************

AU ST EPASHKO, V. S.
TI OPTIMIZATION AND GENERALIZATION OF MODEL SORTING SCHEMES IN
   ALGORITHMS FOR THE GROUP METHOD OF DATA HANDLING.
SO AVTOMATIKA (USSR). VOL.12, NO.4. 36-43. 11.
YR 79.
TR SOV. AUTOM. CONTROL (USA). SAUCBZ. VOL.12, NO.4. 28-33. JULY-AUG.
   1979..

**************************** INSP ****************************
AU SVETALSKIY, B. K.   KOVALCHUK, P. I.
TI A MULTILAYER ALGORITHM OF THE GROUP METHOD OF DATA HANDLING WITH
   SELECTION OF PRIMARY ARGUMENTS.
SO AVTOMATIKA (USSR). VOL.12, NC.4. 31-5. 6.
YR 79.
TR SOV. AUTCM. CONTROL (USA). SAUCBZ. VOL.12, NO.4. 24-7. JULY-AUG.
   1979.

**************************** INSP ****************************
AU IVAKHNENKO, A. G.   BELOZERSKIY, YE. A.   KOZUBOVSKIY, S. P.
TI DISCOVERY OF THE LAWS OF COMPLEX PLANTS WHEN THE OUTPUT VARIABLES ARE
   NOT INLICATED. IIL SELF-ORGANIZATION OF NONAUTONOMOUS SYSTEMS OF
   EQUATIONS.
SO AVTCMATIKA (USSR). VOL.12, NO.4. 15-23. 5.
YR 79.
TR SOV. AUTOM. CONTROL (USA). SAUCBZ. VOL.12, NO.4. 9-17. JULY-AUG.
   1979.

**************************** INSP ****************************
AU CHEBERKUS, V. I.   KUROVSKIY, A. P.
TI SELF-ORGANIZATION OF POINT MODELS FOR PREDICTION OF DISSOLVED OXYGEN
   CONTENT AND SUSPENDED MATTER IN A RIVER.
SO AVTOMATIKA (USSR). VOL.12, NO.5. 85-8. 3.
YR 79.
TR SOV. AUTOM. CONTROL (USA). SAUCBZ. VOL.12, NO.5. 68-71. SEPT.-OCT.
   1979.

**************************** INSP ****************************
AU IVAKHNENKO, A. G.
TI COMBINED SIMULATION-INDUCTION MODELING METHODS FOR ECOLOGICAL
   SYSTEMS.
SO AVTOMATIKA (USSR). VOL.12, NO.5. 8-18. 26.
YR 79.
TR SOV. AUTOM. CONTROL (USA). SAUCBZ. VOL.12, NO.5. 5-12. SEPT.-OCT.
   1979.

**************************** INSP ****************************
AU IKEDA, S.
TI GROUP METHOD OF DATA HANDLING TOWARDS A MODELING OF COMPLEX SYSTEMS.
SO SYST. AND CONTROL (JAPAN). VOL.23, NO.12. 710-17. 20.
YR 79.

**************************** INSP ****************************
AU IVAKHNENKO, A. G.   STEPASHKO, V. S.   KOSTENKO, YU. V.   ZHITORCHUK,
   YU. V. RAO, N.
TI SELF-ORGANIZATION OF COMBINED MODELS FOR PREDICTING CYCLIC PROCESSES
   USING THE CRITERION OF PREDICTION BALANCE.
SO AVTCMATIKA (USSR). VOL.12, NO.2. 8-21. 9.

YR 79.
TR SOV. AUTOM. CONTROL (USA). SAUCBZ. VOL.12, NO.2. 6-17. MARCH-APRIL
   1979..

****************************** INSP ********************************
AU KUZMIN, N. N. KOBOTNIKH, YU. YA. SEDOV, A. N.
TI IDENTIFICATION OF FUNCTIONS WITH ORTHOGONAL PROPERTIES BY THE GROUP
   METHOD CF DATA HANDLING.
SO AVTOMATIKA (USSR). VOL.12, NO.2. 46-54. 3.
YR 79.
TR SOV. AUTOM. CONTROL (USA). SAUCBZ. VOL.12, NO.2. 38-45. MARCH-APRIL
   1979.

****************************** INSP ********************************
AU RYOBE, M. YAMOTO, Y. SHIBUYA, K. SAWARAGI, Y.
TI SYSTEM ANALYSIS OF A CEMENT PLANT BY THE GMSM.
SO SYST. AND CONTROL (JAPAN). VOL.23, NO.10. 587-93. 13.
YR 79.

****************************** INSP ********************************
AU WATARI, H. SCOTT, D. E.
TI A NEW CLASS OF NONLINEAR MULTIVARIATE MODELING ALGORITHMS.
SO 1979 SCSC SUMMER COMPUTER SIMULATION CONFERENCE. TORONTO, CANADA.
   MONTVALE, NJ, USA. XXX980. 86-8. 7. AFIPS. AGU AIAA ISA ET AL. 16-18
   JULY 1979.
YR 79.

****************************** INSP ********************************
AU KONDO, T. TAMURA, H.
TI REVISED GMDE ALGORITHM OF SELF-SELECTING OPTIMAL INTERMEDIATE
   FCLYNOMIALS USING AIC.
SO TRANS. SOC. INSTRUM. AND CONTROL ENG. (JAPAN). VOL.15, NO.4. 466-71.
   7.
YR 79.

****************************** INSP ********************************
AU STONE, R. IVAKHNENKO, A. G. VISOTSKII, V. N. SEMINA, L. P.
TI DISCOVERY OF LAWS FOR COMPLEX PLANTS WHEN OUTPUT VARIABLES ARE NOT
   INDICATED. II. STRUCTURAL IDENTIFICATION.
SO AVTOMATIKA (USSR). VOL.12, NO.1. 25-37. 16.
YR 79.
TR SOV. AUTOM. CONTROL (USA). SAUCBZ. VOL.12, NO.1. 20-30. JAN.-FEB.
   1979.

****************************** INSP ********************************
AU TAMURA, H. KONDO, T.
TI MODELING AND SHORT-TERM PREDICTION OF AIR POLLUTION BY A REVISED
   GMDH.
SO TRANS. SOC. INSTRUM. AND CONTROL ENG. (JAPAN). VOL.15, NO.5. 22-7.
   5.
YR 79.

****************************** INSP ********************************
AU LEONG, H. H. JOHNSON, G. R.
TI MODELLING OF ENERGY CONSUMPTION FOR SPACE HEATING FOR A COMMUNITY VIA
   GMDH APPROACH.
SO PROCEEDINGS OF THE INTERNATIONAL CONFERENCE ON CYBERNETICS AND

SOCIETY. DENVER, CO, USA. NEW YORK, USA. XX8949. 268-73. 5. IEEE.
IEEE. 8-10 OCT. 1979.
YR 79.

******************************** INSP ********************************
AU OONISHI, S. KATO, S. AKITA, K.
TI ESTIMATION OF JUNCTION TEMPERATURE OF THYRISTOR AT NORMAL OPERATING
   CONDITICNS BY GMDH.
SO TRANS. SOC. INSTRUM. AND CONTROL ENG. (JAPAN). VOL.15, NO.4. 560-1.
   5.
YR 79.

******************************** INSP ********************************
AU MURAKAMI, S. KANEYASU, M.
TI A MACBO ECONOMETRIC MODEL OF THE JAPANESE ECONOMY BY GMDH ALGORITHMS.
SO BULL. KYUSHU INST. TECHNOL. (SCL AND TECHNOL.) (JAPAN). NO.38.
   121-30. 4.
YR 79.

******************************** INSP ********************************
AU SAWARAGI, Y. SOEDA, T. TAMURA, H. YOSHIMURA, T. OHE, S. CHUJO,
   Y. ISHIHARA, H.
TI STATISTICAL PREDICTION OF AIR POLLUTION LEVELS USING NON-PHYSICAL
   MODELS.
SO AUTOMATICA (GB). VOL.15, NO.4. 441-51. 15.
YR 79.

******************************** INSP ********************************
AU IKEDA, S. SAWARAGI, Y. JANSSEN, J. M. L. PAU, L. F. STRASZAK, A.
TI A HEURISTIC ECONOMETRIC MODEL BY SELF-ORGANIZATION METHOD.
SO MODELS AND DECISION MAKING IN NATIONAL ECONOMIES. VIENNA, AUSTRIA.
   AMSTERDAM, NETHERLANDS. IX8425. 99-107. 9. NORTH-HOLLAND. IFAC IIASA
   IFORS. 24-27 JAN. 1977.
IN 0 444 85280 8.
YR 79.

******************************** INSP ********************************
AU IVAKHNENKO, A. G.
TI DEVELOPMENT OF MODELS OF OPTIMAL COMPLEXITY USING SELF-ORGANIZATION
   THEORY.
SO INT. J. COMPUT. AND INF. SCI. (USA). VOL.8, NO.2. 111-27. 23.
YR 79.

******************************** INSP ********************************
AU KOVALCHUK, F. I. LAKHNO, Y. S.
TI PREDICTION OF AUTOMOBILE EXHAUST ATMOSPHERIC POLLUTION FIELDS USING
   THE MODEL OF AN INHOMOGENEOUS STRATIFIED MEDIUM.
SO AVTCMATIKA (USSR). VOL.13, NO.6. 31-7. 11.
YR 80.
TR SOV. AUTOM. CONTROL (USA). SAUCBZ. VOL.13, NO.6. 27-32. NCV.-DEC.
   1980..

******************************** INSP ********************************
AU VASILYEV, V. I. KONOVALENKO, V. V. MYSHKO, V. K. KREMEN, N. I.
TI OPTIMIZATION OF THE PARAMETERS OF AN OBJECT USING A STATIC MODEL
   CONSTRUCTED BY THE GROUP METHOD OF DATA HANDLING.
SO AVTOMATIKA (USSR). VOL.13, NO.6. 72-4. 1.

YR 80.
TR SOV. AUTOM. CONTROL (USA). SAUCBZ. VOL.13, NO.6. 63-5. NOV.-DEC.
   1980.

********************************** INSP **********************************

AU IVAKHNENKO, A. G. KROTOV, G. I.
TI SELF-ORGANIZATION OF MODELS WITH VARIABLE COEFFICIENTS FOR OPTIMIZING
   OF THE PLANNING OF WATER RESERVOIRS.
SO AVTOMATIKA (USSR). VOL.13, NO.6. 11-30. 5.
YR 80.
TR SOV. AUTCM. CCNTROL (USA). SAUCBZ. VOL.13, NO.6. 9-26. NOV.-DEC.
   1980..

********************************** INSP **********************************

AU BELOZERSKIY, Y. A. PETUKHOVA, S. A.
TI SELF-ORGANIZATICN OF MODELS OF COMPLEX QUASISTATIONARY PLANTS.
SO AVTOMATIKA (USSR). VOL.13, NO.6. 3-10. 2.
YR 80.
TR SOV. AUTCM. CCNTROL (USA). SAUCBZ. VOL.13, NO.6. 1-8. NOV.-DEC.
   1980.

********************************** INSP **********************************

AU PODKLADNEV, V. YA.
TI USE OF GMDH SELF-ORGANIZATION ALGORITHMS IN THE CONTROL OF CONTINUOUS
   METAL-REFINING PROCESSES.
SO AVTOMATIKA (USSR). VOL.13, NO.5. 25-30. 6.
YR 80.
TR SOV. AUTCM. CCNTROL (USA). SAUCBZ. VOL.13, NO.5. 22-6. SEPT.-OCT.
   1980.

********************************** INSP **********************************

AU YURACHKOVSKIY, YU. P.
TI SELF-ORGANIZATION OF THE PRINCIPAL COMPONENTS IN SIMULATION USING A
   SMALL SAMPLE.
SO AVTOMATIKA (USSR). VOL.13, NO.5. 86-9. 8.
YR 80.
TR SOV. AUTCM. CCNTROL (USA). SAUCBZ. VOL.13, NO.5. 77-80. SEPT.-OCT.
   1980..

********************************** INSP **********************************

AU FOKBASS, V. L. GOLUBEVA, L. V.
TI SELF-ORGANIZATICN OF A MATHEMATICAL MODEL FOR LONG-RANGE PLANNING OF
   THE COST OF COAL MINING.
SO AVTOMATIKA (USSR). VOL.13, NO.3. 19-25. 4.
YR 80.
TR SOV. AUTOM. CCNTROL (USA). SAUCBZ. VOL.13, NO.3. 15-20. MAY-JUNE
   1980.

********************************** INSP **********************************

AU IVAKHNENKC, A. G. KROTOV, G. I. CHEBERKUS, V. I.
TI MULTILAYER ALGORITHM FOR SELF-ORGANIZATION OF LONG-TERM PREDICTIONS
   (ILLUSTRATED BY THE EXAMPLE OF THE LAKE BAYKAL ECOLOGICAL SYSTEM).
SO AVTOMATIKA (USSR). VOL.13, NO.4. 28-47. 9.
YR 80.
TR SOV. AUTOM. CONTROL (USA). SAUCBZ. VOL.13, NO.4. 22-38. JULY-AUG.
   1980..

******************************** INSP ********************************
AU PAK, V. G.
TI ANALYTICAL CONSTRUCTION CF ADAPTIVE NETWORKS REALIZING MULTILAYER
   ECLYNOMIAL ALGORITHMS.
SO AVTCMATIKA (USSR). VOL.13, NO.4. 61-72. 5.
YR 80.
TR SOV. AUTOM. CONTROL (USA). SAUCBZ. VOL.13, NO.4. 50-60. JULY-AUG.
   1980..

******************************** INSP ********************************
AU IVAKHNENKO, A. G. KOPPA, YU. V. LANTAYEVA, D. N. IVAKHNENKO, N.
   A.
TI THE RELATIONSHIP BETWEEN COMPUTER SELF-ORGANIZATION OF MATHEMATICAL
   MODELS AND PATTERN RECOGNITION.
SO AVTOMATIKA (USSR). VOL.13, NO.3. 3-13. 20.
YR 80.
TR SOV. AUTCM. CONTROL (USA). SAUCBZ. VOL.13, NO.3. 1-9. MAY-JUNE 1980..

******************************** INSP ********************************
AU KIKOT, V. S. PATAREU, S. G.
TI PROCESSING CF EXPERIMENTALLY MEASURED RESISTIVITIES OF METALS USING
   GMDE.
SO AVTOMATIKA (USSR). VOL.13, NO.3. 76-9. 5.
YR 80.
TR SOV. AUTCM. CCNTROL (USA). SAUCBZ. VOL.13, NO.3. 67-70. MAY-JUNE
   1980..

******************************** INSP ********************************
AU KLEIN, L. P. MUELLER, I. A. IVAKHNENKO, A. G.
TI MODELING OF THE ECONOMICS OF THE USA BY SELF-ORGANIZATION OF THE
   SYSTEM OF EQUATIONS.
SO AVTOMATIKA (USSR). VOL.13, NO.1. 3-12. 6.
YR 80.
TR SOV. AUTOM. CCNTROL (USA). SAUCBZ. VOL.13, NO.1. 1-8. JAN.-FEB.
   1980.

******************************** INSP ********************************
AU KHOMOVNENKO, B. G. KOLOMIETS, N. G.
TI SELF-ORGANIZATION OF A SYSTEM OF SIMPLE PARTIAL MODELS FOR PREDICTING
   THE WHEAT HARVEST.
SO AVTOMATIKA (USSR). VOL.13, NO.1. 28-37. 11.
IR 80.
TR SOV. AUTOM. CONTROL (USA). SAUCBZ. VOL.13, NO.1. 22-9. JAN.-FEB.
   1980..

******************************** INSP ********************************
AU PONOMARENKO, V. S.
TI SOLUTION OF A CLASS OF MATHEMATICAL PROGRAMMING PROBLEMS USING THE
   GROUP METHOD OF CATA EANDLING (GMDH).
SO AVTCMATIKA (USSR). VOL.13, NO.1. 13-19. 4.
YR 80.
TR SOV. AUTOM. CONTROL (USA). SAUCBZ. VOL.13, NO.1. 9-14. JAN.-FEB.
   1980..

******************************** INSP ********************************
AU IVAKHNENKO, A. G. KOPPA, JU. V. LANTAYOVA, D. IVAKHNENKO, N. A.
TI UNIFICATION OF MATHEMATICAL MODEL SELF-ORGANIZATION THEORY AND THE

        THEORY OF PATTERN RECOGNITION.
SO PROCEECINGS OF THE 5TH INTERNATIONAL CONFERENCE ON PATTERN
    RECOGNITION. MIAMI BEACH, FI, USA. NEW YORK, USA. 2 VOL. XLVI&1382.
    113&41. 23. IEEE. INT. ASSOC. PATTERN RECOGNITION IEEE. 1-4 DEC.
    1980.
YR 80.

********************************* INSP *********************************
AU IKEDA, S. IHARA, J.
TI GROUP METHOD OF DATA HANDLING TOWARDS A MODELING OF COMPLEX SYSTEMS.
    VIII.
SO SYST. AND CONTROL (JAPAN). VOL. 24, NO. 7. 483-91. 10.
YR 80.

********************************* INSP *********************************
AU TAMURA, H. KONDO, T.
TI HEURISTICS FREE GROUP METHOD OF DATA HANDLING ALGORITHM OF GENERATING
    OPTIMAL PARTIAL POLYNOMIALS WITH APPLICATION TO AIR POLLUTION
    PREDICTION.
SO INT. J. SYST. SCI. (GB). VOL. 11, NO. 9. 1095-111. 17.
YR 80.

********************************* INSP *********************************
AU CONISHI, S. AKITA, K.
TI ESTIMATION OF JUNCTION TEMPERATURE OF THYRISTOR AT NORMAL OPERATING
    CONDITIONS AND ITS TEMPERATURE OBSERVATION METHOD.
SO TRANS. SOC. INSTRUM. AND CONTROL ENG. (JAPAN). VOL. 16, NO. 4. 533-8.
    13.
YR 80.

********************************* INSP *********************************
AU IKEDA, S. AN, H.
TI A NONLINEAR MODEL FOR TYPHOON RAINFALL (A GMDH ALGORITHM WITH INPUT
    DATA CLUSTERING).
SO J. SOC. INSTRUM. AND CONTROL ENG. (JAPAN). VOL. 19, NO. 7. 733-7. 7.
YR 80.

********************************* INSP *********************************
AU YOSHIMURA, T. KIYOZUMI, R. SOEDA, T.
TI PREDICTION AT SHORT RANGE OF AIR POLLUTION CONCENTRATION BY THE
    REVISED GMDH USING AIC.
SO J. SOC. INSTRUM. AND CONTROL ENG. (JAPAN). VOL. 19, NO. 7. 725-8. 9.
YR 80.

********************************* INSP *********************************
AU IKEDA, S. IHARA, J.
TI GROUP METHOD OF DATA HANDLING TOWARDS A MODELING OF COMPLEX SYSTEMS.
    VII.
SO SYST. AND CONTROL (JAPAN). VOL. 24, NO. 6. 390-8. 10.
YR 80.

********************************* INSP *********************************
AU IKEDA, S. IHARA, J.
TI GROUP METHOD OF DATA HANDLING TOWARDS A MODELING OF COMPLEX SYSTEMS.
    VI.
SO SYST. AND CONTROL (JAPAN). VOL. 24, NO. 5. 338-46. 9.
YR 80.
********************************* INSP *********************************

AU IHARA, J.
TI GROUP METHCD OF DATA HANDLING TOWARDS A MODELING OF COMPLEX SYSTEMS.
   IV.
SO SYST. AND CONTROL (JAPAN). VCL.24, NO.3. 158-68. 63.
YR 80.

***************************** INSP *********************************
AU TAMURA, H. KONDO, T.
TI NCNLINEAR STEADY STATE MODELING FOR RIVER QUALITY BY A REVISED GMDH.
SO TRANS. SOC. INSTRUM. AND CONTROL ENG. (JAPAN). VOL.16, NO.2. 189-94.
   4.
YR 80.

***************************** INSP *********************************
AU IKEDA, S. IHARA, J.
TI GROUP METHOD OF DATA HANDLING TOWARDS A MODELING OF COMPLEX SYSTEMS.
   V.
SO SYST. AND CCNTROL (JAPAN). VOL.24, NO.4. 264-72. 68.
YR 80.

***************************** INSP *********************************
AU IKEDA, S.
TI GROUP METHOD OF DATA HANDLING TOWARDS A MODELLING CF COMPLEX SYSTEMS.
   II.
SO SYST. AND CONTROL (JAPAN). VOL.24, NO.1. 46-54. 18.
YR 80.

***************************** IN SP *********************************
AU IKEDA, S.
TI GROUP METHCD OF DATA HANDLING TOWARDS A MODELING OF COMPLEX SYSTEMS.
   III.
SO SYST. AND CONTROL (JAPAN). VCL.24, NO.2. 98-106. 48.
YR 80.

***************************** INSP *********************************
AU NAGASAKA, K. KITA, Y. HASHIMOTO, F.
TI IDENTIFICATION OF A MODEL OF GRINDING WHEEL LIFE BY THE GROUP METHOD
   OF DATA HANDLING.
SO WEAR (SWITZERLAND). VOL.58, NO.1. 147-54. 4.
YR 80.

***************************** INSP *********************************
AU BRAVERMAN, V. YA. IVAKHNENKO, N. A.
TI ANALYSIS AND PREDICTION CF LABOR PRODUCTIVITY IN THE ELECTRIC POWER
   CCNSTRUCTION INDUSTRY USING GMDH ALGORITHMS.
SO AVTCMATIKA (USSR). VOL.14, NO.2. 10-15. 5.
YR 81.
TR SOV. AUTOM. CONTROL (USA). SAUCBZ. VOL.14, NO.2. 8-13. MARCH-APRIL
   1981.

***************************** INSP *********************************
AU SVETALSKIY, B. K. KOZUEOVSKIY, S. F. DUBROVIN, I. YA.
TI A MATHEMATICAL MODEL OF THE ECOLOGICAL SYSTEM OF THE SEA OF AZOV.
SO AVTOMATIKA (USSR). VOL.14, NO.2. 16-22. 23.
YR 81.
TR SOV. AUTOM. CONTROL (USA). SAUCBZ. VOL.14, NO.2. 14-20. MARCH-APRIL
   1981..
***************************** INSP *********************************

AU GRISHCHENKO, A. Z. PAVLOV, V. A.
TI ALGORITHMS FOR CONSTRUCTING NONLINEAR CLASS BOUNDARIES IN PATTERN
   RECOGNITION PROBLEMS.
SO AVTCMATIKA (USSR). VOL.14, NO.2. 3-9. 8.
YR 81.
TR SOV. AUTOM. CONTROL (USA). SAUCBZ. VOL.14, NO.2. 1-7. MARCH-APRIL
   1981..

***************************** INSP *********************************
AU IVAKHNENKO, A. G. KBOTOV, G. I. YURACHKOVSKIY, YU. P.
TI AN EXPONENTIAL-HARMONIC ALGORITHM OF THE GROUP METHOD OF DATA
   HANDLING.
SO AVTOMATIKA (USSR). VOL.14, NO.2. 23-9. 9.
YR 81.
TR SOV. AUTOM. CONTROL (USA). SAUCBZ. VOL.14, NO.2. 21-7. MARCH-APRIL
   1981.

***************************** INSP *********************************
AU AKIZUKI, K. ONO, K. SHIGEMASA, T.
TI ON A DESIGN METHOD BY GMIH FOR THE RELATION BETWEEN PROCESS
   PARAMETERS AND CPTIMAL CONTROL PARAMETERS.
SO TRANS. SOC. INSTRUM. AND CCNTROL ENG. (JAPAN). VOL. 17, NO. 6. 696-8.
   7.
YR 81.

***************************** INSP *********************************
AU IVAKHNENKO, A. O. G. BELOZERSKIY, I. A. KOZUBOVSKIY, S. F.
TI SYNTHESIS OF A CONTROL MODEL USING A MULTILAYER GMDH ALGORITHM.
SO AVTOMATIKA (USSR). VOL.14, NO.1. 75-84. 12.
YR 81.
TR SOV. AUTOM. CONTROL (USA). SAUCBZ. VOL.14, NO.1. 69-77. JAN.-FEB.
   1981.

***************************** INSP *********************************
AU IVAKHNENKO, N. A. IVLEV, M. M. SEGAL, V. V.
TI USE OF A MULTICRITERION APPROACH IN THE SELECTION OF FEATURE PAIRS
   AND IN RESTORATION OF A DECISION RULE IN PROBABILISTIC ALGORITHMS OF
   THE GROUP METHOD OF DATA HANDLING.
SC AVTOMATIKA (USSR). VOL. 14, NO. 1. 88-90. 4.
YR 81.
TR SOV. AUTOM. CCNTROL (USA). SAUCBZ. VOL. 14, NO. 1. 82-5. JAN.-FEB.
   1981.

***************************** INSP *********************************
AU SAITO, S. KOBAYASHI, S.
TI MODELING AND PREDICTION OF WATER DEMAND-APPLICATION OF GMDH ALGORITHM
   USING AIC AS TOTAL CRITERICN.
SO TRANS. SOC. INSTRUM. AND CONTROL ENG. (JAPAN). VOL.17, NO.5. 537-44.
   7.
YR 81.

***************************** INSP *********************************
AU NISHIMURA, Y. YONEZAWA, Y.
TI MINIMAL TIME FEEDBACK CONTROL BY USING SWITCHING HYPER SURFACE.
SO TRANS. SOC. INSTRUM. AND CCNTROL ENG. (JAPAN). VOL. 17, NO. 3. 323-9.
   16.
YR 81.
***************************** INSP *********************************

AU RYOBU, M. SAWARAGI, Y.
TI THE ECONOMETRIC MODEL BY GROUP METHOD OF STRUCTURAL MODELING.
SO TRANS. SOC. INSTRUM. AND CONTROL ENG. (JAPAN). VOL.17, NO.1. 79-84.
   8.
YR 81.

******************************** INSP ********************************
AU NISHIKAWA, T. SHIMIZU, S. AKASHI, H.
TI IDENTIFICATION AND FORECASTING IN MANAGEMENT SYSTEM BY USING THE GMDH
   METHOD IN THE CASE OF THE AUTOMOBILE INDUSTRY IN JAPAN.
SO CONTROL SCIENCE AND TECHNOLOGY FOR THE PROGRESS OF SOCIETY.
   PROCEEDINGS OF THE EIGHTH TRIENNIAL WORLD CONGRESS OF THE
   INTERNATIONAL FEDERATION OF AUTOMATIC CONTROL. KYOTO, JAPAN. OXFORD,
   ENGLAND. 7 VOL. LXXIX&3914. 713-18 VOL.1. 6. PERGAMON. 24-28 AUG.
   1981.
RN 0 08 027580 X.
YR 82.

******************************** INSP ********************************
AU YOSHIMURA, T. PANDEY, U. S. TAKAGI, T. SOEDA, T.
TI PREDICTION OF THE PEAK FLOOD USING REVISED GMDH ALGORITHMS.
SO INT. J. SYST. SCI. (GB). VOL.13, NO.5. 547-57. 9.
YR 82.

******************************** INSP ********************************
AU NISHIKAWA, T. SHIMIZU, S.
TI IDENTIFICATION AND FORECASTING IN MANAGEMENT SYSTEMS USING THE GMDH
   METHOD.
SO APPL. MATH. MODELLING (GB). VOL.6, NO.1. 7-15. 7.
YR 82.

******************************** INSP ********************************
AU YOSHIMURA, T. KIYOZUMI, R. NISHINO, K. SOEDA, T.
TI PREDICTION OF AIR POLLUTANT CONCENTRATIONS BY REVISED GMDH ALGORITHMS
   IN TOKUSHIMA PREFECTURE.
SO IEEE TRANS. SYST. MAN AND CYBERN. (USA). VOL.SMC-12, NO.1. 50-6. 18.
YR 82.

Index